Springer Proceedings in Mathematics & Statistics

Volume 388

This book series features volumes composed of selected contributions from workshops and conferences in all areas of current research in mathematics and statistics, including data science, operations research and optimization. In addition to an overall evaluation of the interest, scientific quality, and timeliness of each proposal at the hands of the publisher, individual contributions are all refereed to the high quality standards of leading journals in the field. Thus, this series provides the research community with well-edited, authoritative reports on developments in the most exciting areas of mathematical and statistical research today.

Frederick Hoffman
Editor

Combinatorics, Graph Theory and Computing

SEICCGTC 2020, Boca Raton, USA, March 9–13

Springer

Editor
Frederick Hoffman
Department of Mathematical Sciences
Florida Atlantic University
Boca Raton, FL, USA

ISSN 2194-1009 ISSN 2194-1017 (electronic)
Springer Proceedings in Mathematics & Statistics
ISBN 978-3-031-05374-0 ISBN 978-3-031-05375-7 (eBook)
https://doi.org/10.1007/978-3-031-05375-7

Mathematics Subject Classification: 05-XX, 05-06, 05Cxx, 12-XX, 12-06, 68Rxx, 68R05, 52-XX, 52-06

This Springer imprint is published by the registered company Springer Nature Switzerland AG
The registered company address is: Gewerbestrasse 11, 6330 Cham, Switzerland

Preface I

The Southeastern International Conference on Combinatorics, Graph Theory, and Computing (SEICCGTC) is an international meeting of mathematical scientists, held annually in March, during Spring Break at Florida Atlantic University (FAU) in Boca Raton, Florida. The conference includes a program with plenary lectures by invited speakers, as well as sessions of contributed papers each day. In addition, two or three invitational special sessions are offered each year. A valuable part of the conference is the opportunity afforded for informal conversations about the methods participants employ in their professional work in business, industry, and government and about their current research.

The 51st meeting was held in the newly renovated Student Union building at FAU, March 9–13, 2020. Five distinguished researchers, at various stages of their careers, accepted invitations to attend as plenary speakers at this year's 51st SEICCGTC: Pierre Baldi, University of California, Irvine, USA; Pavol Hell, Simon Fraser University, Canada; Patricia Hersh, University of Oregon, USA; Panos M. Pardalos, University of Florida, USA; and Kai-Uwe Schmidt, Paderborn University, Germany. Dr. Pardalos had to cancel his talk at the last moment, due to illness and the pandemic. Each of the other plenary speakers gave two talks. There were two special sessions this year, one on Research by Women in Graph Theory and its Applications, organized by Leslie Hogben, and one on Extremal Graph Theory, organized by Neal Bushaw. Both were well attended and well received. Plenary and contributed talks covered a wide variety of topics including: new tools for counting and linear programming, using topological methods; graph homomorphism; graphs with loops; highly non-linear functions and coding theory; association schemes; deep learning and its mathematical foundations; extremal graph theory; posets; latin squares; combinatorial games; coloring, connectivity, domination, labeling, and partitioning of graphs; along with associated algorithms and applications.

The coronavirus pandemic of 2020 created some difficulties for our conference. One plenary speaker could not attend, and we experienced 20 cancellations due to the virus outbreak. Several participants had their return travel plans disrupted; a few were quite worried for a day or two. The FAU Department of Mathematical Sciences hosted approximately 150 conference participants and guests, marking another successful meeting of the SEICCGTC! For the most part, the 51st meeting of the SEICCGTC was a great success and conference participants expressed their approval of the overall quality of speakers and programs and the continuous improvements in the technology provided. We have cause to celebrate!

On Tuesday, March 10, 2020, we celebrated the publication of the commemorative book entitled, "50 Years of Combinatorics, Graph theory and Computing," published by CRC/Taylor & Francis. Twenty-nine past plenary speakers and past conference participants contributed 21 chapters for the book, edited by Fan Chung, Ronald Graham, Ronald Mullin, Frederick Hoffman, Douglas West and Leslie Hogben. The Institute of Combinatorics and its Applications held its annual meeting on Wednesday, March 11.

The conference also featured an outdoor reception Monday evening on the Live Oak Pavilion Patio, a sumptuous beachfront banquet Wednesday evening at the Delray Sands Resort, as well an excursion Thursday afternoon to the Flagler Museum in Palm Beach, followed by an informal reception Thursday evening at the brand new Schmidt Family Complex. The social program was capped off by a wonderful Survivors' Party Friday evening, hosted by Aaron Meyerowitz and Andrea Schuver at their home!

This year, we took another step toward elevating the quality of the conference, by agreeing to publish our conference proceedings with Springer Nature, in their PROMS series of hardback conference proceedings. The purpose is to more effectively and efficiently continue to disseminate important advances in the represented disciplines and to ensure that the conference continues to promote better understanding of the roles of modern applied mathematics, combinatorics and computer science; demonstrate the contribution of each discipline to the others; and decrease gaps between the fields, as it did through fifty years of publishing in the journal, *Congressus Numerantium.*

The conference was supported by the Department chair and staff, with technical support by Andrew Gultz. Outside support came from the National Security Agency, Springer Nature, CRC Press/Taylor & Francis, *Algorithms*, and The Institute of Combinatorics and its Applications. Conference coordination and organization was superbly provided by Dr. Maria Provost.

I gratefully acknowledge the support and assistance of Sara Heuss Holliday, Richard Low, Zvi Rosen, Farhad Shahrokhi, and John Wierman in the compilation and reviewing of these Proceedings. We also thank all our referees.

Boca Raton, USA Frederick Hoffman

Preface II

Ratio balancing numbers, introduced here by Jeremy Bartz and his coauthors, are a generalization of balancing numbers, a concept from number theory involving triangular numbers. The authors define the concept and present examples, existence results, and conjectures.

Bohan Qu and Stephen J. Curran show that the number $\beta = (b^{b-1} - 1)/(b - 1)^2$, where $b \geq 3$, has several interesting multiplicative properties. In the base b number system, $\beta = (123\cdots(b-4)(b-3)(b-1))_b$. They show that the digits of the number $K\beta$, for integers K such that $1 \leq K \leq (b-1)^2$, as a number in the base b number system can be generated from an arithmetic sequence reduced modulo $b - 1$ with an appropriate adjustment.

Dennis Davenport and his coauthors report on recent results of their research group on Riordan arrays. They generalize a known row construction of Riordan arrays to a result on the determination of double Riordan arrays.

Timothy Myers constructs the Clifford graph algebra for any windmill graph W(r, m), which consists of m copies of the complete graph K_r adjoined at one common vertex; and for any Dutch windmill graph D_r^m which consists of m copies of the r-cycle graph C_r adjoined at one common vertex. He then applies the construction to give a new proof that these graphs, which possess the friendship property, are precisely the friendship graphs.

Paul Peart and Francois Ramaroson construct and find the values for certain character sums involving quadratic characters. The method is new and employs elliptic curves. Detailed proofs are provided.

In work that originated in an REU at Illinois State University, Joel Jeffries and his coauthors investigate a multigraph G with the underlying structure of a 4-cycle where each edge multiplicity in the set $\{1, 2, 3, 4\}$ is represented. They refer to each of the three such multigraphs as a Stanton 4-cycle. For each such G, they consider λ such that there exists a G-decomposition of ${}^{\lambda}K_n$.

Brigitte Servatius considers the k-plane matroid, which is a matroid on the edge set, I, of a bipartite graph, $H = (A, B; I)$, defined by a counting condition. She shows that $2k$-connectivity of H implies that I is a spanning set for the k-plane matroid on the edge set of the complete bipartite graph on (A, B). For $k = 2$ she explains the connections to rigidity in the plane and to conjectures of Whiteley.

Farhad Shahrokhi derives an upper bound on the trace function of a hypergraph H and gives some applications. For instance, a new upper bound for the VC dimension of H, or $vc(H)$, follows as a consequence and can be used to compute $vc(H)$ in polynomial time provided that H has bounded degeneracy. This was not previously known, and improves computing time in some cases. Another consequence is a general lower bound on the distinguishing transversal number of H that gives rise to applications in domination theory of graphs.

Sarah Heuss Holliday continues work on a question raised in 2017 by Hedetniemi: For which graphs G does the indexed family of open neighborhoods have a system of distinct representatives? In earlier work with collaborators, that question was answered, and necessary conditions and associated parameters were explored. Haenel and Johnson looked over longest paths and cycles. The work here further generalizes and deepens their examinations.

Atif Abueida and Kenneth Roblee examine harmonious labelings of starlike trees. It has been shown using cyclic groups that the disjoint union of an odd cycle on s vertices and starlike trees with the central vertex adjacent to some even t many s-paths is harmonious. They consider the disjoint union of an odd cycle with at least two starlike trees with new notions of harmonious labelings to accommodate the case where $|V| > |E|$.

A mean coloring of a connected graph G of order 3 or more is an edge coloring of G with positive integers such that the mean of the colors of the edges incident with every vertex is an integer. The associated color of a vertex is its chromatic mean. If distinct vertices have distinct chromatic means, then the edge coloring is a rainbow mean coloring of G. In their paper, Ebrahim Salehi and his coauthors investigate rainbow mean colorings of trees.

Peg solitaire is a game in which pegs are placed in every hole but one, and the player jumps over pegs to remove them. In 2011, this game was generalized to graphs. Here, Robert A. Beeler and Aaron D. Gray examine graphs in which any single edge addition changes solvability. They provide necessary and sufficient conditions for solvability for a certain family. They show that infinite subsets of this family are edge critical and determine the maximum number of pegs that can be left on this family with the condition that a jump is made whenever possible. Finally, they give a list of graphs on eight vertices that are edge critical.

A set of vertices, S, in a strongly connected digraph D, is split dominating provided it is: (1) dominating and (2) $D - S$ is trivial or not strongly connected. The split domination number is the minimum cardinality of a split dominating set for that digraph. Sarah Merz and her coauthors show that for any k-regular tournament, the split domination number is at least $(2k+3)/3$ and this bound is tight. They explore properties of regular tournaments with split domination number equal to the lower bound, including sufficient conditions for $\{1\}$-extendability.

David R. Prier and his coauthors give independence and domination results for six chess-like pieces on triangular boards with triangular spaces and triangular boards with hexagonal spaces. The question of independence and domination for these same boards on the surface of a tetrahedron is introduced, and some initial results are given.

A graph has an efficient dominating set if there exists a subset of vertices D such that every vertex in the graph is dominated by exactly one vertex in D. Lyle Paskowitz and his coauthors investigate efficient domination on the stacked versions of each of the eleven Archimedean Lattices, and determine the existence or non-existence of efficient dominating sets on each lattice through integer programming. The proofs of existence are constructive, and the proofs of non-existence are generated by integer programs. They find efficient dominating sets on seven of the stacked lattices and prove that no such sets exist on the other four stacked lattices.

Let G be a graph with vertex set V (G) and edge set E(G). A (p; q)-graph G = $(V;E)$ is said to be AL(k)-traversal if there exist a sequence of vertices $v_1, v_2, ..., v_p$ such that for each $i = 1, 2, ..., p-1$, the distance for v_i and v_{i+1} is equal to k. We call a graph G a k-steps Hamiltonian graph if it has a AL(k)-traversal in G and the distance between v_p and v_1 is k. A graph G is said to be hereditary k-steps hyperhamiltonian if it is k-steps Hamiltonian and for any v in G, the vertex-deleted subgraph G \{v} is also k-steps Hamiltonian. In this paper, Hsin-hao Su and his coauthors investigate subdivision graphs of a wheel graph and $C_4 \times K_2$ to see which are 2-steps Hamiltonian and hereditary non 2-steps Hamiltonian.

Let G be a graph with average degree greater than k–2. Erdős and Sós conjectured that G contains every tree on k vertices as a subgraph. The circumference of the graph G, c(G), is the number of edges on a longest cycle. Gilbert and Tiner proved that if c(G) is at most k, then G contains every tree on k vertices. Here A.M. Heissan and Gary Tiner improve this result and show that the Erdős-Sós conjecture holds for graphs whose circumference is at most $k + 1$.

Yoshimi Egawa and Kenji Kimura consider a relationship between a regular graph and a regular factor of its vertex-deleted subgraph. Katerinis proved that if r is an even integer and k is an integer with $1 \leq k \leq r/2$, and G is an r-regular, r-edge-connected graph of odd order, then $G \setminus \{x\}$ has a k-factor for each $x \in V(G)$. When the result "for each $x \in V(G)$" of Katerinis is replaced "for some $x \in V(G)$," they consider what condition can hold. One main result is: Let r and k be even integers such that $4 \leq k \leq r/2$, and ℓ be a minimum integer such that $\ell \geq r/(r-2k+4)$, and G be an r-regular, 2ℓ-edge-connected graph of odd order. Then, there is some $x \in V(G)$ such that $G \setminus \{x\}$ has a k-factor. Moreover, if $r \geq 4k - 8$, then we can replace 2ℓ-edge-connected with 2-edge-connected.

In his paper, LeRoy B. Beasley gives several definitions of connectedness and extendibility of paths and cycles in directed graphs. He defines sets of digraphs by various types of connectedness or extendibility and gives some containments as well as examples to show proper containment.

Extraconnectivity generalizes the concept of connectivity of a graph but it is more difficult to compute. In his paper, Eddie Cheng and his coauthors compute the g-extraconnectivity of the arrangement graph for small g ($g \leq 6$) with the help of a computer program. In addition, they provide an asymptotic result for general g.

Alan Bickle defines a k-tree as a graph that can be formed by starting with K_{k+1} and iterating the operation of making a new vertex adjacent to all the vertices of a k-clique of the existing graph. When the order $n > k + 1$, a k-path graph is a k-tree with exactly two vertices of degree k. He states a forbidden subgraph characterization for k-paths as k-trees. He characterizes k-trees with diameter $d \geq 2$ based on the k-paths they contain.

In their paper, Marina Skyers and Lee I. Stanley look at representations of the simple random walk, S_n, and show how to effectively rearrange the sequence of terms $S_n/\sqrt{n}$ in order to achieve almost sure convergence to the standard normal on the open interval (0; 1). This is done via a suitable choice of permutation $F : \{0, 1\}^n \to \{0, 1\}^n$. They are interested in optimal rearrangement of the simple random walk. They describe how to minimize the graph-theoretic complexity of these permutations.

M. R. DeDeo analyzes and compares properties of Cayley graphs of permutation groups called transposition graphs, as this family of graphs has better degree and diameter properties than other families of graphs. Cayley graphs of permutation groups generated by transpositions inherit almost all of the properties of the hypercube. In particular, she studies properties of the complete transportation, (transposition) star graph, bubble-sort graph, modified bubble-sort graph and the binary hypercube and uses these properties to determine bounds on the energy of these graphs.

John C. Wierman studies the (4; 8^2) or "bathroom tile," lattice, one of the eleven Archimedean lattices, which are infinite vertex-transitive graphs with edges from the tilings of the plane by regular polygons. The site percolation model retains each vertex of an infinite graph independently with probability $p, 0 \leq p \leq 1$. The site percolation threshold is the critical probability p_c^{site}, above which the subgraph induced by retained vertices contains an infinite connected component almost surely, and below which all components are finite almost surely. Using computational improvements for the substitution method, the upper bound for the site percolation threshold of the (4; 82) lattice is reduced from 0.785661 to 0.749002.

Boca Raton, USA Frederick Hoffman

Contents

Ratio Balancing Numbers

Jeremiah Bartz, Bruce Dearden, Joel Iiams, and Jerry Metzger

Abstract Balancing numbers were introduced by Behera and Panda while investigating when the sum of two triangular numbers is a triangular number. We introduce a variation called ratio balancing numbers which generalizes the sums considered and involves an integral ratio condition. Often ratio balancing numbers retain the familiar properties of balancing numbers. However, a distinct feature of ratio balancing numbers is that they exist in finite numbers for certain choices of parameters. Computational evidence leads us to conjecture that for any integer d, there are choices of parameters which yield finitely many, but at least d, ratio balancing numbers.

Keywords Balancing numbers · Triangular numbers · Recurrence relations

1 Introduction

Behera and Panda [5] defined balancing numbers as positive integers B satisfying

$$1 + 2 + \cdots + (B - 1) = (B + 1) + \cdots + (B + r)$$

for some integer $r \geq 0$. The previous equation is equivalent to

$$T(B - 1) + T(B) = T(B + r) \tag{1}$$

J. Bartz (✉) · B. Dearden · J. Iiams · J. Metzger
University of North Dakota, Grand Forks, ND 58202, USA
e-mail: jeremiah.bartz@und.edu

B. Dearden
e-mail: bruce.dearden@und.edu

J. Iiams
e-mail: joel.iiams@und.edu

J. Metzger
e-mail: jerry.metzger@und.edu

F. Hoffman (ed.), *Combinatorics, Graph Theory and Computing*,
Springer Proceedings in Mathematics & Statistics 388,
https://doi.org/10.1007/978-3-031-05375-7_1

where $T(i) = \frac{i(i+1)}{2}$ is the ith triangular number. The identity $T(5) + T(6) = T(8)$ shows that 6 is a balancing number. Traditionally, 1 is considered the initial balancing number since it satisfies (1). The collection of balancing numbers forms an infinite sequence which appears in The Online Encyclopedia of Integer Sequences [8] as A001109.

Many variations of balancing numbers have been studied [1, 2, 4, 9–13]. That being said, none so far have incorporated sums of the same type as the well known identity $3T(B-1) + T(B) = T(2B-1)$ [6, p. 13]. Motivated to include such sums, we introduce ratio balancing numbers which involve an integral ratio condition. We show that often ratio balancing numbers retain the familiar properties of balancing numbers. However, a distinct feature of ratio balancing numbers is that they exist in finite numbers for certain choices of parameters. Computational evidence leads us to conjecture that for any integer d, there are choices of parameters which yield finitely many, but at least d, ratio balancing numbers.

The paper is organized as follows. Definitions and examples of ratio balancing numbers and related quantities are given in Sect. 2. In Sect. 3, we prove that there are only finitely many ratio balancing numbers for certain choices of parameters and present our conjecture. We derive in Sect. 4 several familiar properties of balancing numbers for ratio balancing numbers when infinitely many exist. Additionally, we present the surprising restriction on jump sizes when generating an infinite class of ratio balancing numbers; the jump size is either 1 or 2.

2 Ratio Balancing Numbers and Related Quantities

Let $p, q, k, w \in \mathbb{Z}$ with $p, q \geq 1$ and $k \geq 0$. We are interested in finding integers B with $B \geq k$ such that

$$1 + \cdots + (B-k) : (B+1) + \cdots + (B+r) :: p : q.$$

This is equivalent to B satisfying

$$qT(B-k) + pT(B) = pT(B+r) \tag{2}$$

where $T(i) = \frac{i(i+1)}{2}$ is the ith triangular number. We are also interested when the two sides of (2) differ by a fixed integer w. This leads to the following definition.

Definition 1 Let $p, q, k, w \in \mathbb{Z}$ with $p, q \geq 1$ and $k \geq 0$. An integer B is called a ratio balancing number with ratio $p : q$, gap k, and weight w, or more simply an $R(p, q, k, w)$-balancing number if $B \geq k$ and

$$q\,T(B-k) + p\,T(B) + w = p\,T(B+r) \tag{3}$$

for some integer $r \geq 0$. We refer to r as the $R(p, q, k, w)$-balancer corresponding to the $R(p, q, k, w)$-balancing number B.

Solving (3) for r and B, respectively, gives

$$r = \frac{-p(2B+1) + \sqrt{4p(p+q)B^2 + 4p(p+q-2qk)B + 4pqk(k-1) + p^2 + 8pw}}{2p} \tag{4}$$

and

$$B = \frac{2pr + q(2k-1) + \sqrt{4p(p+q)r^2 + 8pqkr + q^2 - 8qw}}{2q} \tag{5}$$

where we take the positive square root so that $r \geq 0$ and $B \geq k$. Thus B is an $R(p, q, k, w)$-balancing number with $R(p, q, k, w)$-balancer r implies that the quantity $4p(p+q)B^2 + 4p(p+q-2qk)B + 4pqk(k-1) + p^2 + 8pw$ as well as $4p(p+q)r^2 + 8pqkr + q^2 - 8qw$ are both perfect square. This motivates the next definitions.

Definition 2 Let B be an $R(p, q, k, w)$-balancing number with $R(p, q, k, w)$-balancer r. Define its $R(p, q, k, w)$-Lucas balancing number to be

$$C = \sqrt{4p(p+q)B^2 + 4p(p+q-2qk)B + 4pqk(k-1) + p^2 + 8pw}$$

and its $R(p, q, k, w)$-Lucas balancer $\hat{r}$ to be

$$\hat{r} = \sqrt{4p(p+q)r^2 + 8pqkr + q^2 - 8qw}.$$

We say (B, C) is an $R(p, q, k, w)$-balancing pair and $(r, \hat{r})$ its $R(p, q, k, w)$-balancer pair.

The integral pair (B, C) is an $R(p, q, k, w)$-balancing pair if and only if the following three conditions hold:

1. $B \geq k$;
2. $4p(p+q)B^2 + 4p(p+q-2qk)B + 4pqk(k-1) + p^2 + 8pw$ is a perfect square;
3. $C \equiv p \pmod{2p}$.

The third condition follows from (4). The second condition implies that the $R(p, q, k, w)$-balancing pair (B, C) is a solution to the Pell like-equation

$$y^2 = 4p(p+q)x^2 + 4p(p+q-2qk)x + 4pqk(k-1) + p^2 + 8pw. \tag{6}$$

Multiplying by $p(p+q)$ we see (6) can be expressed as

$$z^2 - p(p+q)y^2 = N(p, q, k, w) \tag{7}$$

where $z = 2p(p+q)x + p(p+q-2qk)$ and

$$N(p, q, k, w) = p^2q(p+q) - 4p^3qk^2 - 8p^2(p+q)w.$$

Equation (7) is useful for studying ratio balancing numbers and is referred to as the $R(p, q, k, w)$-companion equation. In particular, an integral solution (z, y) to the $R(p, q, k, w)$-companion equation corresponds to an $R(p, q, k, w)$-balancing pair (B, C) where

$$B = \frac{z - p(p+q-2qk)}{2p(p+q)}$$

and $C = y$ provided the following four conditions hold:

1. $z \geq p^2(2k+1) + pq$;
2. $y > 0$;
3. $y \equiv p \pmod{2p}$;
4. $z \equiv p(p+q-2qk) \pmod{2p(p+q)}$.

The first three conditions are analogues to those described above for integer pairs (B, C). The last condition is necessary for integral values of (z, y) to yield integral values of (B, C).

Similarly we note that the $R(p, q, k, w)$-balancer pair $(r, \hat{r})$ is a solution to the equation

$$y^2 = 4p(p+q)x^2 + 8pqkx + q^2 - 8qw.$$

Multiplying by $p(p+q)$ and substituting $z = 2p(p+q)x + 2pqk$ yields the $R(p, q, k, w)$-balancer companion equation

$$z^2 - p(p+q)y^2 = 4p^2q^2k^2 - pq^2(p+q) + 8pq(p+q)w.$$

We are also interested in the index of the triangular number appearing on the right hand side of (2) and make the following definition.

Definition 3 The counterbalancer m of an $R(p, q, k, w)$-balancing number B with $R(p, q, k, w)$-balancer r is defined to be $m = B + r$.

Several relationships between the quantities defined above are given in the next result. These follow quickly from applying the definitions to (4) and (5).

Proposition 1 *Suppose (B, C) is an $R(p, q, k, w)$-balancing pair with $(r, \hat{r})$ its associated $R(p, q, k, w)$-balancer pair and m its counterbalancer. Then*

(a) $r = \frac{-p(2B+1)+C}{2p}$;
(b) $\hat{r} = 2qB - 2pr - 2qk + q$;
(c) $\hat{r} = 2(p+q)B - C + p + (1-2k)q$;
(d) $m = \frac{C-p}{2p}$.

Example 1 The identity $7 \cdot T(3) + T(5) + 9 = T(11)$ shows that the number 5 is an $R(1, 7, 2, 9)$-balancing number with balancer 6 and corresponding $R(1, 7, 2, 9)$-Lucas balancing number 23. Since $2 \cdot T(14) + 3 \cdot T(15) = 3 \cdot T(19)$, the number 15 is an $R(3, 2, 1, 0)$-balancing number with balancer 4. Its corresponding $R(3, 2, 1, 0)$-Lucas balancing number is 117. Every positive integer B is an $R(1, 3, 1, 0)$-balancing number with balancer $B - 1$ since $3 \cdot T(B - 1) + T(B) = T(2B - 1)$. The corresponding $R(1, 3, 1, 0)$-Lucas balancing number is $4B - 1$. These last two examples are discussed further in Examples 7 and 6, respectively.

Example 2 If B is an $R(p, q, k, w)$-balancing number, multiplying (3) by any positive integer c shows B is also an $R(cp, cq, k, cw)$-balancing number.

Example 3 Ratio balancing numbers unify many variations of balancing numbers previously studied. The $R(1, 1, k, w)$-balancing numbers are the almost k-gap balancing numbers [1]. In particular, the $R(1, 1, 0, 0)$-, $R(1, 1, 1, 0)$-, and $R(1, 1, k, 0)$-balancing numbers are cobalancing [11], balancing numbers [5], and upper k-gap balancing numbers [2], respectively. The $R(1, 1, 1, -k^2)$-balancing numbers are the k-circular balancing numbers [10]. Lastly, the $R(1, 1, 1, 1)$- and $R(1, 1, 1, -1)$-balancing numbers are the almost balancing numbers of the first and second kind [9], respectively.

3 Counting Ratio Balancing Numbers

In this section, we establish that $R(p, q, k, w)$-balancing numbers, depending on the choice of parameter values, either do not exist, exist in a finite number, or exist in a finite number of infinite classes. The situation where a finite number of $R(p, q, k, w)$-balancing numbers exist is of particular interest; this case does not arise in other variations of balancing numbers previously studied.

From the discussion in Sect. 2, $R(p, q, k, w)$-balancing numbers can be derived from solutions to the $R(p, q, k, w)$-companion equation which satisfy four conditions. Recall that the $R(p, q, k, w)$-companion equation given in (7) is

$$z^2 - Dy^2 = N \tag{8}$$

where $D = p(p + q)$ and $N = N(p, q, k, w)$. From the theory of Pell equations [7], the existence of solutions to (8) depend on the values of D and N. If D is not a perfect square, then (8) either has no solutions or infinitely many solutions which appear in a finite number of infinite classes. The latter situation is explored further in Sect. 4. If D is a perfect square and $N \neq 0$, then (8) has finitely many solutions (possibly none). Since each $R(p, q, k, w)$-balancing pair corresponds to one of these finitely many solutions, we obtain the following.

Theorem 1 *Suppose $p(p + q)$ is a perfect square. If $N(p, q, k, w) \neq 0$, then there are finitely many (possibly none) $R(p, q, k, w)$-balancing numbers.*

Example 4 The numbers 1 and 3 are the only two $R(1, 24, 1, 0)$-balancing numbers. To see this, observe the $R(1, 24, 1, 0)$-companion equation is $z^2 - 25y^2 = 504$

whose solutions in the positive integers (z, y) are $(23, 1)$, $(27, 3)$, and $(127, 25)$. Only the latter two solutions satisfy the four conditions required to yield ratio balancing numbers.

When there are finitely many $R(p, q, k, w)$-balancing numbers, experimental evidence shows that there are most often three or fewer $R(p, q, k, w)$-balancing numbers. There are four $R(1, 2550408, 1, 0)$-balancing numbers, namely 1, 2, 200, and 318801. Moreover, four is the largest observed number of ratio balancing numbers so far for a fixed set of parameters in the finite case. Despite the perceived rarity of balancing numbers in the finite case, we make the following conjecture.

Conjecture 1 Let d be a positive integer. There exists values of p, q, k, and w with $p(p+q)$ a perfect square and $N(p, q, k, w) \neq 0$ which yield at least d $R(p, q, k, w)$-balancing numbers.

The search interval for ratio balancing numbers in the finite case can be made more efficient in some situations with the following theorem. This result provides an upper bound for B for a certain class of ratio balancing numbers. Observe that the condition that $p(p+q)$ is a perfect square is equivalent to p and $p+q$ each being square when $\gcd(p, q) = 1$.

Theorem 2 *Let $k, w \in \mathbb{Z}$ with $k \geq 0$. Suppose $p = a^2$ and $q = b^2 - a^2$ for some positive integers a and b such that $a < b$ and $\gcd(p, q) = 1$. If b does not divide $2k$, then B is a $R(p, q, k, w)$-balancing number only if $B \leq \max\{M_1, M_2\}$ where*

$$M_1 = \frac{4a^2(b^2 - a^2)k(k-1) + a^4 + 8a^2w - t^2}{4abt - 4a^2(2a^2k + (1-2k)b^2)},$$

$$M_2 = \frac{4a^2(b^2 - a^2)k(k-1) + a^4 + 8a^2w - (t+1)^2}{4ab(t+1) - 4a^2(2a^2k + (1-2k)b^2)},$$

and

$$t = \left\lfloor \frac{a(2a^2k + (1-2k)b^2)}{b} \right\rfloor. \tag{9}$$

Proof Recall that (B, C) is a $R(p, q, k, w)$-balancing pair only if (B, C) is a solution to (6) which after substitution becomes

$$y^2 = 4a^2b^2x^2 + 4a^2(2a^2k + (1-2k)b^2)x + 4a^2(b^2 - a^2)k(k-1) + a^4 + 8a^2w.$$

We determine a choice of t which depends on a, b, and k so that the quantity y^2 lies strictly between the consecutive squares $(2abx + t)^2$ and $(2abx + t + 1)^2$ for sufficiently large integers x, hence cannot be a square of an integer. Observe that the inequalities $(2abx + t)^2 < y^2 < (2abx + t + 1)^2$ reduce to

$$4abtx + t^2 < y_0 < 4ab(t+1)x + (t+1)^2$$

where

$$y_0 = 4a^2(2a^2k + (1 - 2k)b^2)x + 4a^2(b^2 - a^2)k(k - 1) + a^4 + 8a^2w.$$

We select t so that

$$4abt < 4a^2(2a^2k + (1 - 2k)b^2) < 4ab(t + 1).$$

From a geometric viewpoint, this choice guarantees that the *balance line*

$$y = 4a^2(2a^2k + (1 - 2k)b^2)x + 4a^2(b^2 - a^2)k(k - 1) + a^4 + 8a^2w$$

lies strictly between the *bounding lines* $y = 4abtx + t^2$ and $y = 4ab(t + 1)x + (t + 1)^2$ for sufficiently large x. In particular, the choice of t in (9) is sufficient under the given hypotheses unless $\frac{a(2a^2k+(1-2k)b^2)}{b}$ is an integer. This occurs exactly when $2a^3k \equiv 0 \pmod{b}$ or equivalently b divides $2k$ under the assumptions above. The upper bound on B follows from observing that the balance line lies strictly between the two bounding lines for x greater than the largest x-coordinate of the intersection points obtained from the bounding lines with the balance line. □

Remark 1 The argument made in the proof of Theorem 2 can be sharpened by considering divisibility conditions and the relative positioning of the balance and bounding lines at $x = 0$. We omit these details for convenience of the reader since the emphasis of the result is demonstrate a technique to establish an upper bound for B.

Example 5 The unique $R(4, 5, 1, 0)$-balancing number is 1. Using the notation of Theorem 2, we have $t = -1$ and balance line is $y = -16x + 16$. The two bounding lines are $y = 0$ and $y = -24x + 1$. From the intersection of the bounding lines with the balance line, we see that any $R(4, 5, 1, 0)$-balancing number B satisfies $B \le \max\{-15/8, 1\} = 1$. The statement follows since $B = k$ with $r = 0$ always satisfies (3) when $w = 0$.

Lastly we consider the degenerate case when $p(p + q)$ is a perfect square and $N(p, q, k, w) = 0$. If additionally $w = 0$, then $N = 0$ exactly when $q = (4k^2 - 1)p$. Consequently, this situation can be completely described combining the next example with comments given in Example 2. We remark that this case is more subtle for general w.

Example 6 Let $k \ge 1$. The $R(1, 4k^2 - 1, k, 0)$-balancing numbers consist of all integers $B \ge k$. To see this, observe that $z = 8k^2B - 8k^3 + 4k^2 + 2k$ and $y = 4kB - 4k^2 + 2k + 1$ are solutions to the $R(1, 4k^2 - 1, k, 0)$-companion equation $z^2 - 4k^2y^2 = 0$ for each integer $B \ge k$ and satisfy the four conditions described in Sect. 2. The corresponding identity in terms of triangular numbers is

$$(4k^2 - 1)T(B - k) + T(B) = T(2kB - 2k^2 + k).$$

4 Functions Generating Ratio Balancing Numbers and Related Results

When $p(p+q)$ is not a perfect square, the standard balancing number techniques can be used to generate balancing numbers from known balancing numbers. From Pell equation theory, integral solutions to (7), if they exist, occur in a finite number of cyclic classes. That is, if (z', y') is a solution corresponding to an $R(p, q, k, w)$-balancing number, then so is (z'', y'') where

$$z'' + y''\sqrt{p(p+q)} = (\alpha + \beta\sqrt{p(p+q)})^j(z' + y'\sqrt{p(p+q)})$$

or equivalently in matrix form

$$V^j : \begin{bmatrix} z'' \\ y'' \end{bmatrix} = \begin{bmatrix} \alpha & p(p+q)\beta \\ \beta & \alpha \end{bmatrix}^j \begin{bmatrix} z' \\ y' \end{bmatrix}. \tag{10}$$

Here $\alpha + \beta\sqrt{p(p+q)}$ is the fundamental solution to $z^2 - p(p+q)y^2 = 1$ and j is the minimal positive integer such that $y'' \equiv p \pmod{2p}$ and $z'' \equiv p(p+q-2qk)$ $\pmod{2p(p+q)}$. We refer to $j = j(p, q, k, w)$ as the *jump size* for $R(p, q, k, w)$-balancing numbers.

Almost balancing numbers, which include balancing numbers, always have a jump size of $j = 1$. Replacing triangular numbers in (1) with general figurate numbers give the polygonal-balancing numbers [4]. Depending on the choice of parameters, polygonal balancing numbers can have arbitrarily large jump sizes [3]. The next theorem shows that the jump sizes for ratio balancing numbers satisfies $j \leq 2$, striking a middle ground between the results for almost and polygonal balancing numbers.

Theorem 3 *Suppose $p(p+q)$ is not a perfect square and (z', y') is a solution to the $R(p, q, k, w)$-companion equation corresponding to an $R(p, q, k, w)$-balancing number. Let $\alpha + \beta\sqrt{p(p+q)}$ be the fundamental solution to $z^2 - p(p+q)y^2 = 1$. Then (z'', y'') is also a solution corresponding to an $R(p, q, k, w)$-balancing number where*

$$\begin{bmatrix} z'' \\ y'' \end{bmatrix} = \begin{bmatrix} \alpha & p(p+q)\beta \\ \beta & \alpha \end{bmatrix}^2 \begin{bmatrix} z' \\ y' \end{bmatrix}. \tag{11}$$

Hence, the jump size j is at most two.

Proof By assumption $z' \geq p^2(2k+1) + pq$, $y' > 0$, $y' \equiv p \pmod{2p}$, and $z' \equiv p(p+q-2qk) \pmod{2p(p+q)}$. For $j = 2$, we see that (11) becomes

$$\begin{bmatrix} z'' \\ y'' \end{bmatrix} = \begin{bmatrix} \alpha^2 + p(p+q)\beta^2 & 2p(p+q)\alpha\beta \\ 2\alpha\beta & \alpha^2 + p(p+q)\beta^2 \end{bmatrix} \begin{bmatrix} z' \\ y' \end{bmatrix}.$$

Clearly, $z'' \geq p^2(2k+1) + pq$ and $y'' > 0$. Since $\alpha^2 - p(p+q)\beta^2 = 1$ and $z' \equiv p(p+q) \pmod{2p}$, we see

$$y'' \equiv 2\alpha\beta z' + (2\alpha^2 - 1)y' \equiv p \pmod{2p}.$$

Again using $\alpha^2 - p(p+q)v\beta^2 = 1$, observe

$$\begin{aligned} z'' &\equiv (2\alpha^2 - 1)z' + 2p(p+q)\alpha\beta y' \pmod{2p(p+q)} \\ &\equiv -4\alpha^2 pqk - p(p+q-2qk) \pmod{2p(p+q)} \\ &\equiv -4(1 + p(p+q)\beta^2)pqk - p(p+q-2qk) \pmod{2p(p+q)} \\ &\equiv p(p+q-2qk) \pmod{2p(p+q)}. \end{aligned}$$

Thus (z'', y'') satisfies the four conditions given in Sect. 2. □

Ratio balancing numbers with $j = 1$ remain of particular interest and can be characterized as follows.

Theorem 4 *Suppose $p(p+q)$ is not a perfect square and (z', y') is a solution to the $R(p, q, k, w)$-companion equation corresponding to an $R(p, q, k, w)$-balancing number. Let $\alpha + \beta\sqrt{p(p+q)}$ be the fundamental solution to $z^2 - p(p+q)y^2 = 1$. Then (z'', y'') is also a solution corresponding to an $R(p, q, k, w)$-balancing number where*

$$\begin{bmatrix} z'' \\ y'' \end{bmatrix} = \begin{bmatrix} \alpha & p(p+q)\beta \\ \beta & \alpha \end{bmatrix}^j \begin{bmatrix} z' \\ y' \end{bmatrix} \tag{12}$$

with $j = 1$ if and only if the following conditions are satisfied:

1. $\alpha + \beta(p+q) \equiv 1 \pmod 2$;
2. $(\alpha - 1)(p+q-2qk) + p(p+q)\beta \equiv 0 \pmod{2(p+q)}$.

Proof It is immediate from the hypotheses that $z'' \geq p^2(2k+1) + pq$ and $y > 0$. For (z'', y'') to corresponding to an $R(p, q, k, w)$-balancing number with $j = 1$, we require $y'' \equiv p \pmod{2p}$ and $z'' \equiv p(p+q-2qk) \pmod{2p(p+q)}$. On the other hand, it follows from (12) and noting $z' \equiv p(p+q) \pmod{2p}$ that $y'' \equiv (\alpha + \beta(p+q))p \pmod{2p}$ and $z'' \equiv \alpha(p+q-2qk) + p(p+q)\beta y' \pmod{2p(p+q)}$. These observations reduce to the stated conditions. □

Observe that the relations $y_i = C_i$ and $z_i = 2p(p+q)B_i + p(p+q-2qk)$ can be expressed as

$$S : \begin{bmatrix} z_i \\ y_i \end{bmatrix} = \begin{bmatrix} 2p(p+q) & 0 \\ 0 & 1 \end{bmatrix} \begin{bmatrix} B_i \\ C_i \end{bmatrix} + \begin{bmatrix} p(p+q) - 2pqk \\ 0 \end{bmatrix} \tag{13}$$

and

$$S^{-1} : \begin{bmatrix} B_i \\ C_i \end{bmatrix} = \begin{bmatrix} \frac{1}{2p(p+q)} & 0 \\ 0 & 1 \end{bmatrix} \begin{bmatrix} z_i \\ y_i \end{bmatrix} + \begin{bmatrix} \frac{-p(p+q)+2pqk}{2p(p+q)} \\ 0 \end{bmatrix}. \tag{14}$$

Using (13) and (14), we can express (10) in terms of $R(p, q, k, w)$-balancing pairs as $J = S^{-1}V^jS$ where j is the jump size. Analogous expressions can be derived for generating $R(p,q,k,w)$-balancer pairs from known ones. The next two subsections utilize the jump size and the maps above to obtain a collection of recurrence relations of $R(p,q,k,w)$-balancing numbers and related sequences.

4.1 One Jump Case

When $j=1$, it is straightforward to see that $J=S^{-1}VS$ is given by

$$\begin{bmatrix} B_{i+1} \\ C_{i+1} \end{bmatrix} = \begin{bmatrix} \alpha & \frac{\beta}{2} \\ 2p(p+q)\beta & \alpha \end{bmatrix} \begin{bmatrix} B_i \\ C_i \end{bmatrix} + \begin{bmatrix} \frac{(p(p+q)-2pqk)(\alpha-1)}{2p(p+q)} \\ (p(p+q)-2pqk)\beta \end{bmatrix}$$

and $J^{-1}=S^{-1}V^{-1}S$ by

$$\begin{bmatrix} B_{i-1} \\ C_{i-1} \end{bmatrix} = \begin{bmatrix} \alpha & -\frac{\beta}{2} \\ -2p(p+q)\beta & \alpha \end{bmatrix} \begin{bmatrix} B_i \\ C_i \end{bmatrix} + \begin{bmatrix} \frac{(p(p+q)-2pqk)(\alpha-1)}{2p(p+q)} \\ -(p(p+q)-2pqk)\beta \end{bmatrix}.$$

Using the techniques used to prove analogous statements in [2, 5], we obtain the following results.

Proposition 2 *Let $((B_i,C_i))_{i\ge 0}$ be a class of $R(p,q,k,w)$-balancing pairs with jump size $j=1$, $((r_i,\hat{r}_i))_{i\ge 0}$ its $R(p,q,k,w)$-balancer pairs, and $(m_i)_{i\ge 0}$ its associated counterbalancers. Then*

(a) $B_{i+1}=2\alpha B_i - B_{i-1} + \frac{(p(p+q)-2pqk)(\alpha-1)}{p(p+q)}$;
(b) $C_{i+1}=2\alpha C_i - C_{i-1}$;
(c) $r_{i+1}=2\alpha r_i - r_{i-1} + \frac{2qk(\alpha-1)}{p+q}$;
(d) $\hat{r}_{i+1}=2\alpha\hat{r}_i - \hat{r}_{i-1}$;
(e) $m_{i+1}=2\alpha m_i - m_{i-1} + \alpha - 1$.

Moreover,

$$\lim_{i\to\infty}\frac{B_{i+1}}{B_i} = \lim_{i\to\infty}\frac{r_{i+1}}{r_i} = \lim_{i\to\infty}\frac{m_{i+1}}{m_i} = \alpha+\sqrt{\alpha^2-1}.$$

4.2 Two Jump Case

For $j=2$, the identity $\alpha^2+p(p+q)\beta^2=2\alpha^2-1$ is used to simply the presentation. In this case, $J=S^{-1}V^2S$ is given by

$$\begin{bmatrix} B_{i+1} \\ C_{i+1} \end{bmatrix} = \begin{bmatrix} 2\alpha^2-1 & \alpha\beta \\ 4p(p+q)\alpha\beta & 2\alpha^2-1 \end{bmatrix} \begin{bmatrix} B_i \\ C_i \end{bmatrix} + \begin{bmatrix} (p(p+q)-2pqk)\beta^2 \\ 2(p(p+q)-2pqk)\alpha\beta \end{bmatrix}$$

and $J^{-1}=S^{-1}V^{-2}S$ by

Table 1 Initial $R(3, 2, 1, 0)$-balancing numbers and associated sequences.

i	0_a	0_b	1_a	1_b	2_a	2_b	3_a	3_b
B	1	15	70	936	4345	58023	269326	3596496
C	9	117	543	7251	33657	449445	2086191	27858339
r	0	4	20	272	1264	16884	78372	1046560
$\hat{r}$	2	34	158	2110	9794	130786	607070	8106622
m	1	19	90	1208	5609	74907	347698	4643056

$$\begin{bmatrix} B_{i-1} \\ C_{i-1} \end{bmatrix} = \begin{bmatrix} 2\alpha^2 - 1 & -\alpha\beta \\ -4p(p+q)\alpha\beta & 2\alpha^2 - 1 \end{bmatrix} \begin{bmatrix} B_i \\ C_i \end{bmatrix} + \begin{bmatrix} (p(p+q) - 2pqk)\beta^2 \\ -2(p(p+q) - 2pqk)\alpha\beta \end{bmatrix}.$$

Proceeding similarly as in the $j = 1$ case, we obtain the following results.

Proposition 3 *Let $((B_i, C_i))_{i\geq 0}$ be a class of $R(p, q, k, w)$-balancing pairs with jump size $j = 2$, $((r_i, \hat{r}_i))_{i\geq 0}$ its $R(p, q, k, w)$-balancer pairs, and $(m_i)_{i\geq 0}$ its associated counterbalancers. Then*

(a) $B_{i+1} = 2(2\alpha^2 - 1)B_i - B_{i-1} + 2(p(p+q) - 2pqk)\beta^2$;
(b) $C_{i+1} = 2(2\alpha^2 - 1)C_i - C_{i-1}$;
(c) $r_{i+1} = 2(2\alpha^2 - 1)r_i - r_{i-1} + 4pqk\beta^2$;
(d) $\hat{r}_{i+1} = 2(2\alpha^2 - 1)\hat{r}_i - \hat{r}_{i-1}$;
(e) $m_{i+1} = 2(2\alpha^2 - 1)m_i - m_{i-1} + 2p(p+q)\beta^2$.

Moreover,

$$\lim_{i\to\infty} \frac{B_{i+1}}{B_i} = \lim_{i\to\infty} \frac{r_{i+1}}{r_i} = \lim_{i\to\infty} \frac{m_{i+1}}{m_i} = 2\alpha^2 - 1 + \sqrt{(2\alpha^2 - 1)^2 - 1}.$$

Example 7 There are two classes of $R(3, 2, 1, 0)$-balancing pairs whose initial terms are $(1, 9)$ and $(15, 117)$, respectively. The initial $R(3, 2, 1, 0)$-balancing numbers and associated sequences are given in Table 1. Here $\alpha = 4$, $\beta = 1$, and $j = 2$. From Proposition 3, we see that two of the recursive relations for each class are $B_{i+1} = 62B_i - B_{i-1} + 6$ and $r_{i+1} = 62r_i - r_{i-1} + 24$. None of these sequences or subsequences appear in The On-line Encyclopedia of Integer Sequences.

References

1. Bartz, J., Dearden, B., Iiams, J.: Almost gap balancing numbers. Integers 18(A79), 1–12 (2018).
2. Bartz, J., Dearden, B., Iiams, J.: Classes of gap balancing numbers. Rocky Mountain J. Math 51(2), 399–411 (2021).
3. Bartz, J., Dearden, B., Iiams, J.: Jump Sizes for Polygonal Balancing Numbers. Australas. J. Combin. (2022) Accepted and to appear.

4. Bartz, J., Dearden, B., Iiams, J.: Polygonal Balancing Numbers I. Integers 22(A54), 1–26 (2022).
5. Behera, A., Panda, G.K.: On the square roots of triangular numbers. Fibonacci Quart. 37, 98–105 (1999).
6. Deza, E., Deza, M.M.: Figurate Numbers. World Scientific, Singapore (2012).
7. Nagell, T.: Introduction to Number Theory. Wiley, New York (1951).
8. OEIS Foundation Inc., The On-Line Encyclopedia of Integer Sequences. http://oeis.org(2020)
9. Panda, G.K., Panda, A.K.: Almost balancing numbers. J. Indian Math. Soc. (N. S.) **82**, 147–156 (2015)
10. Panda, G.K., Panda, A.K.: Circular balancing numbers. Fibonacci Quart. 55, 309–314 (2017).
11. Panda, G.K., Ray, P.K.: Cobalancing numbers and cobalancers. Int. J. Math. Math. Sci. 8, 1189–1200 (2005).
12. Panda, G.K., Rout, S.S.: Gap balancing numbers. Fibonacci Quart. 51, 239–249 (2013).
13. Rout, S.S., Panda, G.K.: *k*-Gap balancing numbers. Period. Math. Hungar. 70, 109–121 (2015).

An Unexpected Digit Permutation from Multiplying in Any Number Base

Bohan Qu and Stephen J. Curran

Abstract We show that the number $\beta = (b^{b-1} - 1)/(b - 1)^2$, where $b \geq 3$, has several interesting multiplicative properties. In the base b number system, we have $\beta = (123 \cdots (b-4)(b-3)(b-1))_b$. We show that the digits of the number $K\beta$, for integers K such that $1 \leq K \leq (b-1)^2$, as a number in the base b number system can be generated from an arithmetic sequence reduced modulo $b - 1$ with an appropriate adjustment. The proof of this result involves an interplay between multiplication of K with β in the base b number system and the formation of an arithmetic sequence associated with the digits of K expressed as a number in the base $b - 1$ number system. We pose several questions related to this result as well.

Keywords Radix representation · Multiplicative properties · Multiplicative structure

1 Introduction

The number 12, 345, 679, whose digits are generated from the sequence of integers from 1 to 9 with the digit 8 omitted, has several interesting multiplicative properties [1–3]. These properties are a special case of the multiplicative properties of the number

$$\beta = \frac{b^{b-1} - 1}{(b-1)^2}$$

expressed as a number in the base b number system. When we represent β in the base b number system, the digits of β are the sequence of integers from 1 to $b - 1$ with

B. Qu
Columbia University, New York, NY 10027, USA
e-mail: b.qu@columbia.edu

S. J. Curran (✉)
University of Pittsburgh at Johnstown, Johnstown, PA 15904, USA
e-mail: sjcurran@pitt.edu

F. Hoffman (ed.), *Combinatorics, Graph Theory and Computing*,
Springer Proceedings in Mathematics & Statistics 388,
https://doi.org/10.1007/978-3-031-05375-7_2

the digit $b - 2$ omitted. For example, in the base $b = 16$ number system, we have $\beta = 12,345,678,9AB,CDF_{16}$ where A through F represent the digits 10 through 15, respectively. One surprising property involving the product $K\beta$ in the base b number system, where K is an integer satisfying $1 \leq K \leq (b-1)^2$, is the following result.

Theorem 1 *Let $b \geq 3$ be an integer, let K be an integer such that $1 \leq K \leq (b-1)^2$, and let K and $b - 1$ be relatively prime. Let d be the unique integer such that $0 < d < b$ and $K + d$ is divisible by $b - 1$. Then the digits of the product $K\beta$ expressed in the base b number system includes each digit $0, 1, \ldots, b - 1$ exactly once, except the digit d which does not appear as a digit in $K\beta$.*

The proof of Theorem 1 will be given at the end of this paper. As an example of Theorem 1, consider $b = 10$, $K = 41$ and $d = 4$. We observe that $K = 41$ and $b - 1 = 9$ are relatively prime, and $K + d = 45$ is divisible by $b - 1 = 9$. Then $K\beta = 41 \times 12,345,679 = 506,172,839$ contains each of the digits 0,1,...,9 exactly once, except the digit $d = 4$ which never appears. As another example, let $b = 16$, $K = 143$ and $d = 7$. We observe that $K = 143$ and $b - 1 = 15$ are relatively prime, and $K + d = 150$ is divisible by $b - 1 = 15$. Then $K\beta = 8F_{16} \times 12,345,678,9AB,CDF_{16} = A2B,3C4,D5E,6F8,091_{16}$ contains each digit $0, 1, \ldots, F$ exactly once, except the digit $d = 7$ which never appears.

In fact, one can generate the digits of $K\beta$ by calculating the terms of an arithmetic sequence reduced modulo $b - 1$ together with an appropriate adjustment. Let $K = (b-1)j + k$, where j and k are integers such that $0 \leq j \leq b - 2$ and $1 \leq k \leq b - 1$. We calculate the digits of $K\beta$ in the base b number system by reducing the sequence of integers $\{ki + j : i = 0, 1, \ldots, b - 2\}$ modulo $b - 1$ and then adding 1 to those values that are greater than or equal to $b - k - 1$. We first introduce some notation in order to state the main theorem of this paper. The proof of the main theorem involves an interplay between the multiplication of K with β in the base b number system and the terms in the sequence of integers $\{ki + j : i = 0, 1, \ldots, b - 2\}$ after they are reduced modulo $b - 1$. We discuss several questions related to this result at the end of this paper.

2 Main Theorem

We begin by introducing the notation needed to state our main theorem.

Definition 1 (1) Let b be an integer such that $b \geq 3$. Let K be an integer such that $1 \leq K \leq (b-1)^2$. Let j and k be the unique integers such that $0 \leq j \leq b - 2$, $1 \leq k \leq b - 1$ and $K = (b-1)j + k$. Let

$$\beta = \frac{b^{b-1} - 1}{(b-1)^2} = \big(1\,2\,3 \ldots (b-4)(b-3)(b-1)\big)_b.$$

(2) For all integers i such that $0 \le i \le b-2$, let $a_{i,j,k}$ be the unique integers such that

$$K\beta = \big((b-1)j+k\big)\beta = \sum_{i=0}^{b-2} a_{i,j,k}\, b^{b-2-i}$$

and $0 \le a_{i,j,k} < b$, for all i such that $0 \le i \le b-2$.
(3) Let $q_{i,j,k}$ and $r_{i,j,k}$ be the unique integers such that $ki + j = (b-1)q_{i,j,k} + r_{i,j,k}$ and $0 \le r_{i,j,k} < b-1$.
(4) Let $q'_{i,j,k}$ and $r'_{i,j,k}$ be the unique integers such that $ki + j = bq'_{i,j,k} + r'_{i,j,k}$ and $0 \le r'_{i,j,k} < b$.
(5) We define the integers $c_{i,j,k}$ as follows:

$$c_{i,j,k} = \begin{cases} r_{i,j,k}, & \text{if } r_{i,j,k} < b-k-1, \text{ and} \\ r_{i,j,k}+1, & \text{if } r_{i,j,k} \ge b-k-1. \end{cases}$$

(6) We define the integers $\epsilon_{i,j,k}$ by letting $\epsilon_{i,j,k} = q_{i,j,k} - q'_{i,j,k}$. Then

$$\epsilon_{i,j,k} = \begin{cases} 0, \text{ if } q_{i,j,k} = q'_{i,j,k}, \text{ and} \\ 1, \text{ if } q_{i,j,k} = q'_{i,j,k} + 1. \end{cases}$$

The significance of the value of $\epsilon_{i,j,k}$ is that it determines when there is a carry 1 in the product $K\beta$ from the b^{b-2-i}'s digit to the $b^{b-2-(i-1)}$'s digit when the product is carried out in the base b number system. When there is a carry 1 from the b^{b-2-i}'s digit to the $b^{b-2-(i-1)}$'s digit, $\epsilon_{i,j,k}$ represents that carry 1. We state the main theorem of this paper.

Theorem 2 *Let b and K be integers such that $b \ge 3$ and $1 \le K \le (b-1)^2$. For all integers i such that $0 \le i \le b-2$, let j, k, $a_{i,j,k}$, $r_{i,j,k}$ and $c_{i,j,k}$ be the integers defined in Definition 1. Then, for all $0 \le i \le b-2$, $a_{i,j,k} = c_{i,j,k}$; and thus*

$$K\beta = \sum_{i=0}^{b-2} c_{i,j,k}\, b^{b-2-i}.$$

I.e., $K\beta = (c_{0,j,k}\, c_{1,j,k} \dots c_{b-2,j,k})_b$ is the representation of the integer $K\beta$ in the base b number system.

We illustrate Theorem 2 with an example in the base 10 number system. We let $K = 41$ and $b = 10$. Then, by Definition 1, we have $K = 41 = (b-1)j + k = 9 \cdot 4 + 5$ where $j = 4$ and $k = 5$. We calculate the remainders $r_{i,4,5}$ of the integers $ki + j = 5i + 4$ upon division by 9 for $i = 0, 1, \dots, 8$ I.e., $r_{i,4,5} \equiv 5i + 4 \pmod 9$ for $i = 0, 1, \dots, 8$. See column 3 of Table 1. For those remainders $r_{i,4,5} < b - k - 1 = 4$, we define $c_{i,4,5} = r_{i,4,5}$. Also, for the remainders $r_{i,4,5} \ge b - k - 1 = 4$, we define $c_{i,3,4} = r_{i,4,5} + 1$. See column 5 of Table 1.

We prove Theorem 2 for the special case when $(b-1)|K$ in Proposition 8. Then we prove Theorem 2 for the general case when $(b-1) \nmid K$ in Theorem 3.

Table 1 The values of the digits in the product $K\beta = 41 \times 12,345,679 = 506,172,839$

i	$5i+4$	$r_{i,4,5} \equiv 5i+4 \pmod 9$	Is $r_{i,4,5} \geq 4$?	$c_{i,4,5}$
0	4	4	True	5
1	9	0	False	0
2	14	5	True	6
3	19	1	False	1
4	24	6	True	7
5	29	2	False	2
6	34	7	True	8
7	39	3	False	3
8	44	8	True	9

3 Other Interesting Results

When the integer k satisfies $1 \leq k \leq b-1$, we view $k\beta$ as a number with $b-1$ digits in the base b number system that begins with the digit 0. This will allow us to generate all products $K\beta$ with $1 \leq K \leq (b-1)^2$ from the products $k\beta$ with $1 \leq k \leq (b-1)/2$. For example, when $b = 10$, we use Theorem 2 to calculate $k\beta$ for $1 \leq k \leq 9/2$. Thus

$$1\beta = 012,345,679; \qquad 2\beta = 024,691,358;$$
$$3\beta = 037,037,037; \text{ and } 4\beta = 049,382,716.$$

Proposition 1 *Let k be an integer such that $1 \leq k \leq (b-1)/2$ and $\gcd(b-1,k) = 1$. Let $K = (b-1)^2 - k = (b-1)(b-2) + (b-k-1)$. Then $a_{i,b-2,b-k-1} = (b-1) - a_{i,0,k}$ for all integers i such that $0 \leq i \leq b-2$. I.e., $K\beta$ is the $(b-1)$'s complement of $k\beta$ in the base b number system.*

From Proposition 1, we recognize $(81-k)\beta$ as the 9's complement of $k\beta$ for $k = 1, 2,$ and 4. Thus

$$80\beta = 81\beta - 1\beta = 999,999,999 - 012,345,679 = 987,654,320;$$
$$79\beta = 81\beta - 2\beta = 999,999,999 - 024,691,358 = 975,308,641 \text{ and}$$
$$77\beta = 81\beta - 4\beta = 999,999,999 - 049,382,716 = 950,617,283.$$

Proposition 2 *Let K be an integer such that $1 \leq K \leq (b-1)^2$ and $\gcd(b-1,K) = 1$. Let j and k be the unique integers such that $K = (b-1)j + k$, $0 \leq j \leq b-2$ and $1 \leq k \leq b-1$. Let j_3 be the unique integer such that $kj_3 \equiv j \pmod{b-1}$ and $1 \leq j_3 \leq b-2$. Then for all integers i such that $0 \leq i \leq b-2$, we have $a_{i,j,k} = a_{i+j_3,0,k}$ where the indices i and $i+j_3$ are taken modulo $b-1$.*

As a consequence of Proposition 2, in the decimal number system, the digits in the products 5β, 7β and 8β are cyclic permutations of the digits in the products 77β, 79β and 80β, respectively, that begin with the digit 0.

$$\begin{aligned}\text{From } 77\beta &= 950, 617, 283, \text{ we obtain } 5\beta = 061, 728, 395.\\ \text{From } 79\beta &= 975, 308, 641, \text{ we obtain } 7\beta = 086, 419, 753.\\ \text{From } 80\beta &= 987, 654, 320, \text{ we obtain } 8\beta = 098, 765, 432.\end{aligned}$$

Furthermore, as a consequence of Proposition 2, we can now generate all products $K\beta$ where $1 \le K \le 81$ and $\gcd(K, 9) = 1$. For example, let $K = (b-1)j + k = 9 \cdot 7 + 5 = 68$ where $j = 7$ and $k = 5$. Then the digits of 68β are a cyclic permutation of the digits of $5\beta = 061, 728, 395$ that begin with the digit $c_{0,7,5} = r_{0,7,5} + 1 = j + 1 = 8$ since $r_{0,7,5} = j = 7 \ge b - k - 1 = 5$. Thus $68\beta = 839, 506, 172$.

Proposition 3 *Let k be an integer such that $1 \le k \le (b-1)/2$ and $d = \gcd(b-1, k) > 1$. Let $K = (b-1)(b-d) - k = (b-1)(b-d-1) + (b-k-1)$. Then for all integers i such that $0 \le i \le b-2$, we have $a_{i,b-d-1,b-k-1} = (b-d) - a_{i,0,k}$. I.e., $K\beta$ is the $(b-d)$'s complement of $k\beta$ in the base b number system.*

As an application of Proposition 3, we consider $60\beta = (7 \cdot 9)\beta - 3\beta = 777, 777, 777 - 037, 037, 037 = 740, 740, 740$.

Proposition 4 *Let K be an integer such that $1 \le K \le (b-1)^2$, $d = \gcd(b-1, K) > 1$ and $(b-1) \nmid K$. Let j and k be the unique integers such that $K = (b-1)j + k$, $0 \le j \le b-2$ and $1 \le k \le b-1$. Let $k_1 = k/d$ and $\ell = (b-1)/d$. Let j_1 and j_2 be the unique integers such that $j = dj_2 + j_1$, $0 \le j_1 < d$ and $0 \le j_2 < \ell$. Let j_3 be the unique integer such that $k_1 j_3 \equiv j_2 \pmod{\ell}$ and $0 \le j_3 < \ell$, Then for all integers i such that $0 \le i \le b-2$, we have $a_{i,j,k} = a_{i+j_3,j_1,k}$ where the indices i and $i + j_3$ are taken modulo $b-1$.*

As a consequence of Proposition 4, the digits of 6β are a cyclic permutation of the digits of $60\beta = (9 \cdot 6 + 6)\beta = 740, 740, 740$ that begins with the digit 0. Thus $6\beta = 074, 074, 074$.

Proposition 5 *Let k be an integer such that $1 \le k < b-1$ and $d = \gcd(b-1, k) > 1$. Let $k_1 = k/d$. Then, for all integers j_1 such that $0 \le j_1 < d$, we have $a_{i,j_1,k} = a_{i,0,k} + j_1$. I.e., we add j_1 to each digit of $k\beta = (a_{0,0,k}a_{1,0,k} \ldots a_{b-2,0,k})_b$ in the base b number system to generate the digits of $\big((b-1)j_1 + k\big)\beta = (a_{0,j_1,k}a_{1,j_1,k} \ldots a_{b-2,j_1,k})_b$.*

As a consequence of Proposition 5, we have

$$\begin{aligned}12\beta &= 9\beta + 3\beta = 111, 111, 111 + 037, 037, 037 = 148, 148, 148;\\ 21\beta &= 18\beta + 3\beta = 222, 222, 222 + 037, 037, 037 = 259, 259, 259;\\ 15\beta &= 9\beta + 6\beta = 111, 111, 111 + 074, 074, 074 = 185, 185, 185 \text{ and}\\ 24\beta &= 18\beta + 6\beta = 222, 222, 222 + 074, 074, 074 = 296, 296, 296.\end{aligned}$$

Furthermore, as a consequence of Proposition 4, we can generate the digits in all of the products $K\beta$ for integers K such that $1 \leq K \leq 81$, $3|K$ and $9 \not| K$. For example, let $K = 48 = 9j + k$ where $j = 5$ and $k = 3$. We write $j = 5 = 3j_2 + j_1 = 3 \cdot 1 + 2$ where $j_1 = 2$ and $j_2 = 1$. Then the digits of 48β is a cyclic permutation on the digits of $\big((b-1)j_1 + k\big)\beta = 21\beta = 259, 259, 259$ that begins with the digit $a_{0,5,3} = r_{0,5,3} = j = 5$ since $r_{0,5,3} = 5 < b - k - 1 = 6$. Thus $48\beta = 592, 592, 592$.

4 Demonstration of Results

We begin by showing that the digits of β in the base b number system from left to right are the terms in the sequence of integers $0, 1, \ldots, b-1$ with the digit $b-2$ omitted from the list.

Proposition 6 *Let b be an integer such that $b \geq 3$. Then*

$$\beta = \sum_{i=1}^{b-2} ib^{b-2-i} \quad + 1.$$

I.e., for all integers i such that $0 \leq i \leq b-3$, we have $a_{i,0,1} = i$. In addition, we have $a_{b-2,0,1} = b-1$.

Proof We apply the summation formula

$$\sum_{i=1}^{n} i\, x^i = \frac{nx^{n+2} - (n+1)x^{n+1} + x}{(x-1)^2}$$

with $x = b^{-1}$ and $n = b-2$ to obtain

$$\begin{aligned} b^{b-2}\sum_{i=1}^{b-2} ib^{-i} &= b^{b-2}\left(\frac{(b-2)b^{-b} - (b-1)b^{-b+1} + b^{-1}}{(b^{-1}-1)^2}\right) \\ &= \frac{b^{b-1} - 1 - (b-1)^2}{(b-1)^2} = \beta - 1 \end{aligned}$$

Hence,

$$\beta = \sum_{i=0}^{b-2} ib^{b-2-i} \quad + 1.$$

□

In the following proposition we determine a summation formula for the product $K\beta$ where K is an integer between 1 and $(b-1)^2$.

Proposition 7 *Let b be an integer such that $b \geq 3$, and let K be an integer such that $1 \leq K \leq (b-1)^2$. Let j and k be the unique integers such that $K = (b-1)j + k$, $0 \leq j \leq b-2$ and $1 \leq k \leq b-1$. Then*

$$K\beta = \sum_{i=0}^{b-2} (ki + j)b^{b-2-i} \quad + k.$$

Proof By Proposition 6, we have

$$\beta = \sum_{i=0}^{b-2} ib^{b-2-i} \quad + 1.$$

Then

$$\begin{aligned} K\beta &= \sum_{i=1}^{b-2} \big(bj + (k-j)\big)ib^{b-2-i} \quad + K \\ &= \sum_{i=0}^{b-3} j(i+1)b^{b-2-i} + \sum_{i=1}^{b-2} (k-j)ib^{b-2-i} \quad + K \\ &= jb^{b-2} + \sum_{i=1}^{b-3} (ki+j)b^{b-2-i} \quad + \big(k(b-2) + j\big) \quad + k \\ &= \sum_{i=0}^{b-2} (ki+j)b^{b-2-i} \quad + k. \end{aligned}$$

□

We replace $ki + j$ in Proposition 7 with the quotients $q'_{i,j,k}$ and remainders $r'_{i,j,k}$ of $ki + j$ upon division by b in Lemma 1. See Definition 1.4.

Lemma 1 *Let K be an integer such that $1 \leq K \leq (b-1)^2$. Let j and k be the unique integers such that $K = (b-1)j + k$, $0 \leq j \leq b-2$ and $1 \leq k \leq b-1$. Let $q'_{i,j,k}$ and $r'_{i,j,k}$ be the unique integers such that $ki + j = bq'_{i,j,k} + r'_{i,j,k}$ and $0 \leq r'_{i,j,k} < b$. Then*

$$K\beta = \sum_{i=0}^{b-3} (q'_{i+1,j,k} + r'_{i,j,k})b^{b-2-i} \quad + (r'_{b-2,j,k} + k).$$

Proof By Proposition 7, we have

$$K\beta = \sum_{i=0}^{b-2} (ki + j)b^{b-2-i} \quad + k.$$

By Definition 1.4, we have

$$
\begin{aligned}
K\beta &= \sum_{i=0}^{b-2}(bq'_{i,j,k} + r'_{i,j,k})b^{b-2-i} \quad + k \\
&= q'_{0,j,k}b^{b-1} + \sum_{i=0}^{b-3} q'_{i+1,j,k}b^{b-2-i} + \sum_{i=0}^{b-3} r'_{i,j,k}b^{b-2-i} \quad + r'_{b-2,j,k} + k \\
&= q'_{0,j,k}b^{b-1} + \sum_{i=0}^{b-3}(q'_{i+1,j,k} + r'_{i,j,k})b^{b-2-i} \quad + (r'_{b-2,j,k} + k).
\end{aligned}
$$

Observe that $j = k \cdot 0 + j = bq'_{0,jk} + r'_{0,jk}$ where $0 \le r'_{0,j,k} < b$. Since $0 \le j \le b-2$, we have $q'_{0,j,k} = 0$ and $r'_{0,j,k} = j$. Hence

$$
K\beta = \sum_{i=0}^{b-3}(q'_{i+1,j,k} + r'_{i,j,k})b^{b-2-i} \quad + (r'_{b-2,j,k} + k).
$$

□

In Proposition 8, we prove the special case of Theorem 2 when $b-1$ divides K.

Proposition 8 *Let K be an integer such that $1 \le K \le (b-1)^2$ and $(b-1)|K$. Let $K_1 = K/(b-1)$. Then the unique integers j and k such that $K = (b-1)j + k$, $0 \le j \le b-2$ and $1 \le k \le b-1$ are $j = K_1 - 1$ and $k = b-1$. Furthermore, for all integers i such that $0 \le i \le b-2$, we have $a_{i,K_1-1,b-1} = K_1$. I.e., $K\beta = (K_1K_1 \dots K_1)_b$ in the base b number system.*

Proof Note that $ki + j = (b-1)i + (K_1 - 1)$. Thus the unique integers $q_{i,j,k}$ and $r_{i,j,k}$ such that $ki + j = (b-1)q_{i,j,k} + r_{i,j,k}$, and $0 \le r_{i,j,k} < b-1$ are $q_{i,j,k} = q_{i,K_1-1,b-1} = i$ and $r_{i,j,k} = r_{i,K_1-1,b-1} = K_1 - 1$. Since $r_{i,K_1-1,b-1} = K_1 - 1 \ge b - k - 1 = 0$, we have $c_{i,K_1-1,b-1} = r_{i,K_1-1,b-1} + 1 = K_1$ for all integers i such that $0 \le i \le b-2$. Next, we observe that

$$
(b-1)\beta = \frac{b^{b-1}-1}{b-1} = \sum_{i=0}^{b-2} b^{b-2-i} = (11\dots1)_b.
$$

Thus

$$
\begin{aligned}
K\beta = K_1(b-1)\beta = K_1\left(\frac{b^{b-1}-1}{b-1}\right) &= \sum_{i=0}^{b-2} K_1 b^{b-2-i} \\
&= (K_1K_1\dots K_1)_b = \sum_{i=0}^{b-2} c_{i,K_1-1,b-1}b^{b-2-i}.
\end{aligned}
$$

□

Lemma 2 *Let K be an integer such that $1 \leq K \leq (b-1)^2$ and $(b-1) \nmid K$. Let j and k be the unique integers such that $K = (b-1)j + k$, $0 \leq j \leq b-2$ and $1 \leq k \leq b-2$. Let $q_{i,j,k}$ and $r_{i,j,k}$ be the unique integers such that $ki + j = (b-1)q_{i,j,k} + r_{i,j,k}$ and $0 \leq r_{i,j,k} \leq b-2$.*

(1) If $r_{i,j,k} < b-k-1$, for some integer i such that $0 \leq i \leq b-3$, then $q_{i+1,j,k} = q_{i,j,k}$ and $r_{i+1,j,k} = r_{i,j,k} + k$.
(2) If $r_{i,j,k} \geq b-k-1$, for some integer i such that $0 \leq i \leq b-3$, then $q_{i+1,j,k} = q_{i,j,k} + 1$ and $r_{i+1,j,k} = r_{i,j,k} + k - b + 1$.

Proof Observe that $(b-1)(q_{i+1,j,k} - q_{i,j,k}) + (r_{i+1,j,k} - r_{i,j,k}) = \big(k(i+1)+j\big) - \big(ki+j\big) = k$. We consider the two cases when either $r_{i,j,k} < b-k-1$ or $r_{i,j,k} \geq b-k-1$.

Case 1 Suppose $r_{i,j,k} < b-k-1$. Since $(b-1)(q_{i+1,j,k} - q_{i,j,k}) = r_{i,j,k} - r_{i+1,j,k} + k$ where $0 \leq r_{i,j,k} < b-k-1$, $0 \leq r_{i+1,j,k} < b-1$ and $1 \leq k \leq b-2$, we have $r_{i,j,k} - r_{i+1,j,k} + k \equiv 0 \pmod{b-1}$ where $-(b-1) < r_{i,j,k} - r_{i+1,j,k} + k < b-1$. Then $(b-1)(q_{i+1,j,k} - q_{i,j,k}) = r_{i,j,k} - r_{i+1,j,k} + k = 0$. Hence, $q_{i+1,j,k} = q_{i,j,k}$ and $r_{i+1,j,k} = r_{i,j,k} + k$.

Case 2 Suppose $r_{i,j,k} \geq b-k-1$. Since $(b-1)(q_{i+1,j,k} - q_{i,j,k}) = r_{i,j,k} - r_{i+1,j,k} + k$ where $b-k-1 \leq r_{i,j,k} < b-1$, $0 \leq r_{i+1,j,k} < b-1$ and $0 < k < b-1$, we have $r_{i,j,k} - r_{i+1,j,k} + k \equiv 0 \pmod{b-1}$ where $0 < r_{i,j,k} - r_{i+1,j,k} + k < 2(b-1)$. Then $(b-1)(q_{i+1,j,k} - q_{i,j,k}) = r_{i,j,k} - r_{i+1,j,k} + k = b-1$. Hence, $q_{i+1,j,k} = q_{i,j,k} + 1$ and $r_{i+1,j,k} = r_{i,j,k} + k - b + 1$. □

Lemma 3 *Let i, j and k be integers such that $0 \leq j \leq b-2$ and $1 \leq k \leq b-2$. For all $1 \leq i \leq b-2$, let $q_{i,j,k}$ and $r_{i,j,k}$ be the unique integers such that $ki + j = (b-1)q_{i,j,k} + r_{i,j,k}$ and $0 \leq r_{i,j,k} < b-1$. For all $1 \leq i \leq b-2$, let $q'_{i,j,k}$ and $r'_{i,j,k}$ be the unique integers such that $ki + j = bq'_{i,j,k} + r'_{i,j,k}$ and $0 \leq r'_{i,j,k} < b$. Then either $q_{i,j,k} = q'_{i,j,k}$ or $q_{i,j,k} = q'_{i,j,k} + 1$.*

Proof From Definitions 1.3 and 1.4, we have $bq'_{i,j,k} + r'_{i,j,k} = ki + j = (b-1)q_{i,j,k} + r_{i,j,k}$. Thus $(b-1)(q_{i,j,k} - q'_{i,j,k}) = q'_{i,j,k} + r'_{i,j,k} - r_{i,j,k}$ where $0 \leq r'_{i,j,k} \leq b-1$ and $0 \leq r_{i,j,k} \leq b-2$. Observe that $bq'_{i,j,k} \leq bq'_{i,j,k} + r'_{i,j,k} = ik + j \leq (b-2)^2 + (b-2) = b^2 - 3b + 2$. Since $b \geq 3$, we have $q'_{i,j,k} \leq b - 3 + 2/b < b-2$. Thus $0 \leq q'_{i,j,k} \leq b-3$. Hence, $q'_{i,j,k} + r'_{i,j,k} - r_{i,j,k} \equiv 0 \pmod{b-1}$ and $-(b-1) < q'_{i,j,k} + r'_{i,j,k} - r_{i,j,k} < 2(b-1)$. Therefore, either $(b-1)(q_{i,j,k} - q'_{i,j,k}) = q'_{i,j,k} + r'_{i,j,k} - r_{i,j,k} = 0$ or $(b-1)(q_{i,j,k} - q'_{i,j,k}) = q'_{i,j,k} + r'_{i,j,k} - r_{i,j,k} = b-1$. Thus, either $q_{i,j,k} = q'_{i,j,k}$ and $q'_{i,j,k} + r'_{i,j,k} = r_{i,j,k}$, or $q_{i,j,k} = q'_{i,j,k} + 1$ and $q'_{i,j,k} + r'_{i,j,k} = r_{i,j,k} + b - 1$. □

Lemma 4 *Let b, β, i, j, k, K, $q_{i,j,k}$, $r_{i,j,k}$, $q'_{i,j,k}$, $r'_{i,j,k}$, $a_{i,j,k}$, $c_{i,j,k}$ and $\epsilon_{i,j,k}$ be the integers defined in Definition 1. Furthermore, suppose that $1 \leq k \leq b-2$ so that $(b-1) \nmid K$. Then $a_{b-2,j,k} = c_{b-2,j,k}$ and*

$$K\beta = \sum_{i=0}^{b-4} (q'_{i+1,j,k} + r'_{i,j,k}) b^{b-2-i} \quad + (q'_{b-2,j,k} + r'_{b-3,j,k} + \epsilon_{b-2,j,k}) b + c_{b-2,j,k}.$$

Furthermore, we have $\epsilon_{b-2,j,k} = q_{b-2,j,k} - q'_{b-2,j,k}$.

Proof From Definition 1.3, $k(b-2)+j = (b-1)q_{b-2,j,k} + r_{b-2,j,k}$ where $0 \le r_{b-2,j,k} < b-1$. Thus $(b-1)(k - q_{b-2,j,k}) = r_{b-2,j,k} + k - j$ where $0 \le j < b-1$, $1 \le k < b-1$ and $0 \le r_{b-2,j,k} < b-1$. Hence, $r_{b-2,j,k} + k - j \equiv 0 \pmod{b-1}$ where $-(b-1) < r_{b-2,j,k} + k - j < 2(b-1)$. Thus, either $(b-1)(k - q_{b-2,j,k}) = r_{b-2,j,k} + k - j = 0$ or $(b-1)(k - q_{b-2,j,k}) = r_{b-2,j,k} + k - j = b-1$. Therefore, either $q_{b-2,j,k} = k$ and $r_{b-2,j,k} = j-k$, or $q_{b-2,j,k} = k-1$ and $r_{b-2,j,k} = j-k+b-1$. First, suppose $q_{b-2,j,k} = k$. Since $r_{b-2,j,k} = j-k$ and $j < b-1$, we have $r_{b-2,j,k} < b-k-1$. Next, suppose $q_{b-2,j,k} = k-1$. Since $r_{b-2,j,k} = j-k+b-1$ and $j \ge 0$, we have $r_{b-2,j,k} \ge b-k-1$.

Therefore, if $q_{b-2,j,k} = k$, then $r_{b-2,j,k} < b-k-1$. Also, if $q_{b-2,j,k} = k-1$, then $r_{b-2,j,k} \ge b-k-1$. Because the only two possible values for $q_{b-2,j,k}$ are k or $k-1$, these implications are equivalences. Therefore, $q_{b-2,j,k} = k$ is equivalent to $r_{b-2,j,k} < b-k-1$. Similarly, $q_{b-2,j,k} = k-1$ is equivalent to $r_{b-2,j,k} \ge b-k-1$.

By Lemma 3, either $q_{b-2,j,k} = q'_{b-2,j,k}$ or $q_{b-2,j,k} = q'_{b-2,j,k} + 1$. We will consider the four cases depending on which of the following two conditions are satisfied: Either $q_{b-2,j,k} = q'_{b-2,j,k}$ or $q_{b-2,j,k} = q'_{b-2,j,k} + 1$, and either $r_{b-2,j,k} < b-k-1$ or $r_{b-2,j,k} \ge b-k-1$. We will deal with each case separately.

In each case, we begin by observing that from Lemma 1 we have

$$K\beta = \sum_{i=0}^{b-3} (q'_{i+1,j,k} + r'_{i,j,k}) b^{b-2-i} \quad + (r'_{b-2,j,k} + k). \tag{1}$$

Then, in each case, we show that we have

$$\begin{aligned} K\beta = \sum_{i=0}^{b-4} (q'_{i+1,j,k} + r'_{i,j,k}) b^{b-2-i} \quad &+ (q'_{b-2,j,k} + r'_{b-3,j,k} + \epsilon_{b-2,j,k}) b \\ &+ c_{b-2,j,k}. \end{aligned} \tag{2}$$

Case 3 We assume $q_{b-2,j,k} = q'_{b-2,j,k}$ and $r_{b-2,j,k} < b-k-1$. Since $r_{b-2,j,k} < b-k-1$ is equivalent to $q_{b-2,j,k} = k$, we have $q_{b-2,j,k} = q'_{b-2,j,k} = k$. From Definitions 1.3 and 1.4, we have

$$\begin{aligned} k(b-2)+j &= bk + r'_{b-2,j,k} \text{ and} \\ k(b-2)+j &= (b-1)k + r_{b-2,j,k}. \end{aligned}$$

Thus $r'_{b-2,j,k} + k = r_{b-2,j,k}$, where $0 \le r_{b-2,j,k} \le b-2$. By Lemma 1, (1) holds. Since $r'_{b-2,j,k} + k = r_{b-2,j,k}$ where $0 \le r_{b-2,j,k} < b$, we have $r'_{b-2,j,k} + k = r_{b-2,j,k} = a_{b-2,j,k}$. Also, because $r_{b-2,j,k} < b-k-1$, we have $c_{b-2,j,k} = r_{b-2,j,k} = a_{b-2,j,k}$. Since $r'_{b-2,j,k} + k = a_{b-2,j,k}$, there is no carry 1 to the b^1's digit. Observe that $\epsilon_{b-2,j,k} = q_{b-2,j,k} - q'_{b-2,j,k} = 0$. Thus (2) holds.

Case 4 We assume $q_{b-2,j,k} = q'_{b-2,j,k}$ and $r_{b-2,j,k} \geq b-k-1$. Since $r_{b-2,j,k} \geq b-k-1$ is equivalent to $q_{b-2,j,k} = k-1$, we have $q_{b-2,j,k} = q'_{b-2,j,k} = k-1$. From Definitions 1.3 and 1.4, we have

$$k(b-2)+j = b(k-1)+r'_{b-2,j,k} \text{ and}$$
$$k(b-2)+j = (b-1)(k-1)+r_{b-2,j,k}.$$

Thus $r'_{b-2,j,k}+k = r_{b-2,j,k}+1$, where $1 \leq r_{b-2,j,k}+1 \leq b-1$. By Lemma 1, (1) holds. Since $r'_{b-2,j,k}+k = r_{b-2,j,k}+1$ where $1 \leq r_{b-2,j,k}+1 < b$, we have $a_{b-2,j,k} = r_{b-2,j,k}+1$. Also, because $r_{b-2,j,k} \geq b-k-1$, we have $c_{b-2,j,k} = r_{b-2,j,k}+1 = a_{b-2,j,k}$. Since $r'_{b-2,j,k}+k = a_{b-2,j,k}$, there is no carry 1 to the b^1's digit. Observe that $\epsilon_{b-2,j,k} = q_{b-2,j,k} - q'_{b-2,j,k} = 0$. Thus (2) holds.

Case 5 We assume $q_{b-2,j,k} = q'_{b-2,j,k}+1$ and $r_{b-2,j,k} < b-k-1$. Since $r_{b-2,j,k} < b-k-1$ is equivalent to $q_{b-2,j,k} = k$, we have $q_{b-2,j,k} = k$ and $q'_{b-2,j,k} = k-1$. From Definitions 1.3 and 1.4, we have

$$k(b-2)+j = b(k-1)+r'_{b-2,j,k} \text{ and}$$
$$k(b-2)+j = (b-1)k+r_{b-2,j,k}.$$

Thus $r'_{b-2,j,k}+k = r_{b-2,j,k}+b$, where $b \leq r_{b-2,j,k}+b \leq 2b-2$. By Lemma 1, (1) holds. Since $r'_{b-2,j,k}+k = r_{b-2,j,k}+b$ where $0 \leq r_{b-2,j,k} \leq b-2$, we have $a_{b-2,j,k} = r_{b-2,j,k}$. Also, because $r_{b-2,j,k} < b-k-1$, we have $c_{b-2,j,k} = r_{b-2,j,k} = a_{b-2,j,k}$. Since $r'_{b-2,j,k}+k = a_{b-2,j,k}+b$, there is a carry 1 to the b^1's digit. Observe that $\epsilon_{b-2,j,k} = q_{b-2,j,k} - q'_{b-2,j,k} = 1$. Thus (2) holds.

Case 6 We assume $q_{b-2,j,k} = q'_{b-2,j,k}+1$ and $r_{b-2,j,k} \geq b-k-1$. Since $r_{b-2,j,k} \geq b-k-1$ is equivalent to $q_{b-2,j,k} = k-1$, we have $q_{b-2,j,k} = k-1$ and $q'_{b-2,j,k} = k-2$. From Definitions 1.3 and 1.4, we have

$$k(b-2)+j = b(k-2)+r'_{b-2,j,k} \text{ and}$$
$$k(b-2)+j = (b-1)(k-1)+r_{b-2,j,k}.$$

Thus $r'_{b-2,j,k}+k = (r_{b-2,j,k}+1)+b$, where $b+1 \leq (r_{b-2,j,k}+1)+b \leq 2b-1$. By Lemma 1, (1) holds. Since $r'_{b-2,j,k}+k = (r_{b-2,j,k}+1)+b$ where $1 \leq r_{b-2,j,k}+1 \leq b-1$, we have $a_{b-2,j,k} = r_{b-2,j,k}+1$. Also, because $r_{b-2,j,k} \geq b-k-1$, we have $c_{b-2,j,k} = r_{b-2,j,k}+1 = a_{b-2,j,k}$. Since $r'_{b-2,j,k}+k = a_{b-2,j,k}+b$, there is a carry 1 to the b^1's digit. Observe that $\epsilon_{b-2,j,k} = q_{b-2,j,k} - q'_{b-2,j,k} = 1$. Thus (2) holds. □

Theorem 3 *Let b, β, i, j, k, K, $q_{i,j,k}$, $r_{i,j,k}$, $q'_{i,j,k}$, $r'_{i,j,k}$, $a_{i,j,k}$, $c_{i,j,k}$ and $\epsilon_{i,j,k}$ be the integers defined in Definition 1. Further assume that $(b-1) \nmid K$ so that $k \neq b-1$. Then $a_{i,j,k} = c_{i,j,k}$ for all integers i such that $0 \leq i \leq b-2$. Furthermore,*

$$K\beta = \sum_{i=0}^{b-2} c_{i,j,k} b^{b-2-i}.$$

Note that in Theorem 3 we do not include the trivial case $(b-1)|K$ which is included in Theorem 2. By Proposition 8, the results of Theorem 2 hold for the case when $(b-1)|K$.

Proof Let n be an integer such that $0 \le n \le b-3$, and consider the equation

$$K\beta = \sum_{i=0}^{n-1} (q'_{i+1,j,k} + r'_{i,j,k}) b^{b-2-i} + (q'_{n+1,j,k} + r'_{n,j,k} + \epsilon_{n+1,j,k}) b^{b-2-n} + \sum_{i=n+1}^{b-2} c_{i,j,k} b^{b-2-i}.$$

We apply Mathematical Induction to this equation on the values of n in reverse order. By Lemma 4, we have

$$K\beta = \sum_{i=0}^{b-4} (q'_{i+1,j,k} + r'_{i,j,k}) b^{b-2-i} + (q'_{b-2,j,k} + r'_{b-3,j,k} + \epsilon_{b-2,j,k}) b + c_{b-2,j,k}.$$

This is the base step of the proof. Let n be an integer with $0 \le n \le b-3$ and suppose

$$K\beta = \sum_{i=0}^{n-1} (q'_{i+1,j,k} + r'_{i,j,k}) b^{b-2-i} + (q'_{n+1,j,k} + r'_{n,j,k} + \epsilon_{n+1,j,k}) b^{b-2-n} + \sum_{i=n+1}^{b-2} c_{i,j,k} b^{b-2-i} \tag{3}$$

is true. We want to show that

$$K\beta = \sum_{i=0}^{n-2} (q'_{i+1,j,k} + r'_{i,j,k}) b^{b-2-i} + (q'_{n,j,k} + r'_{n-1,j,k} + \epsilon_{n,j,k}) b^{b-2-(n-1)} + \sum_{i=n}^{b-2} c_{i,j,k} b^{b-2-i} \tag{4}$$

holds. When $n = 0$, (3) becomes

$$K\beta = (q'_{1,j,k} + r'_{0,j,k} + \epsilon_{1,j,k}) b^{b-2} + \sum_{i=1}^{b-2} c_{i,j,k} b^{b-2-i} \tag{5}$$

and (4) becomes

$$K\beta = \sum_{i=0}^{b-2} c_{i,j,k} b^{b-2-i}. \tag{6}$$

By Lemma 3, either $q_{n,j,k} = q'_{n,j,k}$ or $q_{n,j,k} = q'_{n,j,k} + 1$, and either $q_{n+1,j,k} = q'_{n+1,j,k}$ or $q_{n+1,j,k} = q'_{n+1,j,k} + 1$. We prove the inductive step by considering the eight cases depending on whether we have either $r_{n,j,k} < b - k - 1$ or $r_{n,j,k} \geq b - k - 1$, either $q_{n,j,k} = q'_{n,j,k}$ or $q_{n,j,k} = q'_{n,j,k} + 1$, and either $q_{n+1,j,k} = q'_{n+1,j,k}$ or $q_{n+1,j,k} = q'_{n+1,j,k} + 1$. We first consider the argument for the values of n for which $1 \leq n \leq b - 3$. Then we show how to modify this argument to the case $n = 0$.

By Lemma 4, we have $\epsilon_{b-2,j,k} = q_{b-2,j,k} - q'_{b-2,j,k}$. One of the 8 cases in the inductive step produces a contradiction. Of the remaining 7 legitimate cases, we assume that $\epsilon_{n+1,j,k} = q_{n+1,j,k} - q'_{n+1,j,k}$. In each of the 7 legitimate cases, we show that $\epsilon_{n,j,k} = q_{n,j,k} - q'_{n,j,k}$. This establishes the legitimacy of the assumption that $\epsilon_{n+1,j,k} = q_{n+1,j,k} - q'_{n+1,j,k}$ in the 7 legitimate cases.

Case 7 We assume $r_{n,j,k} < b - k - 1$, $q_{n,j,k} = q'_{n,j,k}$ and $q_{n+1,j,k} = q'_{n+1,j,k}$. By the inductive hypothesis, (3) holds. Note that $\epsilon_{n+1,j,k} = q_{n+1,j,k} - q'_{n+1,j,k} = 0$. By Lemma 2, we have $q_{n+1,j,k} = q_{n,j,k}$. Thus $q'_{n,j,k} = q_{n,j,k} = q_{n+1,j,k} = q'_{n+1,j,k}$. From Definitions 1.3 and 1.4, we have

$$bq'_{n+1,j,k} + r'_{n,j,k} = kn + j = (b-1)q'_{n+1,j,k} + r_{n,j,k}$$

which, in turn, implies that

$$q'_{n+1,j,k} + r'_{n,j,k} = r_{n,j,k} \quad \text{where } 0 \leq r_{n,j,k} \leq b - 2.$$

Since $0 \leq q'_{n+1,j,k} + r'_{n,j,k} + \epsilon_{n+1,j,k} \leq b - 2$, we have

$$a_{n,j,k} = q'_{n+1,j,k} + r'_{n,j,k} + \epsilon_{n+1,j,k} = r_{n,j,k}.$$

Because $r_{n,j,k} < b - k - 1$, we have $c_{n,j,k} = r_{n,j,k} = a_{n,j,k}$. Also, since $q'_{n+1,j,k} + r'_{n,j,k} + \epsilon_{n+1,j,k} = a_{n,j,k}$, there is no carry 1 to the $b^{b-2-(n-1)}$'s digit. Observe that $\epsilon_{n,j,k} = q_{n,j,k} - q'_{n,j,k} = 0$. Hence, (4) holds.

Case 8 We assume $r_{n,j,k} < b - k - 1$, $q_{n,j,k} = q'_{n,j,k}$ and $q_{n+1,j,k} = q'_{n+1,j,k} + 1$. Since

$$q'_{i,j,k} = \left\lfloor \frac{ki + j}{b} \right\rfloor$$

increases as i increases, we have $q'_{n+1,j,k} \geq q'_{n,j,k}$. By Lemma 2, we have $q_{n+1,j,k} = q_{n,j,k}$. Thus $q'_{n+1,j,k} = q'_{n,j,k} - 1$. Hence, $q'_{n+1,j,k} < q'_{n,j,k}$. This contradicts the fact that $q'_{n+1,j,k} \geq q'_{n,j,k}$. Therefore, this case never occurs.

Case 9 We assume $r_{n,j,k} \geq b - k - 1$, $q_{n,j,k} = q'_{n,j,k}$ and $q_{n+1,j,k} = q'_{n+1,j,k}$. By the inductive hypothesis, (3) holds. Note that $\epsilon_{n+1,j,k} = q_{n+1,j,k} - q'_{n+1,j,k} = 0$.

By Lemma 2, we have $q_{n+1,j,k} = q_{n,j,k} + 1$. Thus $q'_{n,j,k} = q_{n,j,k} = q_{n+1,j,k} - 1 = q'_{n+1,j,k} - 1$. From Definitions 1.3 and 1.4, we have

$$b(q'_{n+1,j,k} - 1) + r'_{n,j,k} = kn + j = (b-1)(q'_{n+1,j,k} - 1) + r_{n,j,k}$$

which, in turn, implies that

$$q'_{n+1,j,k} + r'_{n,j,k} = r_{n,j,k} + 1 \quad \text{where } 1 \le r_{n,j,k} + 1 \le b - 1.$$

Since $1 \le q'_{n+1,j,k} + r'_{n,j,k} + \epsilon_{n+1,j,k} \le b - 1$, we have

$$a_{n,j,k} = q'_{n+1,j,k} + r'_{n,j,k} + \epsilon_{n+1,j,k} = r_{n,j,k} + 1.$$

Because $r_{n,j,k} \ge b - k - 1$, we have $c_{n,j,k} = r_{n,j,k} + 1 = a_{n,j,k}$. Also, since $q'_{n+1,j,k} + r'_{n,j,k} + \epsilon_{n+1,j,k} = a_{n,j,k}$, there is no carry 1 to the $b^{b-2-(n-1)}$'s digit. Observe that $\epsilon_{n,j,k} = q_{n,j,k} - q'_{n,j,k} = 0$. Hence, (4) holds.

Case 10 We assume $r_{n,j,k} \ge b - k - 1$, $q_{n,j,k} = q'_{n,j,k}$ and $q_{n+1,j,k} = q'_{n+1,j,k} + 1$. By the inductive hypothesis, (3) holds. Note that $\epsilon_{n+1,j,k} = q_{n+1,j,k} - q'_{n+1,j,k} = 1$. By Lemma 2, we have $q_{n+1,j,k} = q_{n,j,k} + 1$. Thus $q'_{n,j,k} = q_{n,j,k} = q_{n+1,j,k} - 1 = q'_{n+1,j,k}$. From Definitions 1.3 and 1.4, we have

$$bq'_{n+1,j,k} + r'_{n,j,k} = kn + j = (b-1)q'_{n+1,j,k} + r_{n,j,k}$$

which, in turn, implies that

$$q'_{n+1,j,k} + r'_{n,j,k} = r_{n,j,k} \quad \text{where } 0 \le r_{n,j,k} \le b - 2.$$

Since $1 \le q'_{n+1,j,k} + r'_{n,j,k} + \epsilon_{n+1,j,k} \le b - 1$, we have

$$a_{n,j,k} = q'_{n+1,j,k} + r'_{n,j,k} + \epsilon_{n+1,j,k} = r_{n,j,k} + 1.$$

Because $r_{n,j,k} \ge b - k - 1$, we have $c_{n,j,k} = r_{n,j,k} + 1 = a_{n,j,k}$. Also, since $q'_{n+1,j,k} + r'_{n,j,k} + \epsilon_{n+1,j,k} = a_{n,j,k}$, there is no carry 1 to the $b^{b-2-(n-1)}$'s digit. Observe that $\epsilon_{n,j,k} = q_{n,j,k} - q'_{n,j,k} = 0$. Hence, (4) holds.

Case 11 We assume $r_{n,j,k} < b - k - 1$, $q_{n,j,k} = q'_{n,j,k} + 1$ and $q_{n+1,j,k} = q'_{n+1,j,k}$. By the inductive hypothesis, (3) holds. Note that $\epsilon_{n+1,j,k} = q_{n+1,j,k} - q'_{n+1,j,k} = 0$. By Lemma 2, we have $q_{n+1,j,k} = q_{n,j,k}$. Thus $q'_{n,j,k} = q_{n,j,k} - 1 = q_{n+1,j,k} - 1 = q'_{n+1,j,k} - 1$ and $q_{n,j,k} = q'_{n+1,j,k}$. From Definitions 1.3 and 1.4, we have

$$b(q'_{n+1,j,k} - 1) + r'_{n,j,k} = kn + j = (b-1)q'_{n+1,j,k} + r_{n,j,k}$$

which, in turn, implies that

$$q'_{n+1,j,k} + r'_{n,j,k} = r_{n,j,k} + b \quad \text{where} \quad b \le r_{n,j,k} + b \le 2b - 2.$$

Since $b \le q'_{n+1,j,k} + r'_{n,j,k} + \epsilon_{n+1,j,k} \le 2b - 2$, we have

$$a_{n,j,k} = q'_{n+1,j,k} + r'_{n,j,k} + \epsilon_{n+1,j,k} - b = r_{n,j,k}.$$

Because $r_{n,j,k} < b - k - 1$, we have $c_{n,j,k} = r_{n,j,k} = a_{n,j,k}$. Also, since $q'_{n+1,j,k} + r'_{n,j,k} + \epsilon_{n+1,j,k} = a_{n,j,k} + b$, there is a carry 1 to the $b^{b-2-(n-1)}$'s digit. Observe that $\epsilon_{n,j,k} = q_{n,j,k} - q'_{n,j,k} = 1$. Hence, (4) holds.

Case 12 We assume $r_{n,j,k} < b - k - 1, q_{n,j,k} = q'_{n,j,k} + 1$ and $q_{n+1,j,k} = q'_{n+1,j,k} + 1$. By the inductive hypothesis, (3) holds. Note that $\epsilon_{n+1,j,k} = q_{n+1,j,k} - q'_{n+1,j,k} = 1$. By Lemma 2, we have $q_{n+1,j,k} = q_{n,j,k}$. Thus $q'_{n,j,k} = q_{n,j,k} - 1 = q_{n+1,j,k} - 1 = q'_{n+1,j,k}$ and $q_{n,j,k} = q'_{n+1,j,k} + 1$. From Definitions 1.3 and 1.4, we have

$$bq'_{n+1,j,k} + r'_{n,j,k} = kn + j = (b-1)(q'_{n+1,j,k} + 1) + r_{n,j,k}$$

which, in turn, implies that

$$q'_{n+1,j,k} + r'_{n,j,k} + 1 = r_{n,j,k} + b \quad \text{where} \quad b \le r_{n,j,k} + b \le 2b - 2.$$

Since $b \le q'_{n+1,j,k} + r'_{n,j,k} + \epsilon_{n+1,j,k} \le 2b - 2$, we have

$$a_{n,j,k} = q'_{n+1,j,k} + r'_{n,j,k} + \epsilon_{n+1,j,k} - b = r_{n,j,k}.$$

Because $r_{n,j,k} < b - k - 1$, we have $c_{n,j,k} = r_{n,j,k} = a_{n,j,k}$. Also, since $q'_{n+1,j,k} + r'_{n,j,k} + \epsilon_{n+1,j,k} = a_{n,j,k} + b$, there is a carry 1 to the $b^{b-2-(n-1)}$'s digit. Observe that $\epsilon_{n,j,k} = q_{n,j,k} - q'_{n,j,k} = 1$. Hence, (4) holds.

Case 13 We assume $r_{n,j,k} \ge b - k - 1$, $q_{n,j,k} = q'_{n,j,k} + 1$ and $q_{n+1,j,k} = q'_{n+1,j,k}$. By the inductive hypothesis, (3) holds. Note that $\epsilon_{n+1,j,k} = q_{n+1,j,k} - q'_{n+1,j,k} = 0$. By Lemma 2, we have $q_{n+1,j,k} = q_{n,j,k} + 1$. Thus $q'_{n,j,k} = q_{n,j,k} - 1 = q_{n+1,j,k} - 2 = q'_{n+1,j,k} - 2$ and $q_{n,j,k} = q'_{n+1,j,k} - 1$. From Definitions 1.3 and 1.4, we have

$$b(q'_{n+1,j,k} - 2) + r'_{n,j,k} = kn + j = (b-1)(q'_{n+1,j,k} - 1) + r_{n,j,k}$$

which, in turn, implies that

$$q'_{n+1,j,k} + r'_{n,j,k} = (r_{n,j,k} + 1) + b \quad \text{where} \quad b + 1 \le (r_{n,j,k} + 1) + b \le 2b - 1.$$

Since $b + 1 \le q'_{n+1,j,k} + r'_{n,j,k} + \epsilon_{n+1,j,k} \le 2b - 1$, we have

$$a_{n,j,k} = q'_{n+1,j,k} + r'_{n,j,k} + \epsilon_{n+1,j,k} - b = r_{n,j,k} + 1.$$

Because $r_{n,j,k} \geq b-k-1$, we have $c_{n,j,k} = r_{n,j,k}+1 = a_{n,j,k}$. Also, since $q'_{n+1,j,k} + r'_{n,j,k} + \epsilon_{n+1,j,k} = a_{n,j,k} + b$, there is a carry 1 to the $b^{b-2-(n-1)}$'s digit. Observe that $\epsilon_{n,j,k} = q_{n,j,k} - q'_{n,j,k} = 1$. Hence, (4) holds.

Case 14 We assume $r_{n,j,k} \geq b-k-1$, $q_{n,j,k} = q'_{n,j,k}+1$ and $q_{n+1,j,k} = q'_{n+1,j,k}+1$. By the inductive hypothesis, (3) holds. Note that $\epsilon_{n+1,j,k} = q_{n+1,j,k} - q'_{n+1,j,k} = 1$. By Lemma 2, we have $q_{n+1,j,k} = q_{n,j,k}+1$. Thus $q'_{n,j,k} = q_{n,j,k} - 1 = q_{n+1,j,k} - 2 = q'_{n+1,j,k} - 1$ and $q_{n,j,k} = q'_{n+1,j,k}$. From Definitions 1.3 and 1.4, we have

$$b(q'_{n+1,j,k}-1) + r'_{n,j,k} = kn + j = (b-1)q'_{n+1,j,k} + r_{n,j,k}$$

which, in turn, implies that

$$q'_{n+1,j,k} + r'_{n,j,k} = r_{n,j,k} + b \quad \text{where} \quad b \leq r_{n,j,k} + b \leq 2b-2.$$

Since $b+1 \leq q'_{n+1,j,k} + r'_{n,j,k} + \epsilon_{n+1,j,k} \leq 2b-1$, we have

$$a_{n,j,k} = q'_{n+1,j,k} + r'_{n,j,k} + \epsilon_{n+1,j,k} - b = r_{n,j,k} + 1.$$

Because $r_{n,j,k} \geq b-k-1$, we have $c_{n,j,k} = r_{n,j,k}+1 = a_{n,j,k}$. Also, since $q'_{n+1,j,k} + r'_{n,j,k} + \epsilon_{n+1,j,k} = a_{n,j,k} + b$, there is a carry 1 to the $b^{b-2-(n-1)}$'s digit. Observe that $\epsilon_{n,j,k} = q_{n,j,k} - q'_{n,j,k} = 1$. Hence, (4) holds.

Lastly, we need to consider the case when $n=0$. From Definitions 1.3 and 1.4, we have

$$bq'_{0,j,k} + r'_{0,j,k} = k \cdot 0 + j = j = (b-1)q_{0,j,k} + r_{0,j,k} \quad \text{where} \quad 0 \leq j \leq b-2.$$

Thus $q'_{0,j,k} = q_{0,j,k} = 0$ and $r'_{0,jk} = r_{0,j,k} = j$. Hence, the only cases in which the assumption $q'_{0,j,k} = q_{0,j,k}$ applies are to Cases 7, 9, and 10. One can easily adapt the arguments given in Cases 7, 9, and 10 to show that (5) implies (6) when $n=0$. We leave the details of these arguments to the reader. This completes the proof of Theorem 3. □

We observe that Theorem 2 is a consequence of both Proposition 8 and Theorem 3; this completes the proof of Theorem 2. We need the following lemma in order to prove Propositions 1, 3 and 5. See Definition 1 for the definitions of $a_{i,j,k}$, $r_{i,j,k}$, and $c_{i,j,k}$.

Lemma 5 *Let $b \geq 3$ be an integer, let k be an integer such that $1 \leq k \leq (b-1)/2$, and let $d = \gcd(b-1,k)$. For all integers i such that $0 \leq i \leq b-2$, let $a_{i,0,k}$ be the digits in the base b representation of $k\beta$. I.e., $k\beta = (a_{0,0,k}a_{1,0,k}\ldots a_{b-2,0,k})_b$. Then, for all integers i such that $0 \leq i \leq b-2$, we have $a_{i,0,k} \leq b-d$.*

Proof By Theorem 2, we have $a_{i,0,k} = c_{i,0,k}$ for all integers i such that $0 \leq i \leq b-2$. Let $k = k_1 d$ and $b-1 = \ell d$ for some positive integers k_1 and ℓ. Since $r_{i,0,k} \equiv ki \pmod{b-1}$, we have $r_{i,0,k} \equiv (k_1 i)d \pmod{\ell d}$. Since $r_{i,0,k} < b-1 = \ell d$ and d

divides $r_{i,0,k}$, we have $r_{i,0,k} \leq \ell d - d = b - 1 - d$. Thus $a_{i,0,k} = c_{i,0,k} \leq r_{i,0,k} + 1 \leq b - d$. □

Proposition 1 is a special case of Proposition 3 if we remove the restriction $d = \gcd(b-1, k) > 1$ and replace it with $d = \gcd(b-1, k) \geq 1$.

Proof of Propositions 1 and 3 By Lemma 5, we have $a_{i,j,k} \leq b - d$ for all integers i such that $0 \leq i \leq b-2$. Since $(b-d)(b-1)\beta = ((b-d)(b-d)\ldots(b-d))_b$, $K\beta = \big((b-d)(b-1) - k\big)\beta = \big((b-d)(b-d)\ldots(b-d)\big)_b - \big(a_{0,0,k}\, a_{1,0,k} \ldots a_{b-2,0,k}\big)_b$. Thus $K\beta = \big(a_{0,b-d-1,b-k-1}\, a_{1,b-d-1,b-k-1} \ldots a_{b-2,b-d-1,b-k-1}\big)_b$ is the $(b-d)$'s complement of $k\beta = \big(a_{0,0,k}\, a_{1,0,k} \ldots a_{b-2,0,k}\big)_b$ in the base b number system. □

Proof of Proposition 5 By Lemma 5, the digits of $k\beta$ are not larger than $b - d$. Since $0 \leq j_1 < d$, we can add the digits of $j_1(b-1)\beta = \big(j_1 j_1 \ldots j_1\big)_b$ to the digits of $k\beta = \big(a_{0,0,k}\, a_{1,0,k} \ldots a_{b-2,0,k}\big)_b$ to obtain the digits of $\big((b-1)j_1 + k\big)\beta = \big(a_{0,j_1,k}\, a_{1,j_1,k} \ldots a_{b-2,j_1,k}\big)_b$. □

Propositions 2 and 4 are direct results of Theorem 2 applied to each particular case.

Proof of Proposition 2 We observe that $r_{i,j,k} \equiv ki + j \pmod{b-1}$ and $r_{i+j_3,0,k} \equiv k(i+j_3) + 0 \pmod{b-1} \equiv ki + j \pmod{b-1}$. Since $0 \leq r_{i,j,k},\ r_{i+j_3,0,k} < b-1$, we have $r_{i,j,k} = r_{i+j_3,0,k}$. On the one hand, if $r_{i,j,k} = r_{i+j_3,0,k} < b-k-1$, then $c_{i,j,k} = r_{i,j,k} = r_{i+j_3,0,k} = c_{i+j_3,0,k}$. On the other hand, if $r_{i,j,k} = r_{i+j_3,0,k} \geq b - k - 1$, then $c_{i,j,k} = r_{i,j,k} + 1 = r_{i+j_3,0,k} + 1 = c_{i+j_3,0,k}$. In either case, by Theorem 2, $a_{i,j,k} = c_{i,j,k} = c_{i+j_3,0,k} = a_{i+j_3,0,k}$. □

Proof of Proposition 4 First, since $k_1 j_3 \equiv j_2 \pmod{\ell}$, we have $kj_3 = (k_1 d) j_3 \equiv j_2 d \pmod{\ell d} = j_2 d \pmod{b-1}$. We observe that $r_{i,j,k} \equiv ki + j \pmod{b-1}$ and $r_{i+j_3,j_1,k} \equiv k(i+j_3) + j_1 \pmod{b-1} \equiv ki + j \pmod{b-1}$. Since $0 \leq r_{i,j,k},\ r_{i+j_3,j_1,k} < b-1$, we have $r_{i,j,k} = r_{i+j_3,j_1,k}$. On the one hand, if $r_{i,j,k} = r_{i+j_3,j_1,k} < b-k-1$, then $c_{i,j,k} = r_{i,j,k} = r_{i+j_3,j_1,k} = c_{i+j_3,j_1,k}$. On the other hand, if $r_{i,j,k} = r_{i+j_3,j_1,k} \geq b-k-1$, then $c_{i,j,k} = r_{i,j,k} + 1 = r_{i+j_3,j_1,k} + 1 = c_{i+j_3,j_1,k}$. In either case, by Theorem 2, $a_{i,j,k} = c_{i,j,k} = c_{i+j_3,j_1,k} = a_{i+j_3,j_1,k}$. □

Proof of Theorem 1 Since $\gcd(K, b-1) = 1$, we have $\gcd(k, b-1) = 1$. Thus k is a generator of $\mathbb{Z}_{b-1}$. Hence, $\mathbb{Z}_{b-1} = \{ki \pmod{b-1} : i = 0, 1, \ldots, b-2\} = \{ki + j \pmod{b-1} : i = 0, 1, \ldots, b-2\} = \{r_{i,j,k} : i = 0, 1, \ldots, b-2\}$. Thus $(r_{i,j,k} : i = 0, 1, \ldots, b-2)$ is a permutation on the set of integers $\{0, 1, \ldots, b-2\}$. We also observe that $d = b - k - 1$. Since $c_{i,j,k} = r_{i,j,k}$ if $r_{i,j,k} < b-k-1$ and $c_{i,j,k} = r_{i,j,k} + 1$ if $r_{i,j,k} \geq b-k-1$, $(c_{i,j,k} : i = 0, 1, \ldots, b-2)$ is a permutation on the set of integers $\{0, 1, 2, \ldots, b-1\} \setminus \{d\}$. By Theorem 2, $K\beta = \sum_{i=0}^{b-2} c_{i,j,k} b^{b-2-i}$. Hence, $K\beta$ contains each digit $0, 1, 2, \ldots, b-1$ exactly once, except the digit d which does not appear as a digit in $K\beta$. □

5 Questions for Further Investigation

We consider several questions related to the results in this paper. In order to ask our questions, we need to first state Theorem 2 in the following way.

Theorem 4 *Let $b \geq 3$ be an integer, and let*

$$\beta = \frac{b^{b-1} - 1}{(b-1)^2}.$$

Let N be an integer such that $1 \leq N < (b-1)^2$. Let a_0 and a_1 be the digits of N expressed in the $(b-1)$-base number system. I.e., $N = (a_1\, a_0)_{b-1} = a_1(b-1) + a_0$ where $0 \leq a_i < b-1$ for $i = 0$ and 1. Then the digits of $N\beta$ can be constructed from the sequence $\{a_1 + a_0 j \pmod{b-1} : 0 \leq j < b-1\}$ by adding 1 to those values in the sequence that are greater than or equal to $b - 1 - a_0$.

Hence, we may think of the digits of $N\beta$ as being formed indirectly from the sequence $\{a_1 + a_0 j \pmod{b-1} : 0 \leq j < b-1\}$. This interpretation of Theorem 4 allows us to formulate the following questions. We first consider some questions related to the number $\beta_2 = (b^{(b-1)^2} - 1)/(b-1)^3$.

Question 1 Let $b \geq 3$ be an integer, and let

$$\beta_2 = \frac{b^{(b-1)^2} - 1}{(b-1)^3}.$$

Let N be an integer such that $1 \leq N < (b-1)^3$. Let a_0, a_1 and a_2 be the digits of N expressed in the $(b-1)$-base number system. I.e., $N = (a_2\, a_1\, a_0)_{b-1} = a_2(b-1)^2 + a_1(b-1) + a_0$ where $0 \leq a_i < b-1$ for $i = 0$, 1, and 2.

1. Can the digits of $N\beta_2$ be constructed (indirectly) from the sequence $\{a_2 + a_1 j + a_0 j^2 \pmod{b-1} : 0 \leq j < (b-1)^2\}$?
2. Is there a quadratic function $p(j)$ such that the digits of $N\beta_2$ can be constructed (indirectly) from the sequence $\{a_2 + a_1 j + a_0 p(j) \pmod{b-1} : 0 \leq j < (b-1)^2\}$?

We next consider some questions related to the number $\beta_k = (b^{(b-1)^k} - 1)/(b-1)^{k+1}$.

Question 2 Let $b \geq 3$ and $k \geq 3$ be integers, and let

$$\beta_k = \frac{b^{(b-1)^k} - 1}{(b-1)^{k+1}}.$$

Let N be an integer such that $1 \leq N < (b-1)^{k+1}$. Let $a_0, a_1, \ldots, a_k$ be the digits of N expressed in the $(b-1)$-base number system. I.e., $N = (a_k\, a_{k-1} \ldots a_1\, a_0)_{b-1}$

$= a_k(b-1)^k + a_{k-1}(b-1)^{k-1} + \cdots + a_1(b-1) + a_0$ where $0 \le a_i < b-1$ for each $0 \le i \le k$.

1. Can the digits of $N\beta_k$ be constructed (indirectly) from the sequence $\{\sum_{\ell=0}^{k} a_{k-\ell} j^\ell \pmod{b-1} : 0 \le j < (b-1)^k\}$?
2. For each integer $2 \le \ell \le k$, is there a polynomial function $p_\ell(j)$ of degree ℓ such that the digits of $N\beta_k$ can be constructed (indirectly) from the sequence $\{a_k + a_{k-1}j + \sum_{\ell=2}^{k} a_{k-\ell} p_\ell(j) \pmod{b-1} : 0 \le j < (b-1)^k\}$?

References

1. Baidu Baike. http://baike.baidu.com/item/12345679
2. Morris, J.P.: Problem Solving with Calculators. The Arithmetic Teacher 25(7), 24–26 (1978)
3. Simon, W.: Mathematical Magic. Dover Publications, New York (1974)

A & Z Sequences for Double Riordan Arrays

Donovan Branch, Dennis Davenport, Shakuan Frankson, Jazmin T. Jones, and Geoffrey Thorpe

Abstract A Riordan array is an infinite lower triangular matrix that is defined by two generating functions, g and f. The coefficients of the generating function g give the zeroth column and the nth column of the matrix is defined by the generating function gf^n. We shall call f the multiplier function. Similarly, the Double Riordan array is an infinite lower triangular matrix that is defined by three generating functions, g, f_1 and f_2. Where the zeroth column of the Double Riordan array is g, the next column is given by gf_1 and the following column will be defined by gf_1f_2. The remaining columns are found by multiplying f_1 and f_2 alternatively. Thus, for a double Riordan array there are two multiplier functions, f_1 and f_2. It is well known that any Riordan array can be determined by a Z-sequence and an A-sequence. This is the row construction of the array. This is not the case for Double Riordan arrays. In this paper, we show that double Riordan arrays can be determined by two Z-sequences and one A-sequence.

Keywords Riordan array · Double Riordan array · A-sequence · Z-sequence

D. Branch · G. Thorpe
Morehouse College, Atlanta, GA, USA
e-mail: donovan.branch@morehouse.edu

G. Thorpe
e-mail: geoffrey.thorpe@morehouse.edu

D. Davenport (✉) · S. Frankson
Howard University, Washington, DC, USA
e-mail: dennis.davenport@howard.edu

S. Frankson
e-mail: shakuan.frankson@bison.howard.edu

J. T. Jones
Clark Atlanta University, Atlanta, GA, USA
e-mail: jazmin.jones@students.cau.edu

F. Hoffman (ed.), *Combinatorics, Graph Theory and Computing*,
Springer Proceedings in Mathematics & Statistics 388,
https://doi.org/10.1007/978-3-031-05375-7_3

1 Introduction

In 1991, Shapiro, Getu, Woan, and Woodson introduced a group of infinite lower triangular matrices called the Riordan group, see [6]. The elements of the group are defined by two power series g and f, where the coefficients of g give the leftmost column and the ith column is given by the coefficients of $g \cdot f^i$, for $i = 1, 2, 3, \ldots$

Explicitly, the following construction is used to build a Riordan array. Let $g(z) = 1 + \sum_{k=1}^{\infty} g_k z^k$ and $f(z) = \sum_{k=1}^{\infty} f_k z^k$, where $f_1 \neq 0$. Let $d_{n,k}$ be the coefficient of z^n in $g(z)\big(f(z)\big)^k$. Then $D = (d_{n,k})_{n,k\geq 0}$ is a Riordan array and an element of the Riordan group. We write $D = \big(g(z), f(z)\big)$.

Before giving our new results, we will define the Riordan group, state the Fundamental Theorem of Riordan Arrays, and give some examples of elements in the Riordan group. In Sect. 2, we define the Double Riordan Group, state the Fundamental Theorem of Double Riordan Arrays, and prove a result about the A- and Z-sequences of a Double Riordan array. In Sect. 3, we will give new subgroups of the Double Riordan Group.

In our research several sequences were found which are in the *Online Encyclopedia of Integer Sequences (OEIS)* [7]; the A-numbers refer to this source.

Example 1 The identity matrix in the Riordan Group is

$$(1, z) = \begin{bmatrix} 1 & & & & \\ 0 & 1 & & & \\ 0 & 0 & 1 & & \vdots \\ 0 & 0 & 0 & 1 & \\ 0 & 0 & 0 & 0 & 1 \\ & & \cdots & & \end{bmatrix}.$$

Example 2 Pascal's matrix is

$$\left(\frac{1}{1-z}, \frac{z}{1-z}\right) = \begin{bmatrix} 1 & & & & \\ 1 & 1 & & & \\ 1 & 2 & 1 & & \vdots \\ 1 & 3 & 3 & 1 & \\ 1 & 4 & 6 & 4 & 1 \\ & & \cdots & & \end{bmatrix}.$$

Example 3 The Fibonacci matrix with Pascal-like columns and Fibonacci row sums is

$$\big(1, z(1+z)\big) = \begin{bmatrix} 1 & & & & \\ 0 & 1 & & & \\ 0 & 1 & 1 & & \vdots \\ 0 & 0 & 2 & 1 & \\ 0 & 0 & 1 & 3 & 1 \\ & & \cdots & & \end{bmatrix}.$$

Theorem 1 (The Fundamental Theorem of Riordan Arrays) *Let $A(z) = \sum_{k=0}^{\infty} a_k z^k$ and $B(z) = \sum_{k=0}^{\infty} b_k z^k$ and let A and B be the column vectors $A = (a_0, a_1, a_2, \ldots)^T$ and $B = (b_0, b_1, b_2, \ldots)^T$. Then $(g, f)A = B$, if and only if $B(z) = g(z)A(f(z))$.*

Theorem 2 *Let (g, f) and (G, F) be two Riordan arrays. Then the operation $*$, given by $(g, f) * (G, F) = \big(g(z)G\big(f(z)\big), F\big(f(z)\big)\big)$ is matrix multiplication which is an associative binary operation, $(1, z)$ is the identity element, and the inverse of (g, f) is $\left(\frac{1}{g(\overline{f})}, \overline{f}\right)$, where $\overline{f}$ is the compositional inverse of f.*

Using the Fundamental Theorem of Riordan Arrays, we can easily prove many combinatorial identities and find ways to invert those identities. Given any Riordan array, every element of the array, except the element in the zeroth row and zeroth column, can be written as a linear combination of elements in the preceding row starting from the preceding column [5]. In addition, every element in the zeroth column other than the first element can be expressed as a linear combination of all elements of the preceding row [4]. Hence, a Riordan Array can be determined by a column construction (using generating functions) or by a row construction (using A- and Z-sequences). The following theorem tells us how to construct a Riordan array using the rows.

Theorem 3 *Let $D = (d_{n,k})$ be an infinite triangular matrix. Then D is a Riordan matrix if and only if there exists two sequences $A = a_0, a_1, a_2, \ldots$ and $B = b_0, b_1, b_2, \ldots$ with $a_0 \neq 0$ and $b_0 \neq 0$ such that*

$$d_{n+1,k+1} = \sum_{j=0}^{\infty} a_j d_{n,k+j};\ k, n = 0, 1, 2, \ldots \tag{1}$$

$$d_{n+1,0} = \sum_{j=0}^{\infty} b_j d_{n,j};\ n = 0, 1, 2, \ldots \tag{2}$$

The sequences $a_0, a_1, a_2, \ldots$ and $b_0, b_1, b_2, \ldots$, respectively, are called the A-sequence and Z-sequence of the Riordan matrix D.

Theorem 4 *Let $D = \big(g(t), f(t)\big)$ be a Riordan array. Let A be the generating function of the A-sequence and Z the generating function of the Z-sequence. Then*

Fig. 1 Example of a Riordan matrix

$$
\begin{bmatrix}
g & gf & gf^2 & gf^3 & gf^4 & gf^5 & \ldots \\
1 & & & & & & \\
1 & 1 & & & & & \\
2 & 3 & 1 & & & & \\
5 & 9 & 5 & 1 & & & \vdots \\
14 & 28 & 20 & 7 & 1 & & \\
42 & 90 & 70 & 35 & 9 & 1 & \\
132 & 297 & 275 & 154 & 54 & 11 & 1 \\
 & & & \ldots & & &
\end{bmatrix}
$$

$$A(t) = \frac{t}{(\overline{f}(t))} \text{ and } Z(t) = \frac{1}{\overline{f}(t)} \cdot \left(1 - \frac{1}{g(\overline{f}(t))}\right),$$

where $\overline{f}$ is the compositional inverse of f.

See [1] for more information about A- and Z-sequences of Riordan arrays. The following example shows how to construct a Riordan array using the A- and Z-sequences. Note that for uniqueness the element g_0 must be given and it cannot be 0. Also, we assume that all other elements in the first row are 0 (Fig. 1).

Example 4

$$Z : (1, 1) \qquad A : (1, 2, 1) \qquad \text{first row} : 1, 0, 0, \ldots$$

We get the following equations;

$$
\begin{aligned}
g &= 1 + tg + tgf \text{ and} \\
gf &= tg + 2tgf + tgf^2 \implies f = t + 2tf + tf^2.
\end{aligned}
$$

Solving this set of equations we get the following for g and f;

$$
\begin{aligned}
g &= \frac{1 - \sqrt{1 - 4t}}{2t} \\
&= 1 + t + 2t^2 + 5t^3 + 14t^4 + 42t^5 + \cdots \\
&\text{and} \\
f &= \frac{1 - 2t - \sqrt{1 - 4t}}{2t} \\
&= t + 2t^2 + 5t^3 + 14t^4 + 42t^5 + 132t^6 + \cdots
\end{aligned}
$$

2 Double Riordan: A & Z Sequences

In a Riordan array, we use one multiplier function. Suppose alternating rules are used to generate an infinite matrix similar to a Riordan array. For example, suppose we are looking at Dyck paths with bicolored edges only at even heights. For this case, we have two rules; one for rows at even height and the other for rows at odd height.

In general, the set of double Riordan arrays is not closed under multiplication. However, if we require that g be an even function and f_1 and f_2 odd, then there is an analog of *The Fundamental Theorem of Riordan Arrays*, which gives us a binary operation, see [2].

Definition 1 Let $g(t) = 1 + \sum_{k=1}^{\infty} g_{2k}t^{2k}$, $f_1(t) = \sum_{k=0}^{\infty} f_{1,2k+1}t^{2k+1}$, and $f_2(t) = \sum_{k=0}^{\infty} f_{2,2k+1}z^{2k+1}$, where $f_{1,1} \neq 0$ and $f_{2,1} \neq 0$. Then the double Riordan matrix (or array) of g, f_1 and f_2, denoted by $(g;\, f_1,\, f_2)$, has column vectors

$$(g,\ gf_1,\ gf_1f_2, gf_1^2f_2, gf_1^2f_2^2,\ \ldots),$$

The set of all aerated double Riordan matrices is denoted as $\mathcal{DR}$.

Theorem 5 (The Fundamental Theorem of Double Riordan Arrays) *Let* $g(t) = \sum_{k=0}^{\infty} g_{2k}t^{2k}$, $f_1(t) = \sum_{k=0}^{\infty} f_{1,2k+1}t^{2k+1}$, *and* $f_2(t) = \sum_{k=0}^{\infty} f_{2,2k+1}t^{2k+1}$.

Case 1: If $A(t) = \sum_{k=0}^{\infty} a_{2k}t^{2k}$ and $B(t) = \sum_{k=0}^{\infty} b_{2k}t^{2k}$, and $A = (a_0, 0, a_2, 0, \ldots)^T$ and $B = (b_0, 0, b_2, 0, \ldots)^T$ are column vectors. Then
$(g,\, f_1,\, f_2)A = B$ if and only if $B(z) = g(z)A\left(\sqrt{f_1(z)f_2(z)}\right)$.

Case 2: If $A(t) = \sum_{k=0}^{\infty} a_{2k+1}t^{2k+1}$ and $B(t) = \sum_{k=0}^{\infty} b_{2k+1}t^{2k+1}$ with $(g,\, f_1,\, f_2)A = B$, then $B(t) = g(t)\sqrt{f_1/f_2}A(\sqrt{f_1(t)f_2(t)})$.

Using the Fundamental Theorem of Double Riordan Arrays, we can define a binary operation on $\mathcal{DR}$.

Definition 2 Let $(g,\, f_1,\, f_2)$ and $(G,\, F_1,\, F_2)$ be elements of $\mathcal{DR}$. Then: $(g;\, f_1,\, f_2)$ $(G;\, F_1,\, F_2) = (gG(\sqrt{f_1f_2});\, \sqrt{f_1/f_2}F_1(\sqrt{f_1f_2}),\, \sqrt{f_2/f_1}F_2(\sqrt{f_1f_2}))$.

The following theorem is analogous to Theorem 2.

Theorem 6 $(\mathcal{DR}, *)$ *is a group. Where the matrix* $(1; t, t)$ *is the identity and* $((1/g(\bar{h});\, t\bar{h}/f_1(\bar{h}),\, t\bar{h}/f_2(\bar{h}))$ *is the inverse of* $(g;\, f_1,\, f_2)$, *where* $h = \sqrt{f_1f_2}$ *and* $\bar{h}$ *is the compositional inverse of* h.

A Riordan array has one Z- and one A-sequence. In this section, we show that elements in $\mathcal{DR}$ can be written using two Z-sequences and one A-sequence. Our approach is different than the one found by He, see [3]. He's row construction of Double Riordan arrays has one Z-sequence and two A-sequences. For our approach, we split the double Riordan array into two arrays and after compression, we get two Riordan arrays that have the same multiplier function. To show this we will use the following definition.

Definition 3 Let $f(z) \in \mathbb{R}[[z^2]]$, where $\mathbb{R}[[z^2]]$ is the set of even formal power series. Then $* : \mathbb{R}[[z^2]] \to \mathbb{R}[[z]]$ is called a compression of the power series if $*(\sum_{k=0}^{\infty} a_k z^{2k}) = \sum_{k=0}^{\infty} a_k z^k$. If $f(z) \in \mathbb{R}[[z^2]]$, we denote its compression as f^*. We will use f for f^*, when this causes no confusion. If $g(z)$ is odd, then a compression of $g(z)$ is $\left(\frac{g(z)}{z}\right)^*$.

Example 5 Let

$$f(z) = 1 + 2z^2 + 6z^4 + 22z^6 + \cdots [A006318]$$

Then

$$f^*(z) = 1 + 2z + 6z^2 + 22z^3 + \cdots$$

Theorem 7 *Let $D = (d_{n,k}) = (g,\ gf_1,\ gf_1f_2, gf_1^2f_2, gf_1^2f_2^2, \ldots) \in \mathcal{DR}$. Then D is uniquely determined by three sequences $A = (a_0, a_1, a_2, \ldots)$, $Z_0=(b_0, b_1, b_2, \ldots)$, and $Z_1 = (c_0, c_1, c_2, \ldots)$, where all elements in column g except $d_{0,0}$ are found by sequence Z_0, all elements in column gf_1 except $d_{1,1}$ are found by sequence Z_1, and the remaining internal entries are found by sequence A.*

Proof To proceed we split D into two matrices, one made with the columns in the even positions and the other with those in the odd positions. So that,

$$\begin{aligned}(g,\ gf_1,\ gf_1f_2, gf_1^2f_2, gf_1^2f_2^2,\ \ldots) &= (g,\ gf_1,\ g(f_1f_2), gf_1(f_1f_2), g(f_1f_2)^2,\ \ldots) \\ &= D_0 + D_1.\end{aligned}$$

Where

$$D_0 = (g,\ \mathbf{0},\ g(f_1f_2),\ \mathbf{0},\ g(f_1f_2)^2,\ \mathbf{0},\ g(f_1f_2)^3,\ \mathbf{0},\ g(f_1f_2)^4,\ \ldots)$$

and

$$D_1 = (\mathbf{0},\ gf_1,\ \mathbf{0},\ gf_1(f_1f_2),\ \mathbf{0},\ gf_1(f_1f_2)^2,\ \mathbf{0},\ gf_1(f_1f_2)^3,\ \mathbf{0},\ gf_1(f_1f_2)^4,\ \ldots).$$

We now compress the formal power series that determine the columns of both D_0 and D_1, remove the $\mathbf{0}$ columns, and shift the rows of D_1 up one. So that

$$D_1^* = \left(\frac{gf_1}{x},\ \frac{gf_1}{x}(f_1f_2),\ \frac{gf_1}{x}(f_1f_2)^2,\ \frac{gf_1}{x}(f_1f_2)^3,\ \frac{gf_1}{x}(f_1f_2)^4,\ \ldots\right).$$

$$\begin{bmatrix} g & gf_1 & gf_1f_2 & gf_1(f_1f_2) & g(f_1f_2)^2 & gf_1(f_1f_2)^2 & g(f_1f_2)^3 & gf_1(f_1f_2)^3 & \cdots \\ 1 &&&&&&&& \\ 0 & 1 &&&&&&& \\ 1 & 0 & 1 &&&&&& \\ 0 & 3 & 0 & 1 &&&&& \\ 2 & 0 & 3 & 0 & 1 &&&& \vdots \\ 0 & 9 & 0 & 5 & 0 & 1 &&& \\ 5 & 0 & 9 & 0 & 5 & 0 & 1 && \\ 0 & 28 & 0 & 20 & 0 & 7 & 0 & 1 & \\ 14 & 0 & 28 & 0 & 20 & 0 & 7 & 0 & 1 \\ &&&& \cdots &&&& \end{bmatrix}$$

Fig. 2 Example of a Double Riordan matrix

Fig. 3 Even columns

$$\begin{bmatrix} g & gh & gh^2 & gh^3 & gh^4 & \cdots \\ 1 &&&&& \\ 0 &&&&& \\ 1 & 1 &&&& \\ 0 & 0 &&&& \\ 2 & 3 & 1 &&& \vdots \\ 0 & 0 & 0 &&& \\ 5 & 9 & 5 & 1 && \\ 0 & 0 & 0 & 0 && \\ 14 & 28 & 20 & 7 & 1 & \\ && \cdots &&& \end{bmatrix} \Longrightarrow \begin{bmatrix} g & gh & gh^2 & gh^3 & gh^4 & \cdots \\ 1 &&&&& \\ 1 & 1 &&&& \\ 2 & 3 & 1 &&& \vdots \\ 5 & 9 & 5 & 1 && \\ 14 & 28 & 20 & 7 & 1 & \\ &&& \cdots && \end{bmatrix}$$

Note that with compression and shifting the rows of D_0 and D_1, both D_0^* and D_1^* become simple Riordan arrays with the same multiplier function $f_1 f_2$. Hence, they have the same A-sequence. And each Riordan array has a Z-sequence that determines the 0th column.

Example 6 To illustrate this process, consider the following double Riordan array, where $C(x) = \frac{1-\sqrt{1-4x}}{2x}$ is the generating function for the Catalan numbers.

$$(g, f_1, f_2) = \left(C(x^2),\ xC^2(x^2),\ x\right)$$

We have two functions g and gf_1 alternating, with a multiplier $h = f_1 f_2$, see Fig. 2. The matrix will be "split" into two matrices with the same multiplier function h. This results in two Z-sequences, Z_0 and Z_1, and a single A-sequence. The Riordan array of the even columns is defined by the power series g and multiplier function $f_1 f_2$, where the coefficients of g give the zeroth column, the first column is $gf_1 f_2$, the second is $g(f_1 f_2)^2$ and so on. Note that, when constructing the two Riordan matrices, we remove the aeration by compression (Fig. 3).

Fig. 4 Odd columns

$$\begin{bmatrix} k & kh & kh^2 & kh^3 & kh^4 & \\ 0 & & & & & \\ 1 & & & & & \\ 0 & & & & & \\ 3 & 1 & & & & \\ 0 & 0 & & & & \vdots \\ 9 & 5 & 1 & & & \\ 0 & 0 & 0 & & & \\ 28 & 20 & 7 & 1 & & \\ & & \cdots & & & \end{bmatrix} \Longrightarrow \begin{bmatrix} \frac{k}{z} & \frac{kh}{z} & \frac{kh^2}{z} & \frac{kh^3}{z} & \frac{kh^4}{z} & \\ 1 & & & & & \\ 3 & 1 & & & & \vdots \\ 9 & 5 & 1 & & & \\ 28 & 20 & 7 & 1 & & \\ & & \cdots & & & \end{bmatrix}$$

Similarly, when constructing the Riordan matrix for the odd columns, we divide by z to shift the rows up one making the constant term of $k = gf_1$ one (Fig. 4).

Each Riordan array has $h^* = zC^2(z)$ as the multiplier function. Thus the A-sequence of each array is

$$\begin{aligned} A(z) &= \frac{z}{\bar{h^*}(z)} \\ &= \frac{z}{\frac{z}{(1+z)^2}} \\ &= (1+z)^2 = 1 + 2z + z^2 \end{aligned}$$

Thus the A-sequence for each Riordan Array is 1, 2, 1. When moving to the Double Riordan Array we aerate the sequence to get $(1, 0, 2, 0, 1)_2$ as the A-sequence, where the subscript 2 indicates we move up 2 rows instead of one row as we do with single Riordan Arrays.

Using similar calculations for the Z_0 and Z_1 sequences we get,

$$\begin{aligned} Z_0(z) &= \frac{1}{\bar{h^*}(z)}\left(1 - \frac{1}{g^*(\bar{h^*}(z))}\right) \\ &= 1 + z \end{aligned}$$

and

$$\begin{aligned} Z_1(z) &= \frac{1}{\bar{h^*}(z)}\left(1 - \frac{1}{k^*(\bar{h^*}(z))}\right) \\ &= \frac{z^2 + 3z + 3}{1+z} \\ &= 3 + \sum_{n=0}^{\infty}(-1)^n z^{n+2} \end{aligned}$$

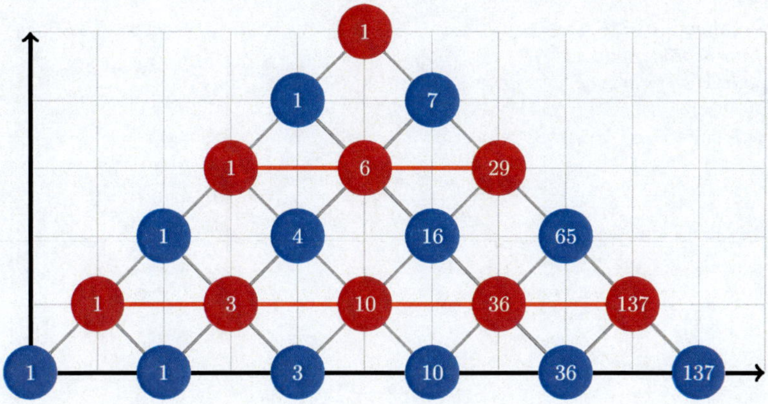

Fig. 5 Schröder path with no level steps at even heights

Hence, the Z-sequences are $1, 1$ and $3, 0, 1, -1, 1, -1, 1, -1, 1, \ldots$. So, for the given double Riordan array they are, $(1, 0, 1)_2$ for the 0th column and $(3, 0, 0, 0, 1, 0, -1, 0, 1, 0, -1, 0, 1, \ldots)_2$ for the 1st column.

Corollary 1 *Let (g, f_1, f_2) be a double Riordan array. Let A, Z_0, and Z_1 be the generating functions for the A and Z sequences respectively. Let $h = f_1 f_2$ and $k = g f_1$. Then*

$$A(t) = \frac{t^2}{(\overline{h}(t))^2},$$

$$Z_0(t) = \frac{1}{(\overline{h}(t))^2}\left(1 - \frac{1}{g(\overline{h}(t))}\right),$$

$$Z_1(t) = \frac{1}{(\overline{h}(t))^2}\left(1 - \frac{f_{1,1}\overline{h}(t)}{k(\overline{h}(t)}\right).$$

Example 7 For the next combinatorial example, we look at Schröder paths with no level steps at even heights, see Fig. 5.

Using the grid, we get the following matrix (Figs. 6 and 7).
The equations are as follows;

$$g - 1 = zg + 2zg - zgh^2 + zgh^3 - zgh^4 + \ldots$$
$$gh = zg + 3zgh + zgh^2 \implies h = z + 3zh + zh^2.$$

Solving these systems of equations we get the following for g and h;

Fig. 6 *Example of $\mathcal{DR}$ matrix for Schröder paths with no level steps at even heights*, where $h = f_1 f_2$ and $k = g f_1$

$$\begin{bmatrix} g & k & gh & kh & gh^2 & kh^2 & gh^3 & kh^3 & gh^4 \\ 1 & & & & & & & & \\ 0 & 1 & & & & & & & \\ 1 & 0 & 1 & & & & & & \\ 0 & 3 & 0 & 1 & & & & & \\ 3 & 0 & 4 & 0 & 1 & & & & \vdots \\ 0 & 10 & 0 & 6 & 0 & 1 & & & \\ 10 & 0 & 16 & 0 & 7 & 0 & 1 & & \\ 0 & 36 & 0 & 29 & 0 & 9 & 0 & 1 & \\ 36 & 0 & 65 & 0 & 38 & 0 & 10 & 0 & 1 \\ & & & & \dots & & & & \end{bmatrix}$$

Fig. 7 Even columns

$$\begin{bmatrix} g & gh & gh^2 & gh^3 & gh^4 \\ 1 & & & & \\ 0 & & & & \\ 1 & 1 & & & \\ 0 & 0 & & & \\ 3 & 4 & 1 & & \\ 0 & 0 & 0 & & \\ 10 & 16 & 7 & 1 & \\ 0 & 0 & 0 & 0 & \\ 36 & 65 & 38 & 10 & 1 \\ & & \dots & & \end{bmatrix} \Longrightarrow \begin{bmatrix} g & gh & gh^2 & gh^3 & gh^4 \\ 1 & & & & \\ 1 & 1 & & & \\ 3 & 4 & 1 & & \\ 10 & 16 & 7 & 1 & \\ 36 & 65 & 38 & 10 & 1 \\ & & \dots & & \end{bmatrix}$$

$$\begin{aligned} g &= \frac{1 - z - \sqrt{5z^2 - 6z + 1}}{2z} \\ &= 1 + z + 3z^2 + 10z^3 + 36z^4 + \cdots [A002212] \\ h &= \frac{1 - 3z - \sqrt{5z^2 - 6z + 1}}{2z} \\ &= z + 3z^2 + 10z^3 + 36z^4 + 137z^5 + 543z^6 + \cdots \\ \bar{h} &= \frac{z}{z^2 + 3z + 1} \\ &= z - 3z^2 + 8z^z - 21z^4 + 55z^z + \cdots [A001906]. \end{aligned}$$

Therefore, the A-sequence and Z-sequence for the matrix composed of the even columns are as follows (Fig. 8);

$$\begin{aligned} A(z) &= \frac{z}{\bar{h}(z)} = \frac{z}{\frac{z}{z^2+3z+1}} = z^2 + 3z + 1 \\ Z(z) &= \frac{1}{\bar{h}}\Big(1 - \frac{1}{g(\bar{h})}\Big) = \frac{z^2 + 3z + 1}{z}\Big(1 - \frac{1}{g(\frac{z}{z^2+3z+1})}\Big) = \frac{z^2 + 3z + 1}{z(z+1)} \\ &= \frac{1}{z}(1 + 2z - z^2 + z^3 - z^4 + z^5 - \cdots). \end{aligned}$$

Fig. 8 Odd columns

$$\begin{bmatrix} k & kh & kh^2 & kh^3 & kh^4 \\ 0 \\ 1 \\ 0 \\ 3 & 1 \\ 0 & 0 \\ 10 & 6 & 1 \\ 0 & 0 & 0 \\ 36 & 29 & 9 & 1 \\ & & \dots \end{bmatrix} \Longrightarrow \begin{bmatrix} \frac{k}{x} & \frac{kh}{x} & \frac{kh^2}{x} & \frac{kh^3}{x} & \frac{kh^4}{x} \\ 1 \\ 3 & 1 \\ 10 & 6 & 1 \\ 36 & 29 & 9 & 1 \\ & & \dots \end{bmatrix}$$

The equations are as follows;

$$k = 1 + 3zk + zkh$$
$$kh = zk + 3zkh + zkh^2 \implies h = z + 3zh + zh^2.$$

Solving these systems of equations we get the following for k and h;

$$\begin{aligned} k &= \frac{1 - 3z - \sqrt{5z^2 - 6z + 1}}{2z^2} \\ &= 1 + 3z + 10z^2 + 36z^3 + 137z^4 + \cdots \ [A002212] \\ h &= \frac{1 - 3z - \sqrt{5z^2 - 6z + 1}}{2z} \\ &= 0 + z + 3z^2 + 10z^3 + 36z^4 + 137z^5 + 543z^6 + \cdots \\ \bar{h} &= \frac{z}{z^2 + 3z + 1} \\ &= z - 3z^2 + 8z^3 - 21z^4 + 55z^5 + \cdots [A001906]. \end{aligned}$$

Therefore, the Z-sequence for the matrix composed of the odd columns is as follows;

$$\begin{aligned} Z(z) &= \frac{1}{\bar{h}}\left(1 - \frac{1}{k(\bar{h})}\right) \\ &= \frac{z^2 + 3z + 1}{z}\left(1 - \frac{1}{k\left(\frac{z}{z^2+3z+1}\right)}\right) \\ &= \frac{3z + 1}{z} = \frac{1}{z}(3z + 1). \end{aligned}$$

Hence, the Z_0-sequence of the Double Riordan matrix is $(1, 0, 2, 0, -1, 0, 1, 0, -1, 0, 1, \ldots)_2$, the Z_1-sequence of the Double Riordan matrix is $(3, 0, 1)_2$ and the A-sequence of the Double Riordan matrix is $(1, 0, 3, 0, 1)_2$.

3 New Subgroups

While studying the Double Riordan group, we found four new subgroups, that we call the Derivative subgroups and the C-Bell Subgroups. The Derivative subgroups are defined by $\{(f', f, cf) \in \mathcal{DR} : c \in \mathbb{R}, c > 0\}$ and $\{(f', cf, f) \in \mathcal{DR} : c \in \mathbb{R}, c > 0\}$. The C-Bell subgroups are defined by $\{(g, czg, f) \in \mathcal{DR} : c \in \mathbb{R}, c > 0\}$ and $\{(g, f, czg) \in \mathcal{DR} : c \in \mathbb{R}, c > 0\}$.

Theorem 8 *Let* $\mathcal{A} = \{(f', f, cf) \in \mathcal{DR} : c \in \mathbb{R}, c > 0\}$. *Then* $\mathcal{A}$ *is a subgroup of* $\mathcal{DR}$.

Proof The identity element is clearly in $\mathcal{A}$, simply let $f = z$ and $c = 1$. Thus, $(f', f, cf) = (1, z, z)$. Now let (f', f, cf) and (g', g, dg) be elements of $\mathcal{A}$. Thus, by definition of multiplication in $\mathcal{DR}$,

$$(f', f, cf) \cdot (g', g, dg) = (f'g'(f\sqrt{c}), \frac{1}{\sqrt{c}}g(f\sqrt{c}), \sqrt{c}dg(f\sqrt{c}))$$

and since the derivative of $\frac{1}{\sqrt{c}}g(f\sqrt{c})$ is $f'g'(f\sqrt{c})$, we have that $\mathcal{A}$ is closed under multiplication. Finally, we need to prove that $\mathcal{A}$ is closed under inverses. Let $(f', f, cf) \in \mathcal{A}$. Then

$$(f', f, cf)^{-1} = \left(\frac{1}{f'(\bar{h})}, \frac{z\bar{h}}{f(\bar{h})}, \frac{z\bar{h}}{cf(\bar{h})}\right),$$

where $h = \sqrt{c}f$.

Claim: The derivative of $\frac{z\bar{h}}{f(\bar{h})}$ is $\dfrac{1}{f'\left(\bar{h}\right)}$.

Note that,

$$h = \sqrt{c}f \implies f(\bar{h}) = \frac{z}{\sqrt{c}}.$$

So that,

$$\frac{z\bar{h}}{f(\bar{h})} = \sqrt{c}\bar{h}.$$

Hence,

$$\frac{d}{dz}\left(\sqrt{c}\bar{h}\right) = \frac{\sqrt{c}}{h'(\bar{h}(z))} = \frac{1}{f'(\bar{h})}.$$

The proof to show that $\{(f', cf, f) \in \mathcal{DR} : c \in \mathbb{R}, c > 0\}$ is a subgroup is similar.

Theorem 9 *Let* $B = \{(g, czg, f) \in \mathcal{DR} : c \in \mathbb{R}, c > 0\}$. *Then* B *is a subgroup of* $\mathcal{DR}$.

Proof Clearly the identity element is in B. Take $g = 1$, $c = 1$ and $f = z$. Then $(g, czg, f) = (1, z, z)$. Now let (g, czg, f) and (G, dzG, F) be elements of B. Then, $(g, czg, f) * (G, dzG, F) = (gG(\sqrt{czgf}), \sqrt{\frac{czg}{f}} d\sqrt{czgf}G(\sqrt{czgf}), L)$

$$= (gG(\sqrt{czgf}), cdzgG(\sqrt{czgf}), L).$$

Thus, B is closed under multiplication since the product is of the form (h, czh, k). Note that the second multiplier function can be any function which we write as L. We now prove that the set is closed under inverses. Let (g, czg, f) be an element of B and $h = \sqrt{czgf}$.

$$\text{Thus, } (g, czg, f)^{-1} = \left(\frac{1}{g\left(\bar{h}\right)}, \frac{z}{cg\left(\bar{h}\right)}, L\right).$$

Hence, B is closed under inverses. Therefore, B is a subgroup of the Double Riordan array.

Similarly we get

$$\{(g, f, czg) \in \mathcal{DR} : c \in \mathbb{R}, c \neq 0\}$$

is a subgroup of $\mathcal{DR}$.

4 Conclusion

An obvious question is, can we have more than 2 multiplier functions? The answer is yes and we call the matrix with k multiplier functions a k-Riordan array. The k-Riordan group is defined similar to the double Riordan group. In the k-Riordan group, we let $g(t) = \sum_{n=0}^{\infty} g_{kn} t^{kn}$ and for each $1 \le i \le k$, $f_i(t) = \sum_{n=0}^{\infty} f_{i,kn+1} t^{kn+1}$, see [3]. For k-Riordan arrays $(g, f_1, f_2, \ldots, f_k)$ and $(G, F_1, F_2, \ldots, F_k)$, multiplication is defined as follows. Let $h(t) = \prod_{i=1}^{k} f_i(t)$. Then

$$(g, f_1, f_2, \ldots, f_k) \cdot (G, F_1, F_2, \ldots, F_k) =$$

$$\left(g(t) \cdot G\left(\sqrt[k]{h(t)}\right), \sqrt[k]{\frac{f_1^k(t)}{h(t)}} \cdot F_1\left(\sqrt[k]{h(t)}\right), \ldots, \sqrt[k]{\frac{f_k^k(t)}{h(t)}} \cdot F_k\left(\sqrt[k]{h(t)}\right)\right).$$

Using our method we can easily find the A- and Z-sequences. Indeed, if D is a k-Riordan array, we split D into k matrices and after compression, we get k Riordan arrays. Columns whose location are congruent to $p \ mod \ k$, where $0 \le p < k$ give the pth matrix. We then compress the matrices. Each of the constructed Riordan arrays will have the same multiplier function $\prod_{i=1}^{k} f_i$. Hence, we get k Z-sequences and one A-sequence.

Acknowledgements This research was made possible by the generous support of the National Security Agency (NSA), Mathematical Association of America (MAA), and National Science Foundation (NSF) grant DUE-1356481.

References

1. Barry, P., *Riordan Arrays: A Primer*, Logic Press, Raleigh, 2016.
2. Davenport, D. E., Shapiro, L. W., and, Woodson, L. C., The Double Riordan Array. *The Electronic Journal of Combinatorics* 18 (2011), 1–16.
3. He, T.-X., Sequence Characterizations of Double Riordan Arrays and Their Compressions, *Linear Algebra and Its Applications* 549 (2018), 176–202.
4. Merlini, D., Rodgers, D. G., Sprugnoli, R., Verri, M. C., On some alternative characterizations of Riordan arrays. *Can. J. Math* 49 (1997), 301–320.
5. Rodgers, D. G., Pascal triangles, Catalan Numbers and renewal arrays. *Discrete Math* 22 (1978), 301–310.
6. Shapiro, L. W., Getu, S., Woan, W., and, Woodson, L. C., The Riordan Group. *Discrete Applied Mathematics* 34 (1991), 229–239.
7. *Sloane's Online Encyclopedia of Integer Sequences*, http://oeis.org/.

Constructing Clifford Algebras for Windmill and Dutch Windmill Graphs; A New Proof of the Friendship Theorem

Timothy Myers

Abstract A Clifford graph algebra $GA(G)$ is a useful structure for studying a simple graph G with n vertices. Such an algebra associates each of its n generators with one of the n vertices of G in a way that depicts the connectivity of G in that any two generators anti-commute or commute depending on whether their corresponding vertices share or do not share an edge. We will construct the Clifford graph algebra for any windmill graph $W(r, m)$, which consist of m copies of the complete graph K_r adjoined at one common vertex; and for any Dutch windmill graph D_r^m which consists of m copies of the r-cycle graph C_r adjoined at one common vertex, then apply this algebraic theory to the class of 3-cycle graphs $F_m = D_3^m$ known as friendship graphs. Specifically, we will use the algebra $GA\big(F_m\big)$ to give a new proof of the fact that those simple graphs which posses the friendship property are precisely the friendship graphs.

Keywords Clifford algebra · Windmill graph · Dutch windmill graph · Friendship graph

Mathematics Subject Classification (2010) Primary 15A66

1 Introduction

The broad goal of this paper is to continue developing applications of Clifford algebras to the subject of algebraic graph theory. This paper will be the third in a recent succession of independent works by Khovanova [11] and Myers [15] which establish some fundamental results in this potential area of study.

Specifically, we will first build a special algebra by selecting a subset of monomials from a basis for an appropriate Clifford algebra with signature so that these monomials will generate a sub-algebra that depicts the connectivity in a simple graph

T. Myers (✉)
Howard University, Washington DC, WA 20059, USA
e-mail: Timyers@howard.edu

F. Hoffman (ed.), *Combinatorics, Graph Theory and Computing*,
Springer Proceedings in Mathematics & Statistics 388,
https://doi.org/10.1007/978-3-031-05375-7_4

such that each pair of generators anti-commute or commute depending on whether their corresponding vertices share or do not share an edge. We will call this sub-algebra a Clifford graph algebra for the given graph G, and denote it as $GA(G)$. In this paper, the graphs of interest will be windmill $W(r, m)$ and Dutch windmill D_m^r graphs. After obtaining an explicit representation for $GA\big(W(r, m)\big)$ and $GA\big(D_m^r\big)$ we will use the algebra $GA\big(F_m\big)$ to give a new proof of the Friendship Theorem.

To appreciate the evolution of the fundamental properties that define any $GA(G)$, we will briefly discuss the history of Clifford algebras and state some intrinsic definitions along the way. This chronicle reveals an ongoing extension of the real numbers to progressively larger embedding algebraic structures; an effort that to this day has spanned more than four centuries.

In 1545 Geralamo Cardano published *Ars Magna*, wherein he used the symbol $\sqrt{-1}$ so describe solutions to quadratic and cubic equations that are unsolvable over $\mathbb{R}$ [20]. Mathematicians such as René Descartes in 1637 continually expressed some disenchantment with the use of the $\sqrt{-1}$ symbol up through the 17th century [6]. Using the notation which Leonard Euler introduced in 1748 [17], in 1831 Carl Gauss de-mystified the $\sqrt{-1}$ symbol by defining the two dimensional field of complex numbers $\mathbb{C}$ as ordered pairs of real numbers subject to the addition and multiplication operations [3] :

$$(a, b) + (c, d) = (a + c, b + d) \text{ and } (a, b)(c, d) = (ac - bd, ad + bc)\ ,$$

wherein the explicit representation $i = (0, 1)$ satisfies $i^2 = -1$, and i along with the real number $1 = (1, 0)$ form a basis which spans $\mathbb{C}$.

In 1843 William Hamilton extended the planar field $\mathbb{C}$ to three dimensional Euclidean space by constructing the four-dimensional division ring $\mathbb{H}$ of quaternions [12]. Similar to the basis for $\mathbb{C}$, the basis for $\mathbb{H}$ consists of the real number 1 and three imaginary units i, j, and k which satisfy $i^2 = j^2 = k^2 = -1$, but these units also anti-commute $ij = -ji$, $jk = -kj$, and $ik = -ki$. In 1876 William Clifford published a work wherein he discussed a class of algebras, called Clifford algebras in his honor, which embedded the exterior product in Grassman's algebra and established generators that posses the squaring and anti-commutativity properties of Hamilton's quaternions [5].

In this work we will define a Clifford algebra as in Definition 1 [2, 13, 14, 19] because it emphasizes the fundamental conditions that relate it to the embedded quaternions. Equipped with a quadratic form, a Clifford algebra with signature is defined as follows.

Definition 1 A real Clifford (*geometric*) algebra of signature (p, q), denoted $\mathbb{G}^{p,q}$, where $p + q = n$, is an associative $\mathbb{R}$-algebra which is generated by the set $S = \{e_1, \ldots, e_n\}$ where the elements in S satisfy the fundamental conditions

$$e_k^2 = \begin{cases} 1 & \text{if } 1 \le k \le p \\ -1 & \text{if } p + 1 \le k \le p + q = n \end{cases}$$

$$e_k e_j = -e_j e_k \text{ for } k \neq j.$$

In particular,
$\mathbb{G}^{(0,n)}$ denotes a geometric algebra where each generator squares to -1,
$\mathbb{G}^n = \mathbb{G}^{(n,0)}$ denotes a geometric algebra where each generator squares to 1.

After Clifford developed his geometric algebras, mathematicians and physicists; namely Sylvester [18], Cartan [4], Weil [22], and Schwinger [16] extended Clifford's algebra into a generalized Clifford algebra, defined as follows (see [9, 21]).

Definition 2 A generalized Clifford algebra is a $\mathbb{C}$-algebra which is generated by the set $S = \{e_1, \ldots, e_n\}$ where the elements in S satisfy the following relations for all $j, k, \ell, m = 1, 2, \ldots, n$

(i) $e_j e_k = \omega_{jk} e_k e_j \ , \ \omega_{jk} e_\ell = e_\ell \omega_{jk} \ , \ \omega_{jk}\omega_{\ell m} = \omega_{\ell m}\omega_{jk}$

(ii) $e_j^{N_j} = \omega_{jk}^{N_j} = \omega_{jk}^{N_k} = 1$ for some $N_j, N_k \in \mathbb{N}$.

To distinguish them from the generalized Clifford algebras in Definition 2, we will refer to the Clifford algebra in Definition 1 as a classical Clifford algebra.

In 2008, T. Khovanova explains how the special case where $\omega_{jk} = \pm 1$ can be used to depict the connectivity between vertices in a finite, simple graph [11]. Hence, Khovanova refers to such a generalized Clifford algebra as a Clifford graph algebra. Although in [11] Khovanova defines a Clifford graph algebra over $\mathbb{C}$, we will alter our definition here from that in [11] by instead defining a Clifford graph algebra over $\mathbb{R}$ with signature (p, q) where $p + q = n$. In this work all graphs will be simple (no multiple edges between any pair of vertices) and finite (finitely many edges and vertices).

Definition 3 A Clifford graph algebra for a simple graph G_n with n vertices $v_1, v_2, \ldots, v_n$, denoted $GA(G_n)$, is an $\mathbb{R}$-algebra with n generators $e'_1, e'_2, \ldots, e'_n$ such that each generator e'_i is paired with exactly one vertex v_i so that the following rules hold

(i) $e'_i e'_j = -e'_j e'_i$ if v_i and v_j are adjacent
$e'_i e'_j = e'_j e'_i$ if v_i and v_j share no edge

(ii)

$$(e'_k)^2 = \begin{cases} 1 & \text{if } 1 \leq k \leq p \\ -1 & \text{if } p + 1 \leq k \leq p + q = n. \end{cases}$$

An objective of this work is to construct a Clifford graph algebra for a given graph G_n, and in particular for windmill and Dutch windmill graphs in a way that is simpler than in the general case where ω_{jk} is an arbitrary complex number. As an additional advantage, the constructive proofs presented here are motivated primarily by the connectivity of G_n. As a first example, a classical Clifford algebra itself can serve as the Clifford graph algebra for the complete graph.

Fig. 1 Schematic depiction of $GA(K_6)$

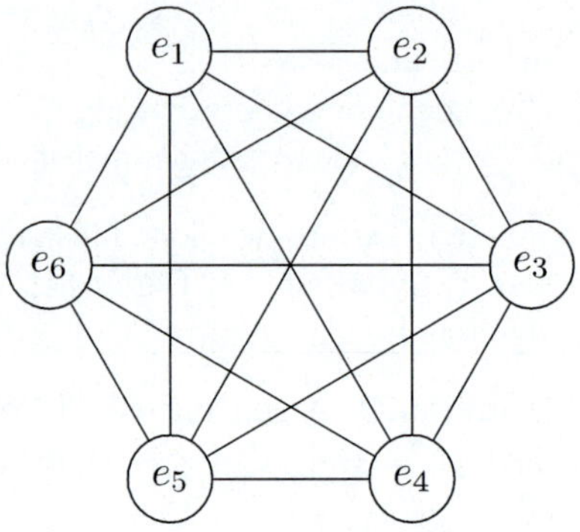

Example 1 Consider the Clifford graph algebra $GA(K_n)$ for the complete graph K_n. Since by definition each pair of vertices in K_n are adjacent (see, for instance, K_6 in Fig. 1), then each pair of distinct generators in the corresponding Clifford graph algebra anti-commute; so in this case any Clifford algebra $\mathbb{G}^{(p,q)}$ with signature (p, q) where $p + q = n$ can serve as the Clifford graph algebra for K_n. Occasionally in this article we will use the underlying graph G_n to illustrate a specific Clifford graph algebra $GA(G_n)$ schematically by labeling each vertex in G_n with its corresponding generator. In particular, the diagram below shows such a schematic representation for $GA(K_6) = \mathbb{G}^{(p,q)}$ such that $p + q = 6$.

If the graph G_n is not complete, the generators in a Classical Clifford algebra will not be able to provide the needed property of commutativity for pairs of vertices that share no edge. As an alternative to constructing a generalized Clifford algebra to serve this purpose for G_n by the process explained in [9], in our case where each $\omega_{jk} = 1$ or $\omega_{jk} = -1$ we will more efficiently prove that we may obtain $GA(G_n)$ directly from a classical Clifford algebra with signature. For convenience, we will choose this underlying algebra to be either $\mathbb{G}^m$ or $\mathbb{G}^{(0,m)}$, where $m > n$. We will construct $GA(G_n)$ by selecting from the basis for either $\mathbb{G}^m$ or $\mathbb{G}^{(0,m)}$ a subset of monomials which satisfies the connectivity conditions of G_n as prescribed in Definition 3; thereby establishing $GA(G_n)$ as a sub-algebra of $\mathbb{G}^m$ or $\mathbb{G}^{(0,m)}$. This method of selection works because every pair of such monomials either commutes or anti-commutes.

As an example of selecting generators for a Clifford graph algebra from a parent algebra, we will construct $GA(G_3)$ from $\mathbb{G}^{(0,3)}$ for each of the four different configurations for G_3 (as presented in [15]).

Example 2 If $n = 3$, the basis B_3 for $\mathbb{G}^{(0,3)}$ contains monomials which can serve as generators for any graph G_3, where

$$B_3 = \left\{ \mathbf{1} \,,\; e_1 \,,\; e_2 \,,\; e_3 \,,\; e_1e_2 \,,\; e_1e_3 \,,\; e_2e_3 \,,\; e_1e_2e_3 \right\}.$$

The Table 1 lists the possible commutativity (c) and anti-commutativity (a) relations between the monomials in B_3.

Table 1 Relations between monomials in B_3

	e_1	e_2	e_3	e_1e_2	e_1e_3	e_2e_3
e_2	a					
e_3	a	a				
e_1e_2	a	a	c			
e_1e_3	a	c	a	a		
e_2e_3	c	a	a	a	a	
$e_1e_2e_3$	c	c	c	c	c	c

Rep.s for $GA(G_3)$ — Generators : — Signature :

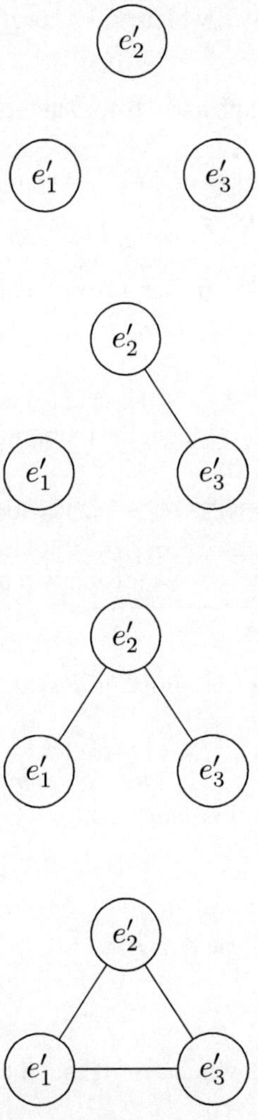

$$e_1' = e_1e_2e_3, \quad e_2' = e_2e_3, \quad e_3' = e_1 \qquad p = 1\ ,\ q = 2$$

$$e_1' = e_1e_2e_3, \quad e_2' = e_1e_2, \quad e_3' = e_1e_3 \qquad p = 1\ ,\ q = 2$$

$$e_1' = e_1, \quad e_2' = e_2, \quad e_3' = e_2e_3 \qquad p = 0\ ,\ q = 3$$

$$e_1' = e_1e_2, \quad e_2' = e_1e_3, \quad e_3' = e_2e_3 \qquad p = 0\ ,\ q = 3$$

Fig. 2 Schematic depictions for $GA(G_3)$

This table shows that the choice of the generators in each of the following possible graphs for G_3 accurately depicts the connectivity of their associated vertices by conforming to property (i) in Definition 3. Also note that the choice of generators satisfies property (ii) in Definition 3. For instance, in the first in Fig. 2 we have that $(e_1')^2 = 1$, $(e_2')^2 = (e_3')^2 = -1$; hence the signature for this graph is $p = 1,\ q = 2$.

2 Preliminaries for a Clifford Graph Algebra

Throughout this work we will assume that the following notations satisfy the stated conditions. The following (i) through (iv) are excerpts from [15] which will be useful in this work.

(i) To avoid trivialities we will always assume that every graph and Clifford algebra considered has at least two vertices or generators.
(ii) The symbols $e_1, \ldots, e_n$ will denote the generators for $\mathbb{G}^n$ or $\mathbb{G}^{(0,n)}$. At times we will indicate this by the notation $\mathbb{G}^n = \langle e_1, \ldots, e_n \rangle$ or $\mathbb{G}^{(0,n)} = \langle e_1, \ldots, e_n \rangle$ [23].
(iii) Indices for vertices of G_n and generators of $GA(G_n)$ are natural numbers, denoted as $i_1, i_2, \ldots, i_n$, which we will assume to satisfy $1 \le i_1 < \cdots < i_r \le n$ where $r \in \mathbb{N}$ and $1 < r \le n$.
(iv) A monomial of the form $e_{i_1} e_{i_2} \cdots e_{i_r}$ where $1 \le i_1 < \cdots < i_r \le n$ is said to have grade r. We will tacitly assume that the symbol $e_{i_1} e_{i_2} \cdots e_{i_r}$ denotes monomial of grade r. We will use the convention that **1** has grade 0.
(v) The monomial of grade n where $\mathbb{G}^n = \langle e_1, \ldots, e_n \rangle$, namely $e_1 e_2 \cdots e_n$ is called the pseudoscalar.
(vi) For convenience, the symbol **1** will denote the multiplicative unit for any classical Clifford algebra.

Proposition 1 ([15]) *If two monomials $e_{j_1} \cdots e_{j_s}$ and $e_{i_1} \cdots e_{i_r}$ share no factor in common, where either r or s is even, they commute.*

Proposition 2 ([15]) *A monomial of even grade $e_{j_1} e_{j_2} \cdots e_{j_{2m-1}} e_{j_{2m}}$ and a monomial $e_{i_1} e_{i_2} \cdots e_{i_r}$ of grade r with exactly one factor in common anti-commute; i.e.*

$$\begin{aligned} &\left(e_{j_1} \cdots e_{j_{2m-1}} e_{j_{2m}}\right)\left(e_{i_1} \cdots e_{i_r}\right) \\ &\qquad = -\left(e_{i_1} e_{i_2} \cdots e_{i_r}\right)\left(e_{j_1} e_{j_2} \cdots e_{j_{2m-1}} e_{j_{2m}}\right). \end{aligned} \tag{1}$$

Definition 4 The symbol $\overline{e}_{2m}$ will denote the monomial $\overline{e}_{2m} = e_2 e_4 \cdots e_{2m}$.

In [15], T. Myers proved the following existence theorem, which asserts that there exists a Clifford graph algebra for any simple graph.

Theorem 3 (Existence of a Clifford Graph Algebra) *Let G_n be a simple graph with n vertices $v_1, \ldots, v_n$. We can always construct a Clifford graph algebra $GA(G_n)$ for G_n as a sub-algebra of $\mathbb{G}^{2^n}$ or $\mathbb{G}^{(0,2^n)}$ by selecting n generators $e'_1, e'_2, \ldots, e'_n$, from the basis of monomials*

$$\{\mathbf{1}\} \cup \left\{e_{i_1} \cdots e_{i_r} \mid 1 \leq i_1 < \cdots < i_r \leq n\right\}$$

for the including algebra $\mathbb{G}^{2^n}$ or $\mathbb{G}^{(0,2^n)}$.

The huge size of the parent algebra in Theorem 3, having 2^n generators, is a consequence of a brief but thorough proof of this theorem. To approach the question of how small the parent Clifford algebra for $GA(G_n)$ can be, we will find a considerably smaller representation for the parent Clifford algebra of particular classes of graphs; and in this paper we will do this for the class of windmill and Dutch windmill graphs. Thus, although Theorem 3 insures that a Clifford graph algebra exists for a given simple graph, the question of choosing the monomials for the generators optimally is still open in theory, and for now depends on the graph.

3 Clifford Algebras for Windmill and Dutch Windmill Graphs

We will now proceed to find and formulate a precise representation for the generators of the Clifford algebras for the windmill and Dutch windmill graphs. This representation will have a closed form and will be efficient in the sense that the monomials selected for generators will be as small in grade as possible; in fact, the generators for all non-central vertices in this formulation will be bivectors. Our exploration will start by defining these classes of graphs and considering some examples.

3.1 Windmill Graphs

Definition 5 ([7]) A windmill graph $W(r, m)$ is a simple graph which consist of m copies of the complete graph K_r adjoined at one common vertex.

Some examples of windmill graphs of the classes $W(4, m)$ and $W(5, m)$ are shown in Figs. 3 and 4.

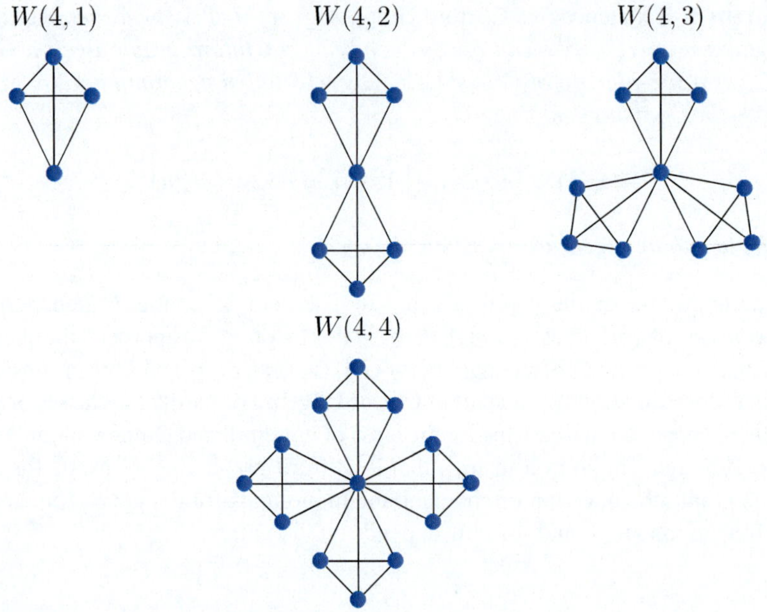

Fig. 3 The class $W(4, m)$: m copies of K_4 adjoined at one common vertex

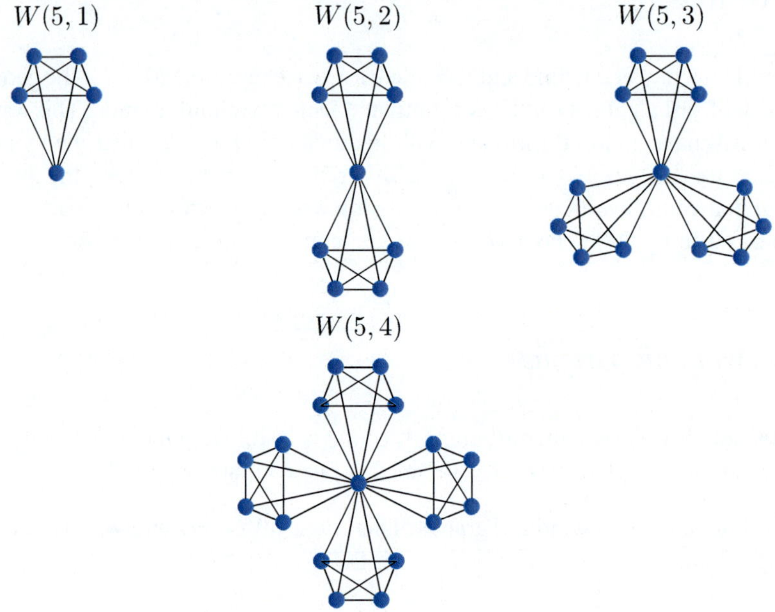

Fig. 4 The class $W(5, m)$: m copies of K_5 adjoined at one common vertex

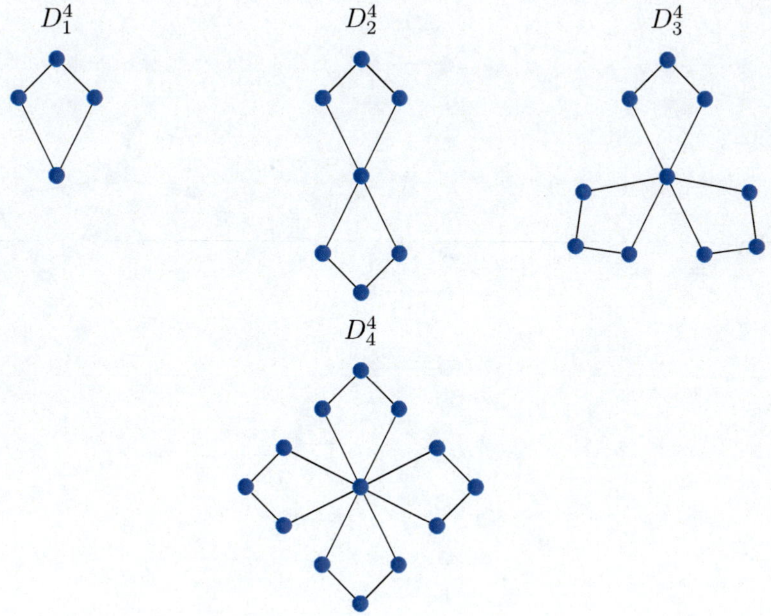

Fig. 5 The class D_m^4 : m C_4-cycles adjoined at exactly one vertex; $3m+1$ vertices

3.2 Dutch Windmill Graphs

Definition 6 ([10]) A Dutch windmill graph D_m^r is a simple graph which consist of m copies of the C_r cycle adjoined at one common vertex.

Some examples of Dutch windmill graphs of the classes D_m^4 and D_m^5 are show in Figs. 5 and 6.

3.3 The Friendship Graph

Our search for $GA\big(W(r,m)\big)$ and $GA\big(D_m^r\big)$ will begin with a graph that is the simplest case of both a windmill and Dutch windmill graph. As expected, the Clifford algebra for this graph will be the simplest considered. We will first mention a special property of this graph.

Definition 7 ([8]) A simple graph with at least 3 vertices has the friendship property if, for any two vertices v_i and v_j, there is exactly one vertex v_k with which each of v_i and v_j share an edge, and we will express this by stating that "v_i and v_j are friends with v_k."

Example 3 The following graph in Fig. 7, a C_3-cycle (triangle) has the friendship property.

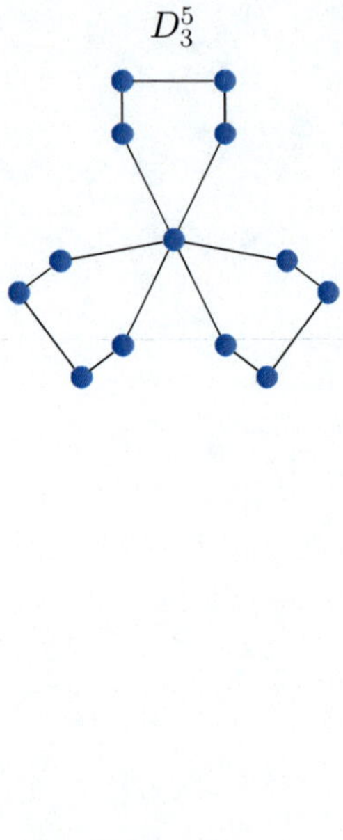

Fig. 6 The class D_m^5 : m C_5-cyles adjoined at exactly one vertex; $4m + 1$ vertices

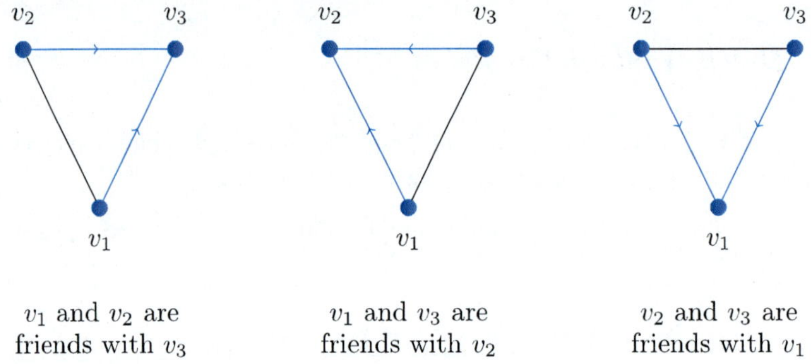

Fig. 7 Illustration of the friendship property

The friendship property prompts the following definition.

Definition 8 ([8])

(i) A friendship graph consists of m C_3-cycles with exactly one common vertex.
(ii) By (i), a friendship graph contains $2m + 1$ vertices and $3m$ edges.

Using standard notation, we will denote a friendship graph of m triangles as F_m. As mentioned, $F_m = W(3, m) = D_3^m$. Figure 8 shows the friendship graphs F_1, F_2, F_3 and F_4 and their notational connections to windmill and Dutch windmill graphs.

From examining any of the graphs in Fig. 8, the following proposition is clear.

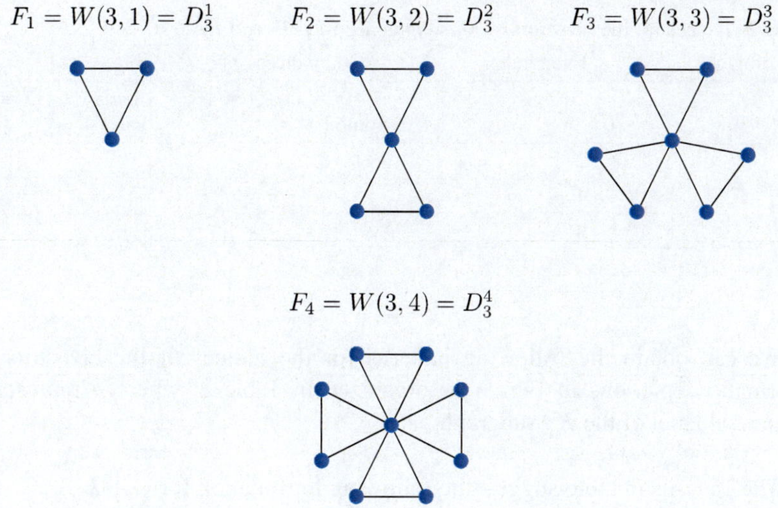

Fig. 8 The friendship graphs F_1, F_2, F_3, and F_4

Proposition 4 ([8]) *Any friendship graph* F_m *has the friendship property (thus the name for this graph).*

In section we will present a new proof of the converse to Proposition 4 that utilizes the Clifford algebra for the friendship graph.

3.4 The Clifford Algebra for the Friendship Graph

Consider the following schematic depictions of the Clifford algebras for F_1, F_2, and F_3 in Fig. 9. The K_3 subgraphs of the friendship graphs are labeled from 1 through 3 in a clockwise direction by increasing indices of the generators, and in Table 2 we denote the numerical label of each K_3 subgraph as n. Note that the properties in Sect. 2 hold for each of these graphs.

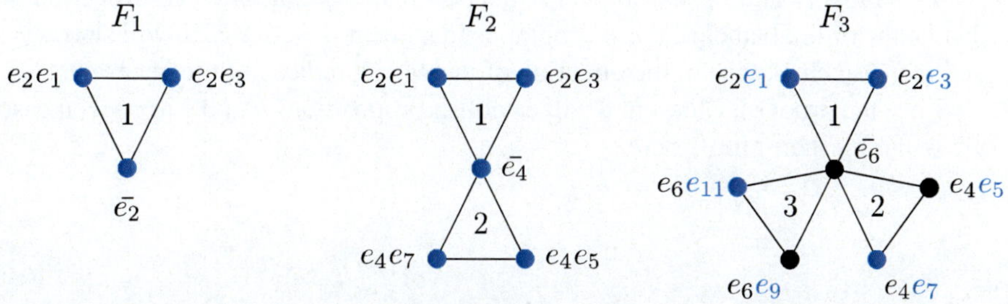

Fig. 9 Schematic depictions of the Clifford graph algebras for F_1, F_2, and F_3

Table 2 Indices for the bivectors of K_3 subgraphs in F_1, F_2 and F_3

K_3 subgraph	Even index	Odd indices	
n	$2n$	$4(n-1)+1 \equiv 1$ mod 4	$4(n-1)+3 \equiv 3$ mod 4
1	2	1	3
2	4	5	7
3	6	9	11

We can obtain the following patterns for the indices of the bivectors in the schematic depictions in Fig. 9, as organized in Table 2, where n represents the numerical label of the K_3 subgraph.

The patterns in table suggest the following formulation for $GA(F_m)$.

Proposition 5 *As a sub-algebra of* $\mathbb{G}^{(0,4m-1)}$ *the Clifford algebra for* F_m *can have the representation*

$$GA(F_m) = \langle e_2e_4 \cdots e_{2m} \text{ and } e_{2n}e_{4n-3}\ ,\ e_{2n}e_{4n-1} \text{ for } n = 1, 2, \ldots, m \rangle.$$

Since F_m is the windmill graph $W(3, m)$ we will establish the proof of Proposition 5 as a special case of a more rigorous development of $GA\big(W(r, m)\big)$ in the proof of Theorem 8.

3.5 The Clifford Algebra for the Class of Windmill Graphs

We will next extend the pattern for labeling odd indexed generators from $GA(F_m)$ to $GA\big(W(r, m)\big)$ by first considering schematic depictions of $W(4, m)$ and $W(5, m)$ for $m = 2, 3, 4$ as shown in Figs. 10 and 11; organize the results in a table, and abstract representations for $GA\big(W(4, m)\big)$ and $GA\big(W(5, m)\big)$ from this information.

Although we will not explicitly number the complete subgraphs in each windmill graph in Figs. 10 and 11, the counter n that occurs in the tables in this subsection will denote the ordinal label of these subgraphs in a given sketch, which will start with 1 and increase clockwise in the direction of increasing indices of the generators.

Since the proof of Theorem 8 will establish Propositions 6 and 7 as special cases, we will omit their proofs here.

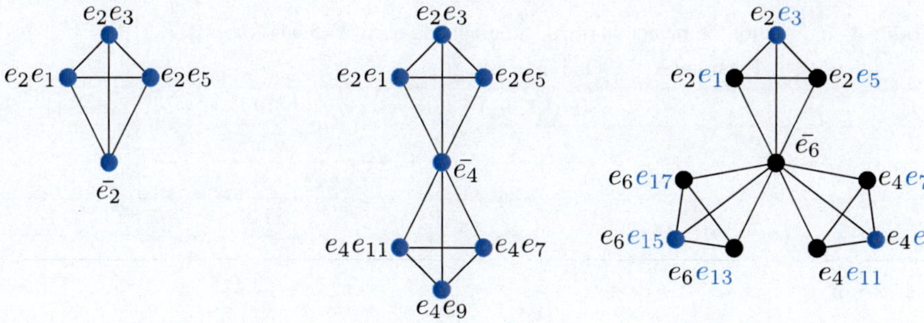

Fig. 10 Schematic depiction of the Clifford graph algebras for $W(4, m)$ for $m = 1, 2, 3$

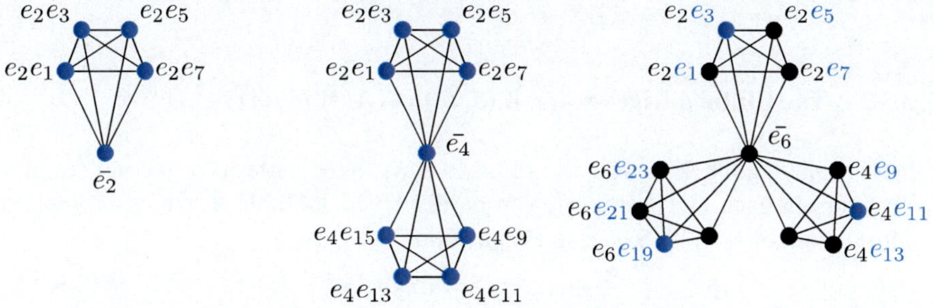

Fig. 11 Schematic depiction of $GA\big(W(5, m)\big)$ for $m = 1, 2$, and 3

3.5.1 The Clifford Algebra for $W(4, m)$: $GA\big(W(4, m)\big) \subset \mathbb{G}^{(0,6m-1)}$

As in Fig. 9, let n denote the K_4 subgraph in each $W(4, m)$ as shown in Fig. 10, with $n = 1$ corresponding to $W(4, 1)$, so that the patterns of the indices of the bivectors of these subgraphs are displayed in Table 3.

The patterns of these indices suggest the following representation for $GA\big(W(4, m)\big)$.

Proposition 6 *As a sub-algebra of* $\mathbb{G}^{(0,6m-1)}$ *the Clifford algebra for* $W(4, m)$ *can have the representation*

Table 3 Indices for the bivectors of K_4 subgraphs in each $W(4, m)$ for $m = 1, 2, 3$

K_4 subgraph	Even index	Odd indices		
n	$2n$	$6(n-1)+1$	$6(n-1)+3$	$6(n-1)+5$
1	2	1	3	5
2	4	7	9	11
3	6	13	15	17

Table 4 Indices for the bivectors of K_5 subgraphs in each $W(5, m)$ for $m = 1, 2, 3$

K_5 subgraph	Even index	Odd indices			
n	$2n$	$8(n-1)+1$	$8(n-1)+3$	$8(n-1)+5$	$8(n-1)+7$
1	2	1	3	5	7
2	4	9	11	13	15
3	6	17	19	21	23

$$GA\big(W(4,m)\big) = \big\langle e_2e_4\cdots e_{2m} \textit{ and } e_{2n}e_{6n-5}\,,\ e_{2n}e_{6n-3}\,,\ e_{2n}e_{6n-1} \quad \textit{for } n = 1, 2, \ldots, m\big\rangle.$$

3.5.2 The Clifford Algebra for $W(5, m)$: $GA\big(W(5, m)\big) \subset \mathbb{G}^{(0,8m-1)}$

With n playing the same role as in Figs. 9 and 10, the patterns of the indices of the bivectors in each $GA\big(W(5, m)\big)$ can be displayed in Table 4, which prompts the formulation of $GA\big(W(5, m)\big)$ in Proposition 7.

Proposition 7 *As a sub-algebra of* $\mathbb{G}^{(0,8m-1)}$, $GA\big(W(5, m)\big)$ *can have the representation*

$$\big\langle e_2e_4\cdots e_{2m} \textit{ and } e_{2n}e_{8n-7}\,,\ e_{2n}e_{8n-5}\,,\ e_{2n}e_{8n-3}\,,\ e_{2n}e_{8n-1} \textit{ for } n = 1, 2, \ldots, m\big\rangle.$$

3.5.3 The Clifford Graph Algebra for $W(r, m)$: $GA\big(W(r, m)\big) \subset \mathbb{G}^{(0,2m(r-1)-1)}$

By combining the tabular information from each of the Tables 2, 3, and 4 we can obtain a formula for labeling in odd indices of the bivectors in the generators for the general class $GA\big(W(r, m)\big)$ as shown in Table 5.

The previous examples for $F_m = W(3, m)$, $W(4, m)$, and $W(5, m)$ motivate the formula for the odd index in each bivector corresponding to the non-central vertices. We can derive this formula

$$2(r-1)(n-1) + j,\ \ j = 1, 3, \ldots, 2(r-2) - 1,\ \ n = 1, \ldots, m \tag{2}$$

by the following combinatorial argument.

Excluding the central vertex, which is colored black in Figs. 10 and 11, $r - 1$ vertices remain. Thus, given any of the non-central $r - 1$ vertices in the first complete K_r sub-graph with an associated bivector having j as the odd index value, to reach

Table 5 Indices for the bivectors of K_r subgraphs for $r = 3, 4, 5$

K_r subgraph size	Number of odd indexed generators per subgraph	Odd indices for generators at cycle n for $n = 1, \ldots, m$	Modular group for odd indices
3	2	$4(n-1)+1$, $4(n-1)+3$,	$\mathbb{Z}_4$
4	3	$6(n-1)+1$, $6(n-1)+3$, $6(n-1)+5$	$\mathbb{Z}_6$
5	4	$8(n-1)+1$, $8(n-1)+3$, $8(n-1)+5$, $8(n-1)+7$	$\mathbb{Z}_8$
r	$r-1$	$2(r-1)(n-1)+j$, for $j = 1, 3, \ldots, 2(r-1)-1$	$\mathbb{Z}_{2(r-1)}$

the corresponding vertex in the next complete sub-graph we must skip over the next $r-1$ vertices. Since each vertex skipped increases the odd index by 2, this skip to the next corresponding vertex is a size of $2(r-1)$, which implies that $2(r-1)-1$ is the highest odd index in the first subgraph. To reach the corresponding vertex in the n-th K_r subgraph for $n = 2, \ldots, m$, this skip size of $2(r-1)$ is repeated $n-1$ times beyond the first complete K_r sub-graph. Therefore, the entire skip size to reach the corresponding vertex at the n-th K_r subgraph is $2(r-1)(n-1)$ beyond the vertex with bivector having a generator with odd index j; and the odd index in the associated bivector is therefore $2(r-2)(n-1)+j$.

Theorem 8 *As a sub-algebra of* $\mathbb{G}^{(0,2m(r-1)-1)}$, *the Clifford graph algebra* $GA\big(W(r,m)\big)$ *for* $W(r,m)$ *can have the representation*

$$\begin{aligned} \big\langle e_2 \cdots e_{2m} \quad & \textit{and} \quad e_{2n} e_{2(r-1)(n-1)+j} \\ \textit{for} \quad n = 1, 2, \ldots, m \quad & \textit{and} \quad j = 1, 3, \ldots, 2(r-1)-1 \quad \big\rangle. \end{aligned} \tag{3}$$

Proof Until we establish this proposition, we will refer to the graph and its associated Clifford algebra as G and $GA(G)$ respectively. We first need to show that the highest index in (3) is at least as large as the total number of generators needed for $GA(G)$. This highest index value will then serve as the number of generators in the parent Clifford algebra. Each complete sub-graph K_r in G contains the central vertex and $r-1$ additional vertices. Since there are m such complete graphs in all in G, then the total number of vertices in G is $m(r-1)+1$.

Note that we only need to compare the highest odd index in (3) with total number of vertices in G since the highest odd index

$$2(r-1)(m-1) + 2(r-1) - 1 = 2m(r-1) - 1$$

occurring in (3) exceeds the highest even index because $r \geq 3$ and $m \geq 1$ imply that $2m(r-1)-1 \geq 2m(2)-1 > 2m$.

Since r and m are at least 3 and 1 respectively it follows that

$$\begin{aligned} rm &\geq 3m \\ rm &\geq m + 2m \geq m + 2 \\ 2mr - 2m - 1 &\geq m(r-1) + 1 \\ 2m(r-1) - 1 &\geq m(r-1) + 1, \end{aligned}$$

and the highest index occurring in (3) is large enough to allow the enumeration of indices in (3); hence the indices of the generators in $\mathbb{G}^{(0,2m(r-1)-1)}$ are large enough for $GA\big(W(r,m)\big)$ to be a sub-algebra of it. In particular, this insures that in Propositions 5, 6, and 7 that the containments $GA(F_m) \subset \mathbb{G}^{(0,4m-1)}$, $GA\big(W(4,m)\big) \subset \mathbb{G}^{(0,6m-1)}$ and $GA\big(W(5,m)\big) \subset \mathbb{G}^{(0,8m-1)}$ are respectively valid.

To study the connectivity in G, we will partition the vertices $v_1, \ldots, v_{m(r-1)+1}$ of G into a singleton containing the central vertex and m subsets of $r-1$ vertices denoted $S_{r,1}, S_{r,2}, \ldots, S_{r,m}$, and enumerate each subset of the vertices contained in each such set as follows. Since there are $r-1$ vertices in each of the m complete graphs K_r excluding the central vertex, naming the central vertex v_r will permit an enumeration from 1 through $r-1$ for the remaining vertices in each set S_r. In this way, we shall enumerate each such set of $r-1$ non-central vertices as:

$$\begin{array}{llllll} S_{r,1} & : v_1, & v_2, & \ldots, & v_{r-1} \\ S_{r,2} & : v_{r+1}, & v_{r+2}, & \ldots, & v_{2r-1} \\ S_{r,3} & : v_{2r+1}, & v_{2r+2}, & \ldots, & v_{3r-1} \\ \vdots & \;\;\vdots & \vdots & \vdots & \vdots \\ S_{r,m} & : v_{(m-1)r+1}, & v_{(m-1)r+2}, & \ldots, & v_{mr-1} \end{array}$$

In general, for each $n = 1, \ldots, m$, the non-central vertices in the set $S_{r,n}$ are given by $v_{r(n-1)+i}$ for $i = 1, 2, \ldots, r-1$. In particular, note that the subscript of the final vertex in this list is $r(n-1) + (r-1) = rn - 1$.

Using this enumeration for the vertices $v_1, \ldots, v_{m(r-1)+1}$, we can establish a correspondence between them and the monomials for $GA(G)$. The key to formulating this correspondence is the relationship $j = 2i - 1$ between the counters i and j, as the following lists for these counters reveal for each cycle $n = 1, \ldots, m$:

$$\text{vertices} \;:\; v_{r(n-1)+i} \qquad \text{for } i = 1,\ 2,\ 3,\ \ldots,\ r-1$$

$$\text{odd indices} \;:\; e_{2(r-1)(n-1)+j} \;\; \text{for } j = 1,\ 3,\ 5,\ \ldots,\ 2(r-1)-1\ .$$

Thus we arrive at the following correspondence between the vertices in G and the monomials in $GA(G)$ which holds for each for each $n = 1, \ldots, m, i = 1, 2, \ldots, r-1$ and $j = 1, 3, \ldots, 2(r-1) - 1$ such that $j = j(i) = 2i - 1$:

$$\text{central vertex} \quad v_r \quad \longleftrightarrow e_2 \cdots e_{2m}$$

$$\text{remaining vertices } v_{(n-1)r+i} \longleftrightarrow e_{2n} e_{2(r-1)(n-1)+j}$$

The monomial $e_2 \cdots e_{2m}$ for the central vertex v_r will anti-commute with any bivector $e_{2n}e_{2(r-1)(n-1)+j}$ for any non-central vertex $v_{(n-1)r+i}$ in any n-th subgraph K_r since these two monomials share precisely the single factor e_{2n} in common. Thus, there is an edge between v_r and any non-central vertex $v_{(n-1)r+i}$ in the n-th subgraph K_r. Since any two bivectors for distinct non-central vertices in the n-th subgraph K_r likewise share exactly the single factor e_{2n}, these will likewise anti-commute, thereby conferring an edge between these vertices. Finally, any non-central vertex $v_{(n_1-1)r+i}$ in the subgraph K_{n_1} with ordinal position n_1 and any non-central vertex $v_{(n_2-1)r+i}$ in the subgraph K_{n_2} with ordinal position n_2 will have corresponding bivectors that commute, since they share no factor in common. Hence, there is no edge between such vertices, which means that K_{n_1} and K_{n_2} share only the central vertex v_r. Therefore $G = W(r, m)$ and the algebra in (3) is a representation for $GA\big(W(r, m)\big)$. □

3.6 *The Clifford Algebra for the Class of Dutch Windmill Graphs*

We will construct $GA(D^r_m)$ in a fashion similar to $GA\big(W(r, m)\big)$ by first determining $GA(D^4_m)$ and $GA(D^5_m)$. Each of the sub-cycle graphs that occur in each Dutch windmill graph in Figs. 12 and 13 will be labeled sequentially by a counter n in a clockwise direction by increasing indices of the generators starting with $n = 1$.

Since the proof of Theorem 11 will establish Propositions 9 and 10 as special cases, we will omit their proofs here.

3.6.1 The Clifford Graph Algebra for D^4_m : $GA(D^4_m) \subset \mathbb{G}^{(0,4m)}$

By letting n denote the labeled number of the K_4 subgraphs in Fig. 12, the indices of these subgraphs can be organized as in Table 6.

Proposition 9 *As sub-algebra of $\mathbb{G}^{(0,4m)}$, the Clifford graph algebra for D^4_m can have the representation*

$$\begin{aligned} &GA(D^4_m) \\ &= \big\langle e_2e_4 \cdots e_{4m} \ \textit{and} \ e_{4(n-1)+2}e_{4(n-1)+1}\ , \\ &\qquad e_{4(n-1)+1}e_{4(n-1)+3},\ \ e_{4(n-1)+4}e_{4(n-1)+3}\ \textit{for} n = 1, 2, \ldots, m\big\rangle. \end{aligned} \tag{4}$$

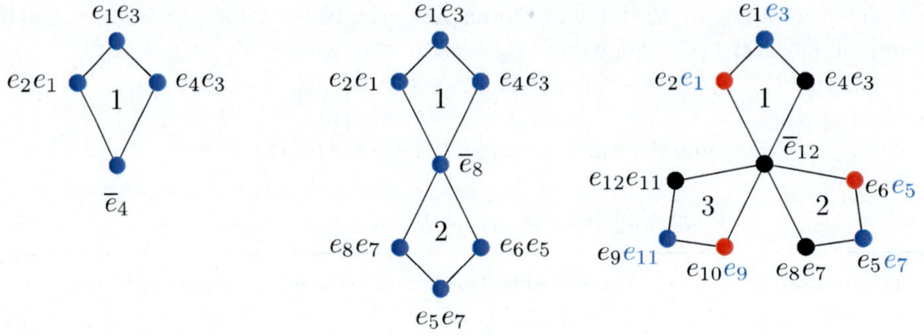

Fig. 12 Schematic depictions of $GA(D_m^4)$ for $m = 1, 2$ and 3

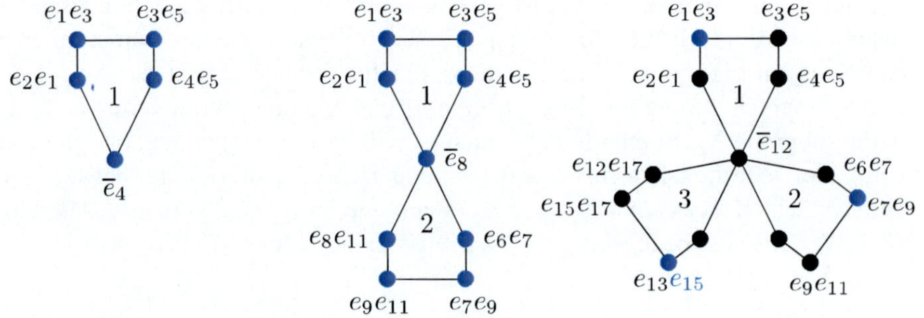

Fig. 13 Schematic depiction of $GA(D_m^5)$ for $m = 1, 2$ and 3

Table 6 Indices for the bivectors of K_4 subgraphs in each D_m^4 graph for $m = 1, 2$ and 3

Cycle	Odd indices		Even indices	
n	$4(n-1)+1$	$4(n-1)+3$	$4(n-1)+2$	$4(n-1)+4$
1	1	3	2	4
2	5	7	6	8
3	9	11	10	12

Note that the number of generators in $\mathbb{G}^{(0,4m)}$ which is the same as the highest occurring index for a generator in (4) is sufficient for labeling the number of vertices $3m+1$ in D_m^4 since $m \geq 1$ implies $4m \geq 3m+1$. Thus $\mathbb{G}^{(0,4m)}$ contains enough generators for the representation in (4) to be a sub-algebra of it.

Also note that the highest even index in (4) exceeds the highest odd index, which will not be the case for $GA(D_m^r)$ when $r \geq 5$.

Table 7 Indices for the bivectors of K_5 subgraphs in each D_m^5 graph for $m = 1, 2$ and 3

Cycle	Odd indices			Even indices	
n	$6(n-1)+1$	$6(n-1)+3$	$6(n-1)+5$	$4(n-1)+2$	$4(n-1)+4$
1	1	3	5	2	4
2	7	9	11	6	8
3	13	15	17	10	12

Table 8 Indices for the bivectors of D_m^r subgraphs for $r = 3, 4, 5$

Cycle size	Number of odd indexed generators per cycle	Odd indices for generators at cycle n for $n = 1, \ldots, m$	Modular group for odd indices
4	2	$4(n-1)+1$, $4(n-1)+3$	$\mathbb{Z}_4$
5	3	$6(n-1)+1$, $6(n-1)+3$, $6(n-1)+5$	$\mathbb{Z}_6$
r	$r-2$	$2(r-2)(n-1)+j$, for $j = 1, 3, \ldots, 2(r-2)-1$	$\mathbb{Z}_{2(r-2)}$

3.6.2 The Clifford Graph Algebra for D_m^5 : $GA(D_m^5) \subset \mathbb{G}^{(0,6m-1)}$

As in previous representations of windmill and Dutch windmill graphs, we will organize the patterns in the indices as in Table 6 (Table 7).

Proposition 10 *As a sub-algebra of* $\mathbb{G}^{(0,6m-1)}$, *the Clifford graph algebra for* D_m^5 *has the representation*

$$GA(D_m^5) = \big\langle e_2 e_4 \cdots e_{4m} \text{ and } e_{4(n-1)+2} e_{6(n-1)+1}\ ,\ e_{6(n-1)+1} e_{6(n-1)+3}\ ,\ e_{6(n-1)+3} e_{6(n-1)+5}\ ,\ e_{4(n-1)+4} e_{6(n-1)+5}\ \text{for } n = 1, 2, \ldots, m \big\rangle.$$

3.6.3 The Clifford Graph Algebra for D_m^r

The Clifford algebra for the general graph Dutch windmill graph D_m^r will become apparent when we arrange the Clifford algebras for D_m^4 and D_m^5 in Table 8.

The previous examples of $GA(D_m^4)$ and $GA(D_m^5)$ help to motivate the formula for the higher of two indices in the bivectors associated with the non-central vertices in each of the m adjoined C_r cycles. We can derive this formula

$$2(r-2)(n-1) + j,\ j = 1, 3, \ldots, 2(r-2) - 1,\ n = 1, \ldots, m \qquad (5)$$

combinatorially as follows. Excluding the central vertex and the final repeated vertex, which are colored black in Figs. 12 and 13, $r-2$ vertices remain. Thus, given any of the first $r-2$ vertices in the first C_r cycle with a bivector having the odd natural number j as the higher of two index values, to reach the corresponding vertex in the next cycle we must skip over the next $r-2$ vertices. Since each vertex skipped increases the odd index by 2, this skip to the next corresponding vertex is a size of $2(r-2)$. To reach a corresponding vertex at cycle n for $n = 2, \ldots, m$, this skip size of $2(r-2)$ is repeated $n-1$ times beyond the first C_r cycle. Therefore, the entire skip size to reach the corresponding vertex at C_r cycle n is $2(r-2)(n-1)$; and the highest odd index occurring in the bivector associated with the corresponding vertex in the n-th C_r cycle is $2(r-2)(n-1)+j$.

The largest odd index in the first C_r cycle obtains the value $2(r-2)-1$ by starting at the first vertex whose bivector has a generator with subscript 1, increasing 1 by $2(r-2)$ to reach the odd index in the corresponding bivector in the second C_r cycle, then subtracting 2 to obtain the highest odd index that repeats in the bivector for the final vertex in the first C_r cycle. Thus, this highest large odd index is $1 + 2(r-2) - 2 = 2(r-2) - 1$.

The formula (5) is the reason why the odd indices occurring in the monomials for $GA(D^r_m)$ are in $\mathbb{Z}_{2(r-2)}$.

Theorem 11 *The Clifford graph algebra for D^m_r has the representation*

$$\begin{aligned} &GA(D^r_m) \\ &= \big\langle e_2 \cdots e_{4m} \quad \textit{and} \quad e_{4(n-1)+2}e_{2(r-2)(n-1)+1}, \\ &\qquad e_{2(r-2)(n-1)+j}e_{2(r-2)(n-1)+j+2}, \quad e_{4(n-1)+4}e_{2(r-2)(n-1)+2(r-2)-1}, \\ &\qquad \textit{for} \quad n = 1, 2, \ldots, m \quad \textit{and} \quad j = 1, 3, \ldots, 2(r-2)-3 \qquad \big\rangle. \end{aligned} \tag{6}$$

and in general $GA(D^r_m) \subset \mathbb{G}^{(0,2m(r-2)-1)}$ if $r \geq 5$, otherwise if $r = 4$, then $GA(D^r_m) \subset \mathbb{G}^{(0,4m)}$.

Proof Let us temporarily reference the graph and the associated algebra in (6) as G and $GA(G)$ respectively. We first need to show that the highest index in (6) surpasses the total number of vertices in G. We already established this for the case $r = 4$ in Sect. 3.6.1, so assume that $r \geq 5$. Note that we only need to compare the highest odd index in (6) with total number of vertices in G since the highest odd index among the monomials in (6), namely $2(r-2)(m-1) + 2(r-2) - 1 = 2m(r-2) - 1$, exceeds $4(m-1) + 4 = 4m$, the highest even index, since

$$\begin{aligned} 2(r-2)(m-1) + 2(r-2) - 1 &= 2(r-2)\big[(m-1)+1\big] \\ &= 2(r-2)m - 1 \geq 2(5-2)m - 1 \\ &= 6m - 1 > 4m. \end{aligned}$$

Like the complete graph K_r, each C_r-cycle contains the central vertex and $r-1$ additional vertices in each of m cycles, so the total number of vertices in G is

$m(r-1)+1$. Since $r \geq 5$, we have that

$$\begin{aligned} r-3 &\geq 2 \\ m(r-3) &\geq 2 \\ 3m+m(r-3) &\geq 3m+2 \\ rm &\geq 3m+2 \\ 2rm-rm &\geq 4m-m+2 \\ 2rm-4m-1 &\geq rm-m+1 \\ 2m(r-2)-1 &\geq m(r-1)+1\,. \end{aligned} \tag{7}$$

Therefore this highest index is large enough to accommodate all of the monomials in (6), and the indices of the generators in $\mathbb{G}^{(0,2m(r-2)-1)}$ are large enough for $GA(D^r_m)$ to be a sub-algebra of it. In particular, this insures that in Proposition 10 the containment $GA(D^5_m) \subset \mathbb{G}^{(0,6m-1)}$ is valid. Note that (7) will not hold in the case where $r=4$ and $m=1$ since then $2m(r-2)-1=3$ but $m(r-1)+1=4$, which is why we require $r \geq 5$ for the inclusion $GA(D^r_m) \subset \mathbb{G}^{(0,2(r-2)m-1)}$.

Now consider the connectivity in G. For each $n=1,..,m$ we will denote a set of $r-1$ vertices as $C_{r,1}, C_{r,2}, \ldots, C_{r,m}$, and enumerate each subset of the vertices $v_1, \ldots, v_{m(r-1)+1}$ contained in each $C_{r,n}$ as follows. Since there are $r-1$ vertices in each such potential C_r cycle excluding the central vertex, naming the central vertex v_r will permit an enumeration from 1 through $r-1$ for the remaining vertices in each potential cycle. In this way, we shall enumerate each such set of $r-1$ non-central vertices as:

$$\begin{array}{llll} C_{r,1}: & v_1, & v_2, & \ldots, & v_{r-1} \\ C_{r,2}: & v_{r+1}, & v_{r+2}, & \ldots, & v_{2r-1} \\ C_{r,3}: & v_{2r+1}, & v_{2r+2}, & \ldots, & v_{3r-1} \\ \vdots & \vdots & \vdots & \vdots & \vdots \\ C_{r,m}: & v_{(m-1)r+1}, & v_{(m-1)r+2}, & \ldots, & v_{mr-1} \end{array}$$

In general, for each $n=1,\ldots,m$, the non-central vertices in the cycle $C_{r,n}$ are given by $v_{r(n-1)+i}$ for $i=1,2,\ldots,r-1$. We shall also refer to the vertex $v_{r(n-1)+1}$ as the initial vertex, $v_{r(n-1)+i}$ for $i=2,\ldots,r-2$ as the middle vertices, and $v_{r(n-1)+(r-1)}$ as the final vertex. In particular, note that the subscript of the final vertex in this list is $r(n-1)+(r-1)=rn-1$.

Using this enumeration for the vertices $v_1,\ldots,v_{m(r-1)+1}$, we can establish a correspondence between them and the monomials for $GA(G)$. The key to formulating this correspondence is the relationship $j+2=2i-1$ between the counters i and j for the middle vertices, as the following lists for these counters reveal for each cycle $n=1,\ldots,m$:

vertices : $v_{r(n-1)+i}$ for $i = 2, 3, 4, \ldots, r-2$

higher : $e_{2(r-1)(n-1)+j}$ for $j + 2 = 3, 5, 7, \ldots, 2(r-2) - 1$.
odd index

Thus we arrive at the following correspondence between the vertices in G and the monomials in $GA(G)$ which holds for each for each $n = 1, \ldots, m, i = 2, \ldots, r-2$ and $j = 1, 3, \ldots, 2(r-2) - 3$ such that $j + 2 = 2i - 1$; i.e. $j = j(i) = 2i - 3$:

central vertex	v_r	$\longleftrightarrow$	$e_2 \cdots e_{2m}$
initial vertex	$v_{(n-1)r+1}$	$\longleftrightarrow$	$e_{4(n-1)+2}e_{2(r-2)(n-1)+1}$
middle vertices	$v_{(n-1)r+i}$	$\longleftrightarrow$	$e_{2(r-2)(n-1)+j}e_{2(r-2)(n-1)+j+2}$
final vertex	v_{nr-1}	$\longleftrightarrow$	$e_{4(n-1)+4}e_{2(r-2)(n-1)+2(r-2)-1}$

Note that for each cycle $C_{r,n}$ the monomial $e_2 \cdots e_{4m}$ for v_r shares exactly one factor $e_{4(n-1)+2}$ with the bivector $e_{4(n-1)+2}e_{2(r-2)(n-1)+1}$ for the initial vertex $v_{(n-1)r+1}$, and it shares precisely one factor $e_{4(n-1)+4}$ with the bivector $e_{4(n-1)+4}e_{2(r-2)(n-1)+2(r-2)-1}$ for the final vertex v_{nr-1}. Thus the monomial for v_r anti-commutes with the bivector for $v_{(n-1)r+1}$ and the bivector for v_{nr-1}, so there is an edge from v_r to $v_{(n-1)r+1}$ and from v_r to v_{nr-1}.

Since the subscripts for the generators in each bivector corresponding to a middle vertex are all odd, whereas the subscripts of the generators in the monomial corresponding to the central vertex are all even, the monomial $e_2 \cdots e_{4m}$ commutes with each such bivector; so there is no edge between v_r and any middle vertex $v_{(n-1)r+i}$ for $i = 1, 3, \ldots, 2(r-2) - 3$.

In addition to the monomial for v_r, the bivector $e_{4(n-1)+2}e_{2(r-2)(n-1)+1}$ for the initial vertex $v_{(n-1)r+1}$ can only anti-commute with the bivector $e_{2(r-2)(n-1)+j}e_{(n-1)r+j+2}$ for a middle vertex $v_{(n-1)r+i}$ when $j = 1$ and consequently $i = 2$; in this case they share the common factor of $e_{2(r-2)(n-1)+1}$. Thus, the initial vertex $v_{(n-1)r+1}$ shares an edge with the central vertex v_r and the single middle vertex $v_{(n-1)r+2}$.

Likewise the bivector $e_{4(n-1)+4}e_{2(r-2)(n-1)+2(r-2)-1}$ for the final vertex v_{nr-1} can only commute with a bivector $e_{2(r-2)(n-1)+j}e_{(n-1)r+j+2}$ for a middle vertex $v_{(n-1)r+i}$ when $j = 2(r-2) - 3$ and subsequently $i = r - 2$; and they commute due to the single common factor $e_{2(r-2)(n-1)+2(r-2)-1}$. Hence, the final vertex v_{nr-1} shares an edge with the central vertex v_r and the middle vertex v_{r-2}.

The bivector $e_{2(r-2)(n-1)+j}e_{2(r-2)(n-1)+j+2}$ for the middle vertex $v_{(n-1)r+i}$ where $j = 2i - 3$ can only anti-commute with the "adjacent" bivectors for middle vertices $v_{(n-1)r+(i-1)}$ and $v_{(n-1)r+(i+1)}$. For instance, the bivector $e_{2(r-2)(n-1)+j}e_{2(r-2)(n-1)+(j+2)}$ for $v_{(n-1)r+i}$ shares the common generator $e_{2(r-2)(n-1)+(j+2)}$ with the bivector $e_{2(r-2)(n-1)+(j+2)}e_{2(r-2)(n-1)+(j+4)}$ for the vertex $v_{(n-1)r+(i+1)}$.

Finally, note that for $1 \le n_1 < n_2 \le m$ each bivector corresponding to $v_{r(n_1-1)+i}$ for $i = 1, 2, \dots, r-1$ must commute with the corresponding bivector for $v_{r(n_2-1)+i}$ because each pair of such bivectors share no common factor. Thus the cycles C_{r,n_1} and C_{r,n_2} share only the vertex v_r. Therefore, the graph associated with $GA(G)$ consists of m C_r-cycles which share exactly one vertex v_r, which is the Dutch windmill graph and so $G = D_m^r$, and $GA(G) = GA(D_m^r)$ as represented in (6). □

Remark 1 Since F_m is the Dutch windmill graph D_m^3, we can also represent $GA(F_m)$ as the special case of $GA(D_m^r)$ where $r = 3$. Note that each C_3 cycle will contain only an initial vertex, final vertex, and the common central vertex; and therefore we can represent $GA(F_m)$ as a sub-algebra of $\mathbb{G}^{(0,4m)}$ as

$$GA(F_m) = \big\langle e_2e_4 \cdots e_{4m} \text{ and } e_{4(n-1)+2}e_{4(n-1)+1}, \\ e_{4(n-1)+4}e_{4(n-1)+3} \text{ for } n = 1, 2, \dots, m\big\rangle. \tag{8}$$

4 The Friendship Theorem

In this section we will present a new proof of the Friendship Theorem. As we do so we will establish, for the benefit of future research, results that explore relationships between $Z\big(GA(G)\big)$ (the center of the algebra $GA(G)$), the adjacency matrix for G, and the parity of the cardinality of a graph with the friendship property. For the remainder of this work we will identify any $GA(G)$ abstractly by its generators using the notation

$$GA(G) = \langle e_1', e_2', \dots, e_n' \rangle,$$

rather than by the monomials which can represent these generators.

4.1 Standard Preliminaries for the Friendship Theorem

The following are standard facts about graphs with the friendship property that will be useful in presenting a new proof of the friendship theorem that uses $GA(F_m)$. We will discuss the brief proofs of these known results because they will provide a theoretical context for and give insight into formulating a new proof of the Friendship Theorem.

Proposition 12 ([1]) *If a simple graph G contains a C_4-cycle subgraph, then G cannot have the friendship property.*

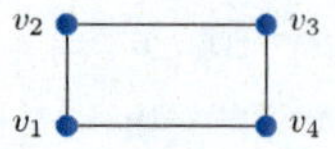

Fig. 14 A single C_4 cycle

Proof Let the C_4 cycle subgraph have vertices v_1, v_2, v_3, and v_4, as shown in Fig. 14. Since, for instance, v_1 and v_3 are friends with both v_2 and v_4 then G cannot have the friendship property. □

Proposition 13 *(i) Let G be a simple graph with the property that any two vertices are friends with at least one vertex. Then each vertex of G has degree of at least 2.*

(ii) If a graph G with the friendship property has exactly three vertices, then G consists of one C_3 cycle, and is a friendship graph.

Proof Let v_i be any vertex in this graph G, and let v_j be a vertex in G distinct from v_i as in Fig. 15. There is a vertex v_k distinct from v_i and v_j with which v_i and v_j are friends. Likewise, there is a vertex v_ℓ distinct from v_i and v_k which is friends with v_i and v_k. Thus v_i has a degree of at least 2, and (i) holds. If G has exactly three vertices, then $v_\ell = v_j$. No additional edges are possible since G is simple. Thus (ii) holds. □

Proposition 14 *Let G be a simple graph consisting of one central vertex which shares an edge with an odd number of vertices such that any two vertices are friends with at least one vertex. Then G contains a C_4 cycle.*

Proof Let v_{2m+2} denote the central vertex, and denote the other vertices as $v_1, \ldots, v_{2m}, v_{2m+1}$ where $m \in \mathbb{N}$. For the sake of illustration, we will assume that the vertices of G are arranged as in Fig. 16. By Proposition 13 the degree of each non-central vertex in G is at least 2, and by assumption each of them share an edge with v_{2m+2}. The only way a C_4 cycle could not occur among vertices v_{2m+2} and v_1 through v_{2m} is if, as Fig. 16 shows, this subgraph consists of C_3 cycles wherein each such cycle contains v_{2m}, v_{2i-1}, and v_{2i} for $i = 1, \ldots, m$. Since v_{2m+1} must share an edge with another vertex other than v_{2m+2}, it will connect with a vertex v_j for some

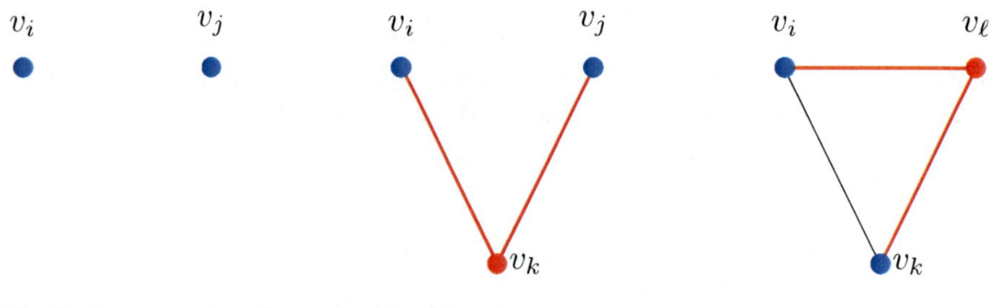

Fig. 15 Representation of a graph with a C_3 cycle

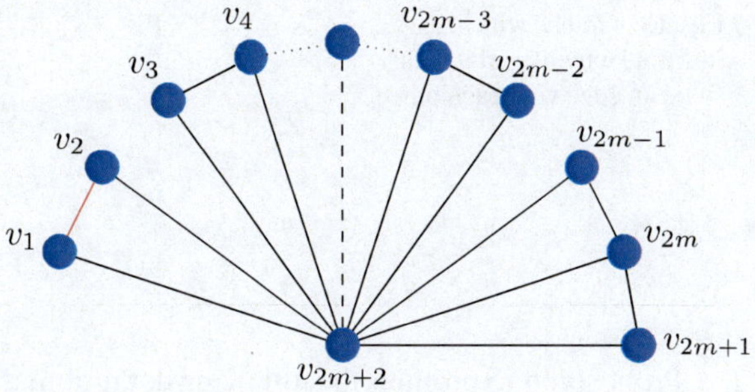

Fig. 16 Representation of a graph with a C_3 cycle

j among $\{1, \ldots, 2m\}$, which will establish a C_4 cycle, such as the cycle between v_{2m+2}, v_{2m+1}, v_{2m}, and v_{2m-1} as in Fig. 16. □

Proposition 15 *Let G be a graph with the friendship property.*

(i) If one vertex is adjoined to G, then this augmented graph cannot have the friendship property.

(ii) If two vertices can be adjoined to G by connecting edges so that the augmented graph has the friendship property; then this extension must be a C_3 cycle which connects to G at one common vertex.

Proof Suppose we can adjoin only one new vertex v_p to G such that the friendship property still holds. By part (i) of Proposition 13, v_p shares an edge with each of at least two vertices v_i and v_j of G which are already friends with a vertex v_k in G, thereby forming a C_4 cycle subgraph in this augmented graph as shown in Fig. 17, which therefore cannot have the friendship property by Proposition 12. Thus, it is not possible to preserve the friendship property by adjoining one vertex.

Now adjoin two new vertices v_p and v_q to G so that the friendship property still holds, and denote this enlarged graph as G'. In order for G' to have the friendship property each of v_p and v_q must share an edge with a vertex v_j in G, so insert these edges. By part (i) of Proposition 13 each of v_p and v_q must share an edge with another vertex. Without loss of generality, suppose that v_p shares an edge with a vertex v_i in G that is different from v_j. Since $v_i \cdot v_j$ are distinct vertices in G, they each share an edge with a vertex v_k of G by the friendship property; which establishes a C_4 cycle subgraph in the enlarged graph between v_p, v_i, v_k, and v_j, thereby implying that G' cannot have the friendship property (see Fig. 18).

Thus, v_p and v_q must share an edge with each other as in Fig. 18, and the augmented graph G' has the friendship property. □

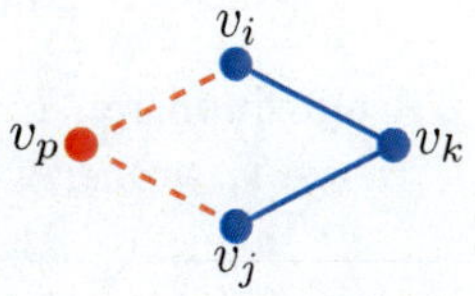

Fig. 17 A graph which cannot have the friendship property

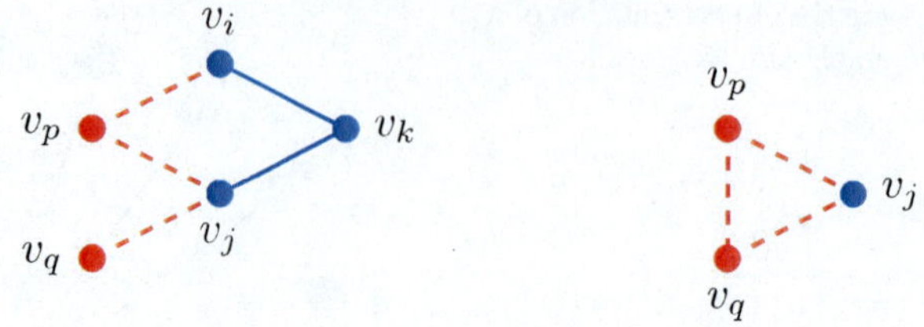

Fig. 18 Graphs which illustrate why v_p and v_q must share an edge with each other

Proposition 15 prompts the following definitions that will be instrumental in developing some properties for proving the Friendship Theorem.

Definition 9 (i) A simple graph G is said to have the quasi-friendship property if any two vertices in G are friends with at least one vertex in G but with the fewest number of common friends possible for this property to hold.
(ii) A simple graph G is friendship property extendable if G has the friendship property and the friendship property still holds when any finite number of disjoint C_3 cycles are adjoined to G so that each such C_3 cycle connects with G at one vertex.
(iii) A simple graph G with the quasi-friendship property is said to be quasi-friendship property extendable if the quasi-friendship property still holds when any finite number of disjoint C_3 cycles are adjoined to G; so that each C_3 cycle connects to G at the one vertex.

Here are some examples and some non-examples of simple graphs with the properties described in Definition 9.

Example 4 (i) A graph with the friendship property has the quasi-friendship property.
(ii) A C_4 cycle with one diagonal is an example of a simple graph which has the quasi-friendship property but not the friendship property. A C_4 cycle with no diagonal does not have the quasi-friendship property since at least one pair of vertices has no common friend. Also, a C_4 cycle with 2 diagonals (K_4) does not have the quasi-friendship property since it has at least one vertex pair that has more common friends than a C_4 cycle with one diagonal.
(iii) The C_3 cycle is the only graph with 3 vertices that is friendship property extendable, and in fact any friendship graph F_m is friendship property extendable by adjoining any finite number (n) of C_3 cycles to the central vertex of F_m to form the enlarged friendship graph F_{m+n}.
(iv) A C_4 cycle with one diagonal is quasi-friendship property extendable by adding C_3 cycles to either of the two vertices with degree 3. Note that such a C_4 cycle is the only graph with 4 vertices which has this property.

Proposition 16 *Every simple graph G with the friendship property is quasi-friendship property extendable.*

Proof First add any finite number of disjoint C_3 cycles to G so that each additional C_3 cycle connects with G at one vertex. Let G' denote this augmented graph. If G' has the property that any two vertices have at least one common friend, then G is quasi-friendship extendable in this case. Otherwise, as a simple graph we will extend G' into a complete graph K_m by adding edges, which has the property that any two vertices have at least one common friend. Therefore, K_m has a subgraph G'' which contains G' having the property that every two vertices have at least one common friend, such that the number of common friends between each such vertex pair is as few as possible. In any case, G is quasi-friendship property extendable. □

4.2 Clifford Algebra Preliminaries for the Friendship Theorem

In [11] T. Khovanova formulates the following result that relates the center of a Clifford Graph algebra with the vertices of the graph. We will present a proof of this fact.

Proposition 17 *Let G_n contain vertices $v_1, \ldots, v_n$, and let the n corresponding generators for $GA(G_n)$ be $e'_1, \ldots, e'_n$. Then the following are equivalent.*

(i) $e'_{i_1} e'_{i_2} \cdots e'_{i_r} \in Z\big(GA(G_n)\big)$.
(ii) *For each $i = 1, \ldots, n$, e'_i anti-commutes with an even number of generators in* $e'_{i_1} e'_{i_2} \cdots e'_{i_r}$.
(iii) *For each $i = 1, \ldots, n$, v_i shares an edge with an even number of vertices in* $\{v_{i_1}, v_{i_2}, \ldots, v_{i_r}\}$.

Proof Let e'_i be any generator of $GA(G_n)$. First assume that the monomial $e'_{i_1} e'_{i_2} \cdots e'_{i_r} \in Z\big(GA(G_n)\big)$, so that in particular,

$$\big(e'_i\big)\big(e'_{i_1} e'_{i_2} \cdots e'_{i_r}\big) = \big(e'_{i_1} e'_{i_2} \cdots e'_{i_r}\big)\big(e'_i\big). \tag{9}$$

If e'_i anti-commutes with an odd number k of generators in $e'_{i_1} e'_{i_2} \cdots e'_{i_r}$, then

$$\begin{aligned}\big(e'_i\big)\big(e'_{i_1} e'_{i_2} \cdots e'_{i_r}\big) &= (-1)^k \big(e'_{i_1} e'_{i_2} \cdots e'_{i_r}\big)\big(e'_i\big) \\ &= -\big(e'_{i_1} e'_{i_2} \cdots e'_{i_r}\big)\big(e'_i\big),\end{aligned} \tag{10}$$

contrary to (9), so e'_i must anti-commute with an even number of generators in $e'_{i_1} e'_{i_2} \cdots e'_{i_r}$.

Conversely, suppose that e'_i anti-commutes with an even number of k generators in $e'_{i_1} e'_{i_2} \cdots e'_{i_r}$. Then the equations in (10) again hold. Since e'_i is any generator of $GA(G_n)$, then any monomial $e'_{j_1} e'_{j_2} \cdots e'_{j_s}$ in the basis for $GA(G_n) = \langle e'_1, e'_2, \ldots, e'_n \rangle$ and hence any $u \in GA(G_n)$ commutes with $e'_{i_1} e'_{i_2} \cdots e'_{i_r}$, which means that $e'_{i_1} e'_{i_2} \cdots e'_{i_r} \in Z\big(GA(G_n)\big)$. Therefore $(i) \Leftrightarrow (ii)$.

To complete this proof, note that e'_i anti-commutes with e'_{i_k} for $k \in \{1, \ldots, r\}$ iff v_i and v_{i_k} share an edge. Thus, the number of generators among $e'_{i_1}, e'_{i_2}, \ldots, e'_{i_r}$ which e'_i anti-commutes with equals the number of vertices among $v'_{i_1}, v'_{i_2}, \ldots, v'_{i_r}$ which v_i shares an edge with. Thus in particular e'_i anti-commutes with an even number of generators in $e'_{i_1}e'_{i_2}\cdots e'_{i_r}$ iff v_i shares an edge with each of an even number of vertices in $\{v_{i_1}, v_{i_2}, \ldots, v_{i_r}\}$, and so $(ii) \Leftrightarrow (iii)$. □

Definition 10 ([11]) A monomial $e'_{i_1}e'_{i_2}\cdots e'_{i_r}$ in $GA(G) = \langle e'_1, \ldots, e_n\rangle$ is central in $GA(G)$ if it satisfies condition (i) in Proposition 17.

The following proposition in [11] relates the notion of a monomial central to $GA(G)$ to the adjacency matrix for G. We will give a brief proof of this fact.

Proposition 18 *Let $\left[a_{ij}\right]_{n\times n}$ be the adjacency matrix for G_n. If the monomial $e'_{i_1}e'_{i_2}\cdots e'_{i_r}$ is central, then the row vectors in rows $i_1, \ldots, i_r$ and column vectors in columns $i_1, \ldots, i_r$ in $\left[a_{ij}\right]_{n\times n}$ must each sum to a vector whose components are even entries.*

Proof Choose any $j \in \{1, \ldots, n\}$ in the adjacency matrix, and consider the j-th column in corresponding to v_j and e'_j. Among each row i_k for $k = 1, \ldots, r$ in this column there is a 1 iff there is an edge between v_j and v_{i_k} iff e_j and e_{i_k} anti-commute (by Proposition 17); otherwise there is a 0. Since $e'_{i_1}e'_{i_2}\cdots e'_{i_r}$ is central, e_j anti-commutes with an even number of generators in this monomial, and so there are an even number of entries among the rows i_k for $k = 1, \ldots, r$ that equal 1, and the remaining entries in these rows are 0. Since any column in $\left[a_{ij}\right]_{n\times n}$ has this property, then the row vectors of this matrix in rows $i_1, \ldots, i_r$ must sum to a vector whose components are even entries. □

We will now use these results of T. Khovanova to develop some properties about the pseudoscalar in Lemmas 19, 20, and Corollary 21 that are the key to proving The Friendship Theorem by means of a Clifford graph algebra.

Lemma 19 *Let G be a simple graph with $2m + 1$ vertices for some $m \in \mathbb{N}$ which is friendship property extendable. The following are true.*

(i) The pseudoscalar $e'_1\cdots e'_{2m+1}$ is central in $GA(G)$ where $GA(G) = \langle e'_1, \ldots, e'_{2m+1}\rangle$.
(ii) Each vertex in G has even degree of at least 2.

Proof We will proceed by induction on m. If $m = 1$, Proposition 13 implies that $G = F_3$, and so $e'_1e'_2e'_3$ is central since any arbitrary generator e'_i among e'_1, e'_2, e'_3 will anti-commute with the other two since the vertex corresponding to e'_i shares exactly one edge with each of the other two vertices in F_3. As discussed in Example 4 this graph is the only friendship property extendable graph with 3 vertices.

Now assume that G is any arbitrary graph which is friendship property extendable. Denote the vertices of G as $v_1, \ldots, v_{2m+1}$ and let $GA(G) = \langle e'_1, \ldots, e'_{2m+1}\rangle$ for some $m \in \mathbb{N}$ such that v_k corresponds to e'_k. Since G is friendship property extendable, we will adjoin two new vertices v_{2m+2} and v_{2m+3} to G at a vertex v_j of G in a C_3 cycle such that the friendship property still holds, and let G' denote this augmented graph. Note that G' is an arbitrary friendship extendable graph with $2(m+1)+1$ vertices.

Denote the generators in $GA(G')$ corresponding to v_{2m+2} and v_{2m+3} as e'_{2m+2} and e'_{2m+3}. By the friendship property there is a vertex v_j in G with which each of v_{2m+2} and v_{2m+3} share an edge. Assume that $e'_1 \cdots e'_{2m+1}$ is central in $GA(G)$. Let $e'_i \in \{e'_1, \ldots, e'_{2m+1}, e'_{2m+2}, e'_{2m+3}\}$. If $e'_i \in \{e'_1, \ldots, e'_{2m+1}\}$, then

$$(e'_i)(e'_1 \cdots e'_{2m+1}) = (e'_1 \cdots e'_{2m+1})(e'_i) \tag{11}$$

since $e'_1 \cdots e'_{2m+1}$ is central in $GA(G)$. If $e'_i \neq e'_j$ then

$$(e'_i)(e'_{2m+2}e'_{2m+3}) = (e'_{2m+2}e'_{2m+3})(e'_i) \tag{12}$$

since in this case v_i shares no edge with v_j, and so in this case (11) and (12) imply

$$(e'_i)(e'_1 \cdots e'_{2m+1}e'_{2m+2}e'_{2m+3}) = (e'_1 \cdots e'_{2m+1}e'_{2m+2}e'_{2m+3})(e'_i). \tag{13}$$

If $e'_i = e'_j$, then

$$(e'_i)(e'_{2m+2}e'_{2m+3}) = (-1)^2(e'_{2m+2}e'_{2m+3})(e'_i) = (e'_{2m+2}e'_{2m+3})(e'_i),$$

and Eq. (13) holds in this case as well. Finally, if $e'_i = e_{2m+2}$ or $e'_i = e'_{2m+3}$ then $(e'_i)(e'_{2m+2}e'_{2m+3}) = -(e'_{2m+2}e'_{2m+3})(e'_i)$. Also, $e'_ie'_j = -e'_je'_i$ since v_i and v_j share and edge. However, since e'_i shares no edge with the remaining generators in $GA(G)$ (excluding e_j), then e'_i commutes with the product of these remaining generators and so

$$\begin{aligned}(e'_i)(e'_1 \cdots e'_{2m+1}e'_{2m+2}e'_{2m+3}) &= (-1)^2(e'_1 \cdots e'_{2m+1}e'_{2m+2}e'_{2m+3})(e'_i)\\ &= (e'_1 \cdots e'_{2m+1}e'_{2m+2}e'_{2m+3})(e'_i).\end{aligned}$$

In any case, $e'_1 \cdots e'_{2m+1}e'_{2m+2}e'_{2m+3} = e'_1 \cdots e'_{2m+1}e'_{2m+2}e'_{2(m+1)+1}$ is central in $GA(G')$; so by the principle of mathematical induction, (i) is true.

By Proposition 13 the degree of each vertex in G is at least 2, and part (ii) of Proposition 17 insures that each such degree must be even; and therefore (ii) holds. □

Of course, Lemma 19 is only useful if a graph that is friendship property extendable has $2m+1$ vertices. The following lemma and corollary will show that this is the case.

Lemma 20 *Let G be a simple graph with an even number of at least 4 vertices that is quasi-friendship property extendable. Let $GA(G) = \langle e'_1, \ldots, e'_{2m}\rangle$ be the Clifford algebra for G where e'_k corresponds to v_k for $k = 1, \ldots, 2m$. Then the pseudoscalar $e'_1 \cdots e'_{2m}$ is not central in $GA(G)$.*

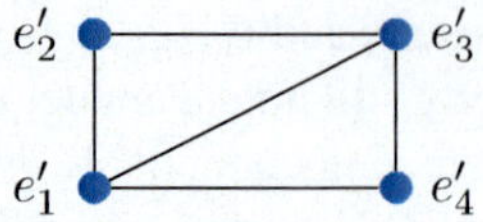

Fig. 19 Initial step in the inductive proof that the Clifford algebra a graph prescribed by Lemma 20 cannot have a central pseudo-scalar

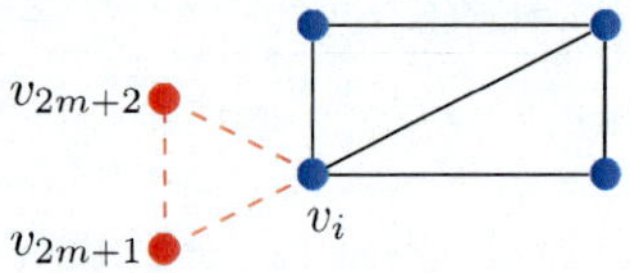

Fig. 20 General part of the inductive proof that the Clifford algebra a graph prescribed by Lemma 20 cannot have a central pseudoscalar

Proof We will proceed by induction on m where $2m$ is the number of vertices in G such that $m \geq 2$. Let $m = 2$, and let G be the C_4 cycle with one diagonal as the schematic depiction of $GA(G)$ shown in Fig. 19. Recall from Example 4 that this C_4 is the only quasi-friendship extendable graph with 4 vertices. Then

$$e'_1(e'_1e'_2e'_3e'_4) = -(e'_1e'_2e'_3e'_4)e'_1 \,.$$

Now let $m \in \mathbb{N}$ such that G is any quasi-friendship extend-able graph with $2m$ vertices and assume that there is some $k \in \mathbb{N}$ such that

$$e'_k(e'_1 \cdots e'_{2m}) = -(e'_1 \cdots e'_{2m})e'_k \,. \tag{14}$$

Since G is quasi-friendship extend-able we will adjoin two more vertices v_{2m+1} and v_{2m+2} to G such that this augmented graph G' has the property that any two vertices in G' are friends with the fewest possible number of other vertices in G'. Thus by Definition 9 these additional vertices are only friends with exactly one vertex v_i in G, and in fact v_{2m+1}, v_{2m+2}, and v_i form a C_3 cycle adjoined to G at v_i. Thus G' is an arbitrary quasi-friendship property extend-able graph with $2(m+1)$ vertices. Let e'_{2m+1} and e'_{2m+2} correspond to v_{2m+1} and v_{2m+2} (see Fig. 20).

If $i \neq k$ then v_k shares no edge with v_{2m+1} or v_{2m+2} so that $e'_k e'_{2m+1} = e'_{2m+1}e'_k$ and $e'_k e'_{2m+2} = e'_{2m+2}e'_k$. If $i = k$ then v_{2m+1} and v_{2m+2} are friends with e'_k so that $e'_k e'_{2m+1} = -e'_{2m+1}e'_k$ and $e'_k e'_{2m+2} = -e'_{2m+2}e'_k$. In any case,

$$e'_k(e'_{2m+1}e'_{2m+2}) = (e'_{2m+1}e'_{2m+2})e'_k \,. \tag{15}$$

By combining (14) and (15) we obtain

$$e'_k(e'_1 \cdots e'_{2m}e'_{2m+1}e'_{2(m+1)}) = -(e'_1 \cdots e'_{2m}e'_{2m+1}e'_{2(m+1)})e'_k .$$

Therefore by the principle of mathematical induction (14) is true for all $m \in \mathbb{N}$; and so the pseudoscalar $e'_1 \cdots e'_{2m}$ is not central in $GA(G)$. □

Corollary 21 *A simple graph G with an even number of at least* 4 *vertices cannot have the friendship property.*

Proof Suppose there is some $m_0 \in \mathbb{N}$ for which some such graph G has the friendship property. By Proposition 16, G is quasi-friendship property extendable. By Lemma 20 the pseudoscalar $e'_1 \cdots e'_{2m_0}$ is not central in $GA(G)$ where $GA(G) = \langle e'_1, \ldots, e'_{2m_0}\rangle$. Thus by part (iii) of Proposition 17 there is some $j \in \{1, \ldots, 2m_0\}$ for which v_j shares an edge with an odd number of vertices in G. By Proposition 14 G contains a C_4 cycle subgraph, and thus cannot have the friendship theorem by Proposition 12, contrary to the assumption that G has the friendship property. Therefore every graph G with an even number of at least 4 vertices cannot have the friendship property. □

4.3 Proof of the Friendship Theorem

Before proving a feature theorem in this work, we will summarizes the important properties, developed in the previous section, that hold for a graph with the friendship property.

Theorem 22 *Let G be a graph with the friendship property. Then the following are true.*

(i) G has $2m + 1$ *vertices for some* $m \in \mathbb{N}$.
(ii) Each vertex of G has even degree of at least 2.
(iii) Adjoining a C_3 *cycle to exactly one vertex of G is the only way to preserve the friendship property by adding* 2 *vertices (and necessary edges) to G.*
(iv) By Proposition 18 all of the row vectors and all of the column vectors in the adjacency matrix for G must each sum respectively to a row vector and column vector with only even entries.

Proof Note that (iii) is true by Proposition 15. By Corollary 21 (i) is true since a graph G with the friendship property must have an odd number of edges, so there is some $m \in \mathbb{N}$ such that G has $2m + 1$ vertices. Condition (ii) is then true by Lemma 19.

Finally, (iv) holds by Proposition 18 since the pseudoscalar $e'_1 \cdots e'_{2m+1}$ in $GA(G) = \langle e'_1, \ldots, e'_{2m+1}\rangle$ is central by Lemma 19. □

Theorem 23 (The Friendship Theorem) *Let G be a simple graph with the friendship property. Then G is a friendship graph.*

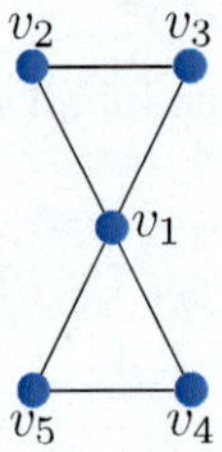

	v_1	v_2	v_3	v_4	v_5
v_1	0	1	1	1	1
v_2	1	0	1	0	0
v_3	1	1	0	0	0
v_4	1	0	0	0	1
v_5	1	0	0	1	0

Fig. 21 A graph satisfying the inductive assumption for the friendship theorem

Proof Recall from Corollary 21 that it is not possible for a graph with an even number of vertices to have the friendship property, so we will only consider graphs that have an odd number of vertices.

If G has 3 vertices, Proposition 13 implies G is a friendship graph in this case where $m = 1$. We will proceed by induction on m. Assume that for any $m \in \mathbb{N}$ such a graph G having $2m + 1$ vertices is a friendship graph, which thus has $3m$ edges. To simplify the notation, we will select $m = 2$, since this choice will include all of the considerations in a proof with more formal statements. G then has the following graph and adjacency matrix (see Fig. 21). □

As illustrated by the adjacency matrix for G in Fig. 21, the adjacency matrix $A = [a_{ij}]_{n\times n}$ (n is odd) for a friendship graph has the following features which follow directly form the definition of a friendship graph. We will keep these conditions in mind as an aid in detecting an inconsistency in the ensuing proof by contradiction.

Adjacency Matrix of a Friendship Graph.

(i) For each $i = 1, \ldots, n$, $a_{ii} = 0$.

(ii) Given any two row vectors $\mathbf{r}_{i_1}$ and $\mathbf{r}_{i_2}$ in A, there is exactly one value of $j \in \{1, \ldots, n\}$ for which the j-th column entry of $\mathbf{r}_{i_1}$ and $\mathbf{r}_{i_2}$ is a 1. That is $a_{i_1 j} = a_{i_2 j} = 1$ for exactly one $j \in \{1, \ldots, n\}$. Given any two column vectors $\mathbf{c}_{j_1}$ and $\mathbf{c}_{j_2}$ in A, there is exactly one value of $i \in \{1, \ldots, n\}$ for which the i-th row entry of $\mathbf{c}_{j_1}$ and $\mathbf{c}_{j_2}$ is a 1. That is $a_{ij_1} = a_{ij_2} = 1$ for exactly one $i \in \{1, \ldots, n\}$.

(iii) Each entry in $\mathbf{r}_1$ is a 1 except for the first; that is $a_{1j} = 1$ for $j = 2, \ldots, n$.

Each entry in $\mathbf{c}_1$ is a 1 except for the first; that is $a_{i1} = 1$ for $i = 2, \ldots, n$.

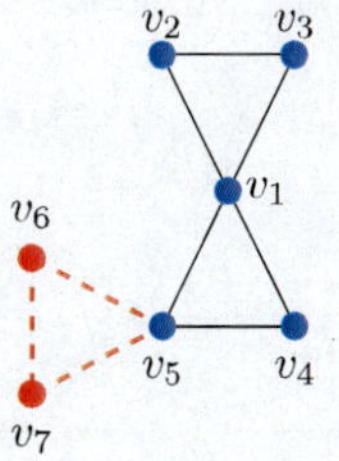

	v_1	v_2	v_3	v_4	v_5	v_6	v_7
v_1	0	1	1	1	1	0	0
v_2	1	0	1	0	0	0	0
v_3	1	1	0	0	0	0	0
v_4	1	0	0	0	1	0	0
v_5	1	0	0	1	0	1	1
v_6	0	0	0	0	1	0	1
v_7	0	0	0	0	1	1	0

Fig. 22 A contradiction results if vertices v_6 and v_7 do not share one edge with the vertex v_1

(iv) For each $i = 2, \dots, n$ and each $j = 2, \dots, n$, $\mathbf{r}_i$ and $\mathbf{c}_j$ can have only two entries that equal 1.

Condition (i) holds because a friendship graph is simple, condition (ii) is equivalent to the friendship property, and (iii) and (iv) hold because there is precisely one central vertex where the C_3 cycles which comprise the friendship graph coincide.

For the inductive step in this proof, recall from Proposition 15 that we cannot preserve the friendship property by adjoining only one vertex to G, but we can adjoin 2 vertices to a vertex of G with a C_3 cycle which, as mentioned in Example 4, makes a friendship graph friendship property extendable if this cycle adjoins to G at the central vertex. We will now prove that this must be the case.

Suppose that this C_3 cycle adjoins at a vertex other than v_1, the central vertex of G. Without loss of generality, we will choose v_5 to be the non-central vertex in G with which the pair of vertices v_6 and v_7 form a C_3 cycle. The adjacency matrix and graph for this augmentation of G are as shown in Fig. 22.

Note that the column vectors $\mathbf{c}_2$ and $\mathbf{c}_5$ contradict property (ii) because there is no value of i for which the i-th row entry for each of these columns is a 1, which means that v_2 and v_6 are friends with no vertex in the augmented graph, which thus does not have the friendship property. If we try to make v_2 and v_6 friends by adding an edge between any two vertices in this augmented graph, there will be at least one vertex with an odd degree, which will form a C_4 cycle subgraph that cannot be eliminated by adding additional edges, as shown in Fig. 23; thereby violating the friendship property by Proposition 12.

Therefore, the adjoining C_3 cycle cannot connect to G at a non-central vertex such as v_5, so the adjoining C_3 cycle must adjoin with G at its central vertex v_1 in order to preserve the friendship property as shown in Fig. 24, which is a friendship graph obtained by increasing m by 1. Therefore by the principle of mathematical induction any simple graph with the friendship property is a friendship graph.

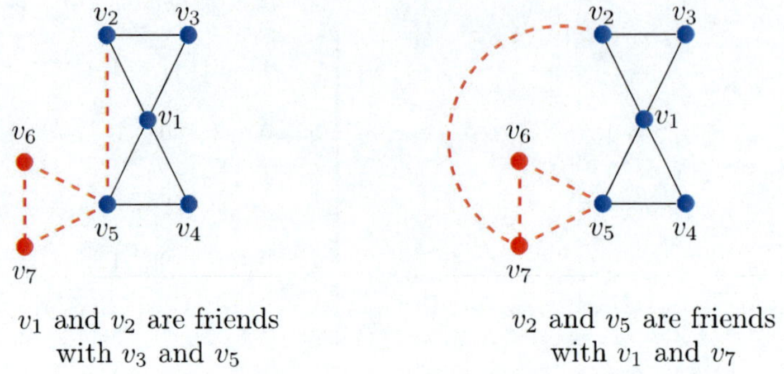

Fig. 23 A contradiction results if vertices v_6 and v_7 do not share one edge with the vertex v_1

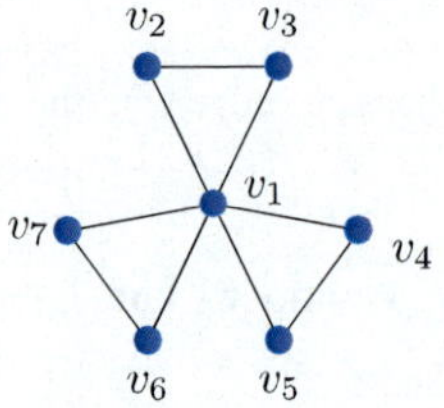

	v_1	v_2	v_3	v_4	v_5	v_6	v_7
v_1	0	1	1	1	1	1	1
v_2	1	0	1	0	0	0	0
v_3	1	1	0	0	0	0	0
v_4	1	0	0	0	1	0	0
v_5	1	0	0	1	0	0	0
v_6	1	0	0	0	0	0	1
v_7	1	0	0	0	0	1	0

Fig. 24 Vertices v_6 and v_7 must each share an edge with v_1

As part (iv) of Theorem 22 asserts, the sum of all the row vectors and column vectors of this adjacency matrix sums to a vector with only even entries.

5 Concluding Remarks

A generalized Clifford algebra in the case where $\omega_{jk} = \pm 1$ can serve as a useful tool for studying graphs. In this work, we demonstrated how to construct such an algebra for windmill and Dutch windmill graphs by selecting monomials from a parent Clifford algebra.

As discussed in Sect. 2, constructing a Clifford graph algebra by choosing monomials of minimal grade from a parent Clifford algebra may eventually suggest a way to generalize this construction process for large classes of graphs. Because decision trees are useful as a predictive tool in machine learning, developing a Clifford algebra for trees may be helpful in this endeavor.

References

1. A. Abdollahi, S. Janbaz, M. Oboudi, *Graphs Cospectral with a Friendship Graph or its Complement*, Transaction on Combinatorics, **2** No. 4, 37-52, 2013
2. E. Baylis, *Clifford (Geometric) Algebras with Applications with applications in Physics, Mathematics, and Engineering*, Birkhäuser, Boston, MA, 1996
3. E. Bell, *The Development of Mathematics*, McGraw-Hill Book Company, NY, 1945
4. E. Cartan, *Les groupes bilinéaires et les Systèmes de nombers complexes*, Annes de la faculté des sciences de Toulouse, **12 No.1**, 1898
5. W. Clifford, *Applications of Grassman's Extensive Algebra*, American Journal of Mathematics, **1 No.4** , 350-358, 1878
6. R. Descartes, *La Géométrie, 1637. Rpt. in The Geometry of René Descartes with a Facsimilie of the First Edition*, Dover Books, Inc., 1954
7. J.A. Gallian *Dynamic Survey* **DS6 Graph Labeling** , Electronic Journal of Combinatorics **DS6, 1-58**, January 2007,
8. C. Huneke *The Friendship Theorem*, The American Mathematical Monthly **109, No.2, 192-194**, February 2002,
9. R. Jagannathan *On Generalized Clifford Algebras and their Physical Applications*, retrieved from arXiv:1005.4300v1 [math.ph], May 2010,
10. M. Kanna, R. Kumar, R. Jagadeesh *Computation of Topological Indices of Dutch Windmill Graph*, Open Journal of Discrete Mathematics, **6, 74-81**, 2016,
11. T. Khovanova *Clifford Algebras and Graphs*, retrieved from arXiv:0810.3322v1 [math.CO], Oct. 2008,
12. E. Kramer *The Nature and Growth of Modern Mathematics*, Princeton University Press, Princeton, NJ, 1981,
13. A. Macdonald, *An Elementary Construction of the Geometric Algebra*, Advances in Applied Clifford Algebras, **12 No.1** , 1-6, 2002
14. A. Macdonald, *Linear and Geometric Algebra*, Createspace Independent Publishing Platform, 2011
15. T. Myers, *Constructing Clifford Graph Algebras from Classical Clifford Algebras*, Congressus Numerantium, **233**, 223-243, 2019
16. J. Schwinger, *Unity Operator Bases*, *Proc. Natl Acad Sci USA Apr* **46 No. 41** , 578-579, 1960
17. I. Stewart, *Significant Figures : The Lives and Works of Great Mathematicians*, Basic Books, NY, 2017
18. J. Sylvester, *The Collected Mathematics Papers of James Joseph Sylvester*, Cambridge University Press, **5 No.3**, 1909
19. R. Scott, S. Staples, *Dynamic Random Walks on Clifford Algebras*, Advances in Pure and Applied Mathematics, **1 No.1**, 2010
20. B. Van der Waerden, *A History of Algebra*, Springer verlag, 1985,
21. A. Vourdas, Quantum Systems with Finite Hilbert Space, Reports on Progress in Physics **67 No. 267.**, 2004,
22. H. Weil, *Quantenmachanik und Gruppentheorie*, Zeitschrift für Physik, **46**, 1-46, 1927
23. Y. wu, P. Hilton, *A Course in Modern Algebra*, John Wiley & Sons, 1974

Finding Exact Values of a Character Sum

Paul Peart and Francois Ramaroson

Abstract Let F_p be a field with p elements, where p is a positive prime. For x in F_p the quadratic character χ is defined as follows: If x is a nonzero square, then $\chi(x) = 1$; if x is a non-square, then $\chi(x) = -1$; $\chi(0) = 0$. Note that x is a square in F_p if and only if there exists a in F_p such that $x = a^2$. Let $f(x) = x^2 + bx + c$ and $g(x) = x^2 + \widetilde{b}x + \widetilde{c}$ be two irreducible polynomials in $F_p[x]$. (That is, $\chi(b^2 - 4c) = \chi(\widetilde{b}^2 - 4\widetilde{c}) = -1$). We will also assume that the resultant of $f(x)$ and $g(x)$ is nonzero in an algebraic closure of F_p. That is $\operatorname{Re} s(f, g) = \prod_{(\alpha,\beta): f(\alpha)=0 \text{ and } g(\beta)=0} (\alpha - \beta) \neq 0$, where the product is taken over all α and β in the algebraic closure for which $f(\alpha) = 0$ and $g(\beta) = 0$. It is easy to show that the above no common roots condition is equivalent to $\operatorname{Re} s(f(x), g(x)) = (c - \widetilde{c})^2 + (b - \widetilde{b})(b\widetilde{c} - \widetilde{b}c) \neq 0$. We now form the character sum W_p given by $W_p = \sum_{x \in F_p} \chi(f(x)g(x))$. We present a new method for computing W_p when $b^2 - 4c \neq \widetilde{b}^2 - 4\widetilde{c} \bmod p$. Our method involves counting points from $F_p \times F_p$ that are on a specified elliptic curve.

Keywords Quadratic character · Elliptic curve

1 Introduction

Let F_p be a field with p elements, where p is a positive prime. For x in F_p the quadratic character χ is defined by

$$\chi(x) = \begin{cases} 1, & \text{if } x \text{ is a square in } F_p \text{ and } x \neq 0 \\ -1, & \text{if } x \text{ is not a square in } F_p \text{ and } x \neq 0 \\ 0, & \text{if } x = 0 \end{cases}$$

P. Peart (✉) · F. Ramaroson
Department of Mathematics, Howard University, Washington D.C. 20059, USA
e-mail: ppeart@howard.edu

F. Hoffman (ed.), *Combinatorics, Graph Theory and Computing*,
Springer Proceedings in Mathematics & Statistics 388,
https://doi.org/10.1007/978-3-031-05375-7_5

Note that x is a square in F_p if and only if there exists a in F_p such that $x = a^2$. Throughout this paper, we will use F_p^* to denote the set of nonzero elements of F_p. That is, $F_p^* = F_p - \{0\}$.

Let $f(x) = x^2 + bx + c$ and $g(x) = x^2 + \widetilde{b}x + \widetilde{c}$ be two irreducible polynomials in $F_p[x]$. (That is, $\chi(b^2 - 4c) = \chi(\widetilde{b}^2 - 4\widetilde{c}) = -1$). We will also assume that the resultant of $f(x)$ and $g(x)$ is nonzero in an algebraic closure of F_p. That is

$$\mathrm{Re}\,s(f,\, g) = \prod_{(\alpha,\,\beta):f(\alpha)=0 \text{ and } g(\beta)=0} (\alpha - \beta) \neq 0$$

where the product is taken over all α and β in the algebraic closure for which $f(\alpha) = 0$ and $g(\beta) = 0$. It is easy to show that the above no common roots condition is equivalent to

$$\mathrm{Re}\,s(f(x),\, g(x)) = (c - \widetilde{c})^2 + \left(b - \widetilde{b}\right)\left(b\widetilde{c} - \widetilde{b}c\right) \neq 0$$

We now form the character sum W_p given by

$$W_p = \sum_{x \in F_p} \chi\left(f(x)g(x)\right)$$

It is well known (see for example Perel'muter [4]) that W_p satisfies

$$\left|W_p\right| < 2\sqrt{p}$$

In this paper, we prove that $W_p = N_p - p - 1$, where N_p is the number of points on a specified elliptic curve. Further, when the parameters in the elliptic curve satisfy certain conditions, we show how to obtain the exact numeric value of W_p for infinitely many primes. Let α be the cardinality of the set $\{x \in F_p : \chi(f(x)) = \chi(g(x)) = 1\}$. This means that α is the number of times that $\chi(f(x))$ and $\chi(g(x))$ are simultaneously squares in F_p. We will prove that

$$W_p = 4\alpha - p + 2$$

In general, a closed form expression for α in terms of p and the coefficients of $f(x)$ and $g(x)$, has not been determined. However, in this paper, we will show how to efficiently compute α in many cases.

We define the following sets.
$S = \left\{x \in F_p : \chi(f(x)) = +1\right\}$, $T = \left\{x \in F_p : \chi(f(x)) = -1\right\}$,
$U = \left\{x \in F_p : \chi(g(x)) = +1\right\}$, $V = \left\{x \in F_p : \chi(g(x)) = -1\right\}$.
The following facts result from the irreducibility of f and g.

(1) $\sum\limits_{x \in F_p} \chi(f(x)) = \sum\limits_{x \in F_p} \chi(g(x)) = -1$.

(2) $S \cup T = F_p$ and $S \cap T = \emptyset$, so $|S| + |T| = p$.

(3) $U \cup V = F_p$ and $U \cap V = \emptyset$, and so $|U| + |V| = p$.

(4) $|T| = |S| + 1$ and $|V| = |U| + 1$.

(5) $|S| = |U| = \frac{p-1}{2}$ and $|T| = |V| = \frac{p+1}{2}$.

(6) If $\alpha = |S \cap U|$, $\beta = |S \cap V|$, $\gamma = |T \cap U|$, and $\delta = |T \cap V|$, then

$\alpha = \delta - 1$, $\beta = \gamma = \frac{p+1-2\delta}{2}$.

(7) $W_p = \sum_{x \in F_p} \chi(f(x)g(x)) = 4\delta - p - 2 = 4\alpha - p + 2$, and $W_p + p - 2 = 0 \bmod 4$.

(8) Let $f(x) = x^2 + bx + c$ and $g(x) = x^2 + \widetilde{b}x + \widetilde{c}$ with $k_1 = b^2 - 4c$ and

$k_2 = \widetilde{b}^2 - 4\widetilde{c}$ non-squares in F_p. Also let $k_3 = 2(\widetilde{b} - b)$, and let L_p be the number of points from $F_p^* \times F_p^*$ on the $uv - curve$:
$u^2v - uv^2 + k_3uv - k_2u + k_1v = 0$. Then $\alpha = \frac{1}{4}L_p$.

(9) Assume that $k_1 \neq k_2$ and let N_p be the number of points from $F_p \times F_p$ on

the $xy - curve : y^2 = x^3 + (8k_2 + k_3^2 - 4k_1)x^2 + 8k_2(2k_2 - 2k_1 + k_3^2)x + 16k_2^2k_3^2$.

Then this $xy - curve$ is an elliptic curve over F_p and $N_p = L_p + 3$.

(10) $W_p = N_p - 3 - p + 2 = N_p - p - 1$.

(11) $\frac{p-2-2\sqrt{p}}{4} < \alpha < \frac{p-2+2\sqrt{p}}{4}$ and $|W_p| < 2\sqrt{p}$

(12) If $k_3 = 0$ and $k_1 = 2k_2$, the elliptic curve in (9) becomes $y^2 = x^3 - 16k_2^2x$.

Let $a = -16k_2^2 (\mathrm{mod}\ p)$, so that the elliptic curve becomes
$y^2 = h(x) = x^3 + ax$ with a nonzero. Note that the condition $k_1 = 2k_2$ requires that $\chi(2) = 1$. According to Theorem 6.2.1, p. 190 in [1],

$N_p = p + \sum_{x=0}^{p-1} \chi(h(x))$. Further, if $p \equiv 1 (\mathrm{mod}\ 4)$, we can take $p = m^2 + n^2$,

where m and n are integers with $m \equiv -\chi(2) (\mathrm{mod}\ 4)$ and $n \equiv mg^{(p-1)/4} (\mathrm{mod}\ p)$, where g is a generator of F_p^*. Now we define $l(a)$ by $a \equiv g^{l(a)} (\mathrm{mod}\ p)$ with $0 \leq l(a) \leq p - 1$. Then

$$\sum_{x=0}^{p-1} \chi(h(x)) = \left\{ \begin{array}{l} 2m(-1)^{(p-1)/4}, \text{ if } l(a) \equiv 0 (\mathrm{mod}\ 4) \\ 2n(-1)^{(p-1)/4}, \text{ if } l(a) \equiv 1 (\mathrm{mod}\ 4) \\ 2m(-1)^{(p+3)/4}, \text{ if } l(a) \equiv 2 (\mathrm{mod}\ 4) \\ 2n(-1)^{(p+3)/4}, \text{ if } l(a) \equiv 3 (\mathrm{mod}\ 4) \end{array} \right\}$$

2 Proofs

Proof (1): Fix θ in $F_p^* = \{x \in F_p : x \neq 0\}$ with θ non-square. Since $b^2 - 4c$ is a non-square in F_p, then $b^2 - 4c = \theta t^2$ for some t in F_p.
$\sum_{x\in F_p} \chi(x^2 + bx + c) = \sum_{x\in F_p} \chi\left(\left(x + \frac{b}{2}\right)^2 - \frac{b^2-4c}{4}\right) = \sum_{y\in F_p} \chi\left(y^2 - \theta t^2\right)$
$= \sum_{y\in F_p} \chi\left(t^2\right) \chi\left(\frac{y^2}{t^2} - \theta\right) = \sum_{z\in F_p} \chi\left(z^2 - \theta\right)$, independent of t.
Consider the double sum:

$$\begin{aligned}
\sum_{w\in F_p} \sum_{z\in F_p} \chi\left(z^2 - \theta w^2\right) &= \sum_{z\in F_p} \chi\left(z^2\right) + \sum_{w\in F_p^*} \sum_{z\in F_p} \chi\left(z^2 - \theta w^2\right) \\
&= p - 1 + \sum_{w\in F_p^*} \chi\left(-\theta w^2\right) + \sum_{w\in F_p^*} \sum_{z\in F_p^*} \chi\left(z^2 - \theta w^2\right) \\
&= p - 1 + (p-1)\chi\left(-\theta\right) + \sum_{w\in F_p^*} \sum_{z\in F_p^*} \chi\left(z^2 - \theta w^2\right) \\
&= p - 1 + (p-1)\chi\left(-\theta\right) + (p-1) \sum_{z\in F_p^*} \chi\left(z^2 - \theta\right) \\
&= (p-1)\left(1 + \chi(-\theta) + \sum_{z\in F_p^*} \chi\left(z^2 - \theta\right)\right) \\
&= (p-1)\left(1 + \chi(-\theta) + \sum_{z\in F_p} \chi\left(z^2 - \theta\right) - \chi(-\theta)\right) \\
&= (p-1)\left(1 + \sum_{z\in F_p} \chi\left(z^2 - \theta\right)\right).
\end{aligned}$$

Here is another evaluation of the same double sum.
$\sum_{w\in F_p} \sum_{z\in F_p} \chi\left(z^2 - \theta w^2\right) = \sum_{w\in F_p} \sum_{z\in F_p} \left(\chi \circ N_{F_{p^2}/F_p}\right)\left(z + w\sqrt{\theta}\right)$
$= \sum_{w\in F_p} \sum_{z\in F_p} \psi\left(z + w\sqrt{\theta}\right)$
where $N_{F_{p^2}/F_p} : F_{p^2} \to F_p$ is the norm map, and $\psi = \chi \circ N_{F_{p^2}/F_p}$ is a non-trivial character of F_{p^2}. But the last sum is equal to $\sum_{y\in F_{p^2}} \psi(y) = 0$. Comparing the two values of the same double sum, we get $(p-1)\left(1 + \sum_{z\in F_p} \chi\left(z^2 - \theta\right)\right) = 0$. That is

$$\sum_{z\in F_p} \chi\left(z^2-\theta\right) = -1.$$

Proof (2) and **Proof (3)**: Since f and g have no roots in F_p, for every $x \in F_p$, $\chi(f(x)) = -1$ or $+1$.

Proof (4): This follows immediately from (1).

Proof (5): (5) follows from (2), (3), and (4).

Proof (6): From (2), (3), and (5), $\alpha+\beta = |S| = \frac{p-1}{2}$, $\gamma+\delta = |T| = \frac{p-1}{2}$, $\alpha + \gamma = |U| = \frac{p-1}{2}$, $\beta+\delta = |V| = \frac{p+1}{2}$. In matrix form, we get the system of equations

$$\begin{bmatrix} 1\,1\,0\,0 \\ 0\,0\,1\,1 \\ 1\,0\,1\,0 \\ 0\,1\,0\,1 \end{bmatrix} \begin{bmatrix} \alpha \\ \beta \\ \gamma \\ \delta \end{bmatrix} = \begin{bmatrix} (p-1)/2 \\ (p+1)/2 \\ (p-1)/2 \\ (p+1)/2 \end{bmatrix}.$$

After applying elementary row operations, the system becomes

$$\begin{bmatrix} 1\,1\,0\,0 \\ 0\,1\,0\,1 \\ 0\,0\,1\,1 \\ 0\,0\,0\,0 \end{bmatrix} \begin{bmatrix} \alpha \\ \beta \\ \gamma \\ \delta \end{bmatrix} = \begin{bmatrix} (p-1)/2 \\ (p+1)/2 \\ (p+1)/2 \\ 0 \end{bmatrix}$$

Solving for α, β, and γ in terms of δ, we get $\alpha = \delta - 1$, $\beta = \gamma = (p+1-2\delta)/2$. Based on the fact that α, β, γ, and δ are nonnegative integers, we see that δ is a positive integer in the set $\{1, 2, ..., (p+1)/2\}$.

Proof (7): With reference to (6), $\mathbf{W}_p = \alpha+\delta-\gamma-\beta = \alpha+(\alpha+1) - 2\frac{p+1-2\delta}{2} = 2\alpha+1-p-1+2\delta = 2\alpha-p+2(\alpha+1) = 4\alpha-p+2 = 4(\delta-1)-p+2 = 4\delta-p-2$. α is a nonnegative integer, so W_p+p-2 is divisible by 4.

Proof (8): Let $f(x) = x^2+bx+c = \gamma^2$ and $g(x) = x^2+\widetilde{b}x+\widetilde{c} = \phi^2$, that is, $\chi(f(x)) = \chi(g(x)) = 1$. Then $(x+\frac{b}{2})^2-\gamma^2 = \frac{b^2}{4}-c \Leftrightarrow (2x+b)^2-4\gamma^2 = b^2-4c = k_1$, and $(2x+\widetilde{b})^2-4\phi^2 = \widetilde{b}^2-4\widetilde{c} = k_2$. So, $(2x+b-2\gamma)(2x+b+2\gamma) = k_1$ and $(2x+\widetilde{b}-2\phi)(2x+\widetilde{b}+2\phi) = k_2$. Let $\alpha_1 = 2x+b-2\gamma$, $\alpha_2 = 2x+b+2\gamma$, $\beta_1 = 2x+\widetilde{b}-2\phi$, $\beta_2 = 2x+\widetilde{b}+2\phi$. The system of equations becomes

$$\alpha_1\alpha_2 = k_1, \quad \beta_1\beta_2 = k_2, \quad \frac{\alpha_1+\alpha_2}{2} - b = \frac{\beta_1+\beta_2}{2} - \widetilde{b} = 2x,$$
$$\alpha_1 \neq 0, \alpha_2 \neq 0, \ \beta_1 \neq 0, \ \beta_2 \neq 0.$$

Substituting for $\alpha_2 = \frac{k_1}{\alpha_1}$ and $\beta_2 = \frac{k_2}{\alpha_2}$ in the last equation, we get

$$\frac{1}{2}(\alpha_1 + \frac{k_1}{\alpha_1}) - b = \frac{1}{2}(\beta_1 + \frac{k_2}{\beta_1}) - \widetilde{b} \Leftrightarrow$$
$$\alpha_1^2\beta_1 - \alpha_1\beta_1^2 + 2(\widetilde{b} - b)\alpha_1\beta_1 - k_2\alpha_1 + k_1\beta_1 = 0.$$

With $u = \alpha_1$ and $v = \beta_1$, we get

$$u^2v - uv^2 + k_3uv - k_2u + k_1v = 0, \ (u, v) \in F_p^* \times F_p^* \qquad (*)$$

This uv-equation when solved over $F_p^* \times F_p^*$ has the same number of solutions as the system $x^2 + bx + c = \gamma^2$, $x^2 + \widetilde{b}x + \widetilde{c} = \phi^2$ solved for (x, γ, ϕ) over $F_p \times F_p^* \times F_p^*$. Also since $(-\gamma)^2 = (p - \gamma)^2$ and $(-\phi)^2 = (p - \phi)^2$, Whenever (x, γ, ϕ) is a solution to the system, so are $(x, \gamma, -\phi)$, $(x, -\gamma, \phi)$, $(x, -\gamma, -\phi)$. So, corresponding to every x in F_p for which $\chi(f(x)) = \chi(g(x)) = 1$, there are exactly four solutions of the system. So,

$$\begin{aligned}\alpha &= \left|\{x \in K_p : \chi_p(x) = \chi_p(x) = 1\}\right| \\ &= \frac{1}{4}\left|\{(u, v) \in F_p^* \times F_p^* : u^2v - uv^2 + k_3uv - k_2u + k_1v = 0\}\right|\end{aligned}$$

Proof (9): We will transform the equation

$$u^2v - uv^2 + k_3uv - k_2u + k_1v = 0, \ (u, v) \in F_p \times F_p \qquad (**)$$

into Weierstrass form $y^2 = x^3 + A_2x^2 + A_1x + A_0$ using Nagell's algorithm as described in Connell [2, p. 116]. We note that $(**)$ has exactly one additional solution namely $(0, 0)$, over $(*)$. We begin by comparing $(**)$ to

$$s_1u^3 + s_2u^2v + s_3uv^2 + s_4v^3 + s_5u^2 + s_6uv + s_7v^2 + s_8u + s_9v = 0$$

We get

$$s_1 = s_4 = s_5 = s_7 = 0, \ s_2 = 1, \ s_3 = -1, \ s_6 = k_3, \ s_8 = -k_2 \neq 0, \ s_9 = k_1 \neq 0$$

Step 1: Since $s_9 \neq 0$.
Step 2: We let $u = \frac{U}{W}$, $v = \frac{V}{W}$, and multiply through by W^3. We get the homogeneous equation $H_3 + H_2W + H_1W^2 = 0$, where $H_3(U, V) = U^2V - UV^2$, $H_2(U, V) = k_3UV$, $H_1(U, V) = -k_2U + k_1V$. The point P with (u, v)-coordinates $(0, 0)$ has projective coordinates $(U, V, W) = (0, 0, 1)$. The tangent line at P given by $H_1 = 0 \Longleftrightarrow -k_2U + k_1V = 0$, meets the curve in the point $Q = (-e_2s_9, e_2s_8, e_3)$, where $e_i = H_i(s_9, -s_8)$, $i = 2, 3$. $e_2 = H_2(k_1, k_2) = k_1k_2k_3$ and $e_3 = H_3(k_1, k_2) = k_1^2k_2 - k_1k_2^2 = k_1k_2(k_1 - k_2)$. We note that e_2, and e_3 are not both zero, since this would require that the two quadratics f and g are the same. We will consider the case in which $e_3 \neq 0$ (that is $k_1 \neq k_2$). This is the case in which the discriminants of the the two quadratic functions are unequal non-squares in F_p^*. Also in this case,

Q is not at infinity. With the following change of coordinates, Q goes to the origin $(\widetilde{U}, \widetilde{V}, \widetilde{W}) = (0, 0, 1)$, and the tangent at P is $-k_2\widetilde{U} + k_1\widetilde{V} = 0$:
$U = \widetilde{U} - \frac{s_9 e_2}{e_3}\widetilde{W} = \widetilde{U} - B\widetilde{W}, \quad V = \widetilde{V} + \frac{s_8 e_2}{e_3}\widetilde{W} = \widetilde{V} - A\widetilde{W}, \quad W = \widetilde{W}$ where $A = \frac{k_2 k_3}{k_1 - k_2}$, and $B = \frac{k_1 k_3}{k_1 - k_2}$.
Returning to affine coordinates, let $u' = \frac{\widetilde{U}}{\widetilde{W}}$, and $v' = \frac{\widetilde{V}}{\widetilde{W}}$.
Step 3: The equation in terms of u' and v' becomes $f_3' + f_2' + f_1' = 0$ where $f_3'(u', v') = (u')^2 v' - u'(v')^2$, $f_2'(u', v') = -A(u')^2 + B(v')^2 - k_3 u'v'$, $f_1'(u', v') = (AB - k_2)u' + (k_1 - AB)v'$. Setting $v' = tu'$, the equation $f_3' + f_2' + f_1' = 0$ becomes $(u')^2 f_3'(1, t) + u' f_2'(1, t) + f_1'(1, t) = 0$. Now let $\phi_i(t) = f_i'(1, t)$. Then

$$\phi_1(t) = (AB - k_2) + (k_1 - AB)t, \quad \phi_2(t) = -A + Bt^2 - k_3 t, \quad \phi_3(t) = t - t^2$$

Therefore, the equation becomes $\phi_3(u')^2 + \phi_2 u' + \phi_1 = 0$, and thus $u' = \frac{-\phi_2 \pm \sqrt{\delta}}{2\phi_3}$, $v' = tu'$, where $\delta = \phi_2^2 - 4\phi_1\phi_3$. We note that the values of t for which $\delta = 0$, are the slopes of the tangent lines to the curve that pass through Q. One such slope is $t_0 = \frac{k_2}{k_1}$. So, $t - t_0$ is a factor of δ, and if we let $t = t_0 + \frac{1}{\tau}$, then $\rho = \tau^4\delta$ is a cubic polynomial in τ. In fact

$$\begin{aligned}\delta(t) = &\frac{k_3^2}{(1 - t_0)^2}(t^2 - (1 - t_0)t - t_0)^2 \\ &-4\left[\frac{k_3^2 t_0}{(1 - t_0)^2} - k_1 t_0 + \left(\frac{k_2}{t_0} - \frac{k_3^2 t_0}{(1 - t_0)^2}\right)t\right](t - t^2)\end{aligned}$$

and

$$\begin{aligned}\rho = \tau^4\delta\left(t_0 + \frac{1}{\tau}\right) &= 4t_0\,(k_2 - k_1)\,\tau^3 + \left(k_3^2 + 8k_2 - 4k_1\right)\tau^2 \\ &+\frac{4k_1^2 - 4k_1k_2 - 2k_3^2 k_1}{k_1 - k_2}\tau + \left(\frac{k_1 k_3}{k_1 - k_2}\right)^2\end{aligned}$$

Comparing to $\rho = c\tau^3 + d\tau^2 + e\tau + k$ in Connell [2, p. 117], we have $c = 4t_0\,(k_2 - k_1)$, $d = k_3^2 + 8k_2 - 4k_1$, $e = \frac{4k_1^2 - 4k_1k_2 - 2k_3^2 k_1}{k_1 - k_2}$, $k = \left(\frac{k_1 k_3}{k_1 - k_2}\right)^2$. We note that $c \neq 0$, which is required for $(**)$ to represent an elliptic curve. Finally, we make the substitutions $\tau = \frac{x}{c}$, and $\rho = \frac{y^2}{c^2}$ to get the Weierstrass equation

$$\begin{aligned}y^2 &= x^3 + dx^2 + cex + c^2 k \\ &= x^3 + \left(k_3^2 + 8k_2 - 4k_1\right)x^2 + 8k_2\left(k_3^2 + 2k_2 - 2k_1\right)x + 16k_2^2 k_3^2 \qquad (***)\end{aligned}$$

We note that the right side of $(***)$ can be factored, and we get

$$y^2 = (x + 4k_2)\left(x^2 + \left(k_3^2 + 4k_2 - 4k_1\right)x + 4k_2 k_3^2\right)$$

We will now show that the elliptic curve $(***)$ has exactly two more points over $F_p \times F_p$ than $(**)$. We note that on $(**)$, $u = 0 \Leftrightarrow v = 0$, so that $(0, 0)$ is the only point with $u = 0$ or $v = 0$. Note also that the point $(u, v) = (-B, -A)$ is on $(**)$. It is easily verified that $v - t_0 u = 0 \Leftrightarrow (u, v) = (0, 0)$ or $(u, v) = (-B, -A)$ .So, starting with a point (u, v) on $(**)$ with $u \neq -B$ and $u \neq 0$, set $u' = u + B$ and $v' = v + A$ and $t = \frac{v'}{u'} = \frac{v+A}{u+B}$ and compute x from the equation

$$t_0 + \frac{c}{x} = t \Leftrightarrow x = \frac{c}{t - t_0}.$$

This gives

$$x = \frac{(u + B)\,c}{v + A - t_0 u - t_0 B} = \frac{(u + B)\,c}{v - t_0 u}$$

We note that x is well defined since $t = t_0$ if and only if $u = 0$. Also $x \neq 0$. Next, compute $\delta = \phi_2^2 - 4\phi_1\phi_3$. Then $y^2 = \frac{x^4\delta}{c^2} \Leftrightarrow y = \pm\frac{x^2}{c}\sqrt{\delta}$. If $\delta = 0$, then (u, v) on $(**)$ corresponds to $(x, 0)$ on $(***)$. If $\delta \neq 0$,then (u, v) corresponds to $\left(x, \frac{x^2}{c}\sqrt{\delta}\right)$ or $\left(x, -\frac{x^2}{c}\sqrt{\delta}\right)$. To make the correspondence one-to-one, note that if $t \neq 0$ and $t \neq 1$,then $u' = \frac{-\phi_2 \pm \sqrt{\delta}}{2\phi_3}$, and $v' = tu'$. This means that the pair of points $\left(\frac{-\phi_2+\sqrt{\delta}}{2\phi_3} - B,\ t\frac{-\phi_2+\sqrt{\delta}}{2\phi_3} - A\right)$ and $\left(\frac{-\phi_2-\sqrt{\delta}}{2\phi_3} - B,\ t\frac{-\phi_2-\sqrt{\delta}}{2\phi_3} - A\right)$ are on $(**)$ and corresponds to the pair $\left(x, \frac{x^2}{c}\sqrt{\delta}\right)$ and $\left(x, -\frac{x^2}{c}\sqrt{\delta}\right)$ on $(***)$. For the one-to-one correspondence, we take $\left(\frac{-\phi_2+\sqrt{\delta}}{2\phi_3} - B,\ t\frac{-\phi_2+\sqrt{\delta}}{2\phi_3} - A\right) \longleftrightarrow \left(x, \frac{x^2}{c}\sqrt{\delta}\right)$and $\left(\frac{-\phi_2-\sqrt{\delta}}{2\phi_3} - B,\ t\frac{-\phi_2-\sqrt{\delta}}{2\phi_3} - A\right) \longleftrightarrow \left(x, -\frac{x^2}{c}\sqrt{\delta}\right)$. If $t = 0$, then $x = 4\,(k_1 - k_2)$ and the equation $\phi_3(u')^2 + \phi_2 u' + \phi_1 = 0$ for u' becomes $-Au' + AB - k_2 = 0$. If $A \neq 0$,then $u' = B - \frac{k_2}{A}$. This gives one solution on$(**)$, namely $(u, v) = \left(-\frac{k_2}{A}, -A\right)$, and two solutions $(x, y) = (4(k_1 - k_2), \pm 4k_1k_3)$ on $(***)$. Note that $A = 0 \Leftrightarrow k_3 = 0 \Leftrightarrow B = 0$. If $t = 0$ and $A = 0$, then there is no solution on $(**)$ and one solution $(4(k_1 - k_2), 0)$ on $(***)$. We note that when $k_3 = 0$, $(***)$ becomes $y^2 = x(x + 4k_2)(x - 4k_1 + 4k_2)$. Now $t = 1 \Leftrightarrow x = \frac{c}{1-t_0} \Leftrightarrow x = -4k_2 \Rightarrow y = 0$. So, $t = 1$ gives one solution $(-4k_2, 0)$ on $(***)$ When $t = 1$, there is no solution on $(**)$, since $\phi_3(1) = \theta_2(1) = 0$ and $\theta_1(1) = k_1 - k_2 \neq 0$ We now consider the case when $t = t_0$. It is easily verified that $t = t_0 \Leftrightarrow u' = B \Leftrightarrow u = 0$. So, $t = t_0$ gives one solution $(u, v) = (0, 0)$ on $(**)$ and none on $(***)$. Finally, when $u = -B \neq 0$, we have $u' = 0 \Rightarrow \theta_1 = 0 \Rightarrow t = \frac{k_2 - AB}{k_1 - AB}$. Let $t_1 = \frac{k_2 - AB}{k_1 - AB}$. Then, since $k_2 \neq k_1$, and k_1 and k_2 are non-squares, $t_1 \neq 0, 1$, and $t_1 = t_0 \Leftrightarrow A = B = 0$. On $(**)$, when $u = -B \neq 0$, the resulting quadratic equation in v gives the distinct roots $v = -A$ and $v = -\frac{k_1}{B}$. So, for $t = t_1$, we get two solutions $(-B, -A)$ and $\left(-B, -\frac{k_1}{B}\right)$ on $(**)$.

Let Λ be the set of all solutions on $(**)$. We express Λ as the union of two disjoint sets Λ_1 and Λ_2 , where

$$\Lambda_1 = \left\{(u,v) \in \Lambda : u \neq 0,\ (u,v) \neq \left(-B, \frac{-k_1}{B}\right),\ t \neq 0,\ t \neq 1\right\},$$

$$\Lambda_2 = \left\{\begin{array}{c}(u,v) \in \Lambda : u = 0, (u,v) = \left(-B, \frac{-k_1}{B}\right),\ t = 0 \\ \text{(that is } (u,v) = \left(-\frac{k_2}{A}, -A\right)),\ t = 1\end{array}\right\}$$

Also, let Ω be the set of all solutions on $(***)$, and express Ω as the union of the disjoint sets Ω_1 and Ω_2, where

$$\Omega_1 = \{(x,y) \in \Omega : x \neq 0,\ t \neq 0,\ t \neq 1\}$$
$$\Omega_2 = \{(x,y) \in \Omega : x = 0,\ t = 0,\ t = 1\}$$

We will show that $card(\Lambda_1) = card(\Omega_1)$ by describing a bijection Ψ between Λ_1 and Ω_1, and that $card(\Omega) = card(\Lambda) + 2$. The bijection Ψ between Λ_1 and Ω_1 is given by

$$\Psi(u,v) = \left(x, \frac{x^2}{c}\sqrt{\delta}\right),\ \text{if } u = \frac{-\phi_2 + \sqrt{\delta}}{2\phi_3} - B,\ \text{where } t = \frac{v+A}{u+B}, u \neq -B,$$
$$t = t_1 \text{ when } u = -B,\ \text{and } x = \frac{c}{t - t_0}$$

$$\Psi(u,v) = \left(x, -\frac{x^2\sqrt{\delta}}{c}\right),\ \text{if } u = \frac{-\phi_2 - \sqrt{\delta}}{2\phi_3} - B$$

$$\Psi^{-1}(x,y) = \left(\frac{-\phi_2 + \sqrt{\delta}}{2\phi_3} - B,\ t\left(\frac{-\phi_2 + \sqrt{\delta}}{2\phi_3}\right) - A\right),\ \text{if } y = \frac{x^2\sqrt{\delta}}{c}.$$

$$\Psi^{-1}(x,y) = \left(\frac{-\phi_2 - \sqrt{\delta}}{2\phi_3} - B,\ t\left(\frac{-\phi_2 - \sqrt{\delta}}{2\phi_3}\right) - A\right),\ \text{if } y = -\frac{x^2\sqrt{\delta}}{c}.$$

$\sqrt{\delta}$ is the smallest integer l in F_p for which $\delta = l^2$. Note that, when $(u,v) = (-B,-A)$,we take $t = t_1 = (k_2 - AB)/(k_1 - AB)$, and then $\phi_1 = 0$ and exactly one of $-\phi_2 + \sqrt{\delta}$ or $-\phi_2 - \sqrt{\delta}$ is 0. So, $\Psi(-B,-A) = \left(\frac{c}{t_1 - t_0}, \frac{c\sqrt{\delta}}{(t_1 - t_0)^2}\right)$ if $-\phi_2 + \sqrt{\delta} = 0$ and $\Psi(-B,-A) = \left(\frac{c}{t_1 - t_0}, \frac{-c\sqrt{\delta}}{(t_1 - t_0)^2}\right)$ if $-\phi_2 - \sqrt{\delta} = 0$

Case(1): $k_3 = 0$.
In this case, $A = B = 0,\ t_1 = t_0 = k_2/k_1$, and $\Lambda_2 = \{(0,0)\}$, and $\Omega_2 = \{(0,0), (4(k_1 - k_2), 0), (-4k_2, 0)\}$.
Case(2): $k_3 \neq 0$
In this case, $A \neq 0$ and $B \neq 0$, $t_1 \neq t_0$,

$$\Lambda_2 = \left\{(0,0), \left(-B, -\frac{k_1}{B}\right), \left(-\frac{k_2}{A}, -A\right)\right\},$$

and

$$\Omega_2 = \{(0, 4k_2k_3), (0, -4k_2k_3), (4(k_1 - k_2), 4k_1k_3), (4(k_1 - k_2), -4k_1k_3), (-4k_2, 0)\}$$

Proof (10): We have from (7) that $W_p = 4\alpha - p + 2$. From (8), $L_p = 4\alpha$. From (9), $N_p = L_p + 3$. So, $W_p = L_p - p + 2 = N_p - 3 - p + 2 = N_p - p - 1$.

Proof (11): Let L_p, M_p, and N_p be the number of points on $(*)$, $(**)$, $(***)$ respectively. We have proved that $L_p = 4\alpha$, $M_p = L_p + 1$, and that $N_p = L_p + 3 = M_p + 2$. Now, according to Hasse's Theorem, (see Silverman [5, p. 138])

$$|N_p - p - 1| < 2\sqrt{p} \Leftrightarrow |W_p| < 2\sqrt{p}$$

Since, $N_p = 4\alpha + 3$, we get

$$|4\alpha + 3 - p - 1| < 2\sqrt{p} \Leftrightarrow \frac{p - 2 - 2\sqrt{p}}{4} < \alpha < \frac{p - 2 + 2\sqrt{p}}{4}$$

Proof (12): This follows immediately from Theorem 6.2.1, p. 190 in [1]. A proof that $p = m^2 + n^2$ when $p \equiv 1 (\bmod 4)$ can be found in [3, p. 95]

3 Counting Points on $(* * *)$

First, using the transformation

$$x = X - \frac{A_2}{3}, \quad y = Y \quad \text{where } A_2 = k_3^2 + 8k_2 - 4k_1$$

we convert $(* * *)$ to the Weierstrass form

$$Y^2 = X^3 + a_1X + a_0$$

Let E be the set of points from $F_p \times F_p$ on this curve. Obviously, $|E| = N_p =$ the number of points on $(* * *)$. It is well known that the points in E together with the point at infinity O form an additive Abelian group in which O is the identity, and the inverse of $P = (X, Y)$ is $-P = (X, -Y)$. Addition is defined as follows: Let $P = (X_1, Y_1)$ and $Q = (X_2, Y_2)$ be two distinct points in E with $Q \neq -P$, then $P + Q = (X_1, Y_1) + (X_2, Y_2) = (X_3, Y_3)$, where $X_3 = \lambda^2 - X_1 - X_2$, $Y_3 = \lambda(X_1 - X_3) - Y_1$, $\lambda = \frac{Y_2 - Y_1}{X_2 - X_1}$. If $P = Q$, the operation is called **point doubling** and we write $2P = (X_3, Y_3)$ with $X_3 = \lambda^2 - 2X_1$, $Y_3 = \lambda(X_1 - X_3) - Y_1$, $\lambda = \frac{3X_1^2 + a_1}{2Y_1}$. Of course, $2(X, 0) = O$. If $d \in K_p$, the point denoted by dP is the point obtained by performing $d - 1$ point additions of P. There is an efficient algorithm for computing dP. This algorithm is called the **Double-and-Add Algorithm** (see Silverman, p. 364). Now, let $E' = E \cup \{O\}$. According to Hasse's Theorem (see Silverman [5, p. 138])

$$\left| \left| E' \right| - 1 - p - 1 \right| < 2\sqrt{p}$$

So, $\left| E' \right|$ is in the Hasse interval $\left(p + 2 - 2\sqrt{p},\ p + 2 + 2\sqrt{p}\right)$. Also, from (8) and (9) above, $\left| E' \right| = 0 \bmod 4$. So, in our search for $\left| E' \right|$, we only need to consider those integers in the Hasse interval that are divisible by 4. Let m be such an integer and suppose that for a point P in E, m is the only integer for which $mP = O$. Then according to Hasse's theorem $\left| E' \right| = m$, and then, we get $N_p = m - 1$. There are several efficient algorithms for finding m. The basic algorithm is the so-called Baby-Step-Giant-Step (BSGS) algorithm (Silverman [6, p. 382]). Several improvements to BSGS have been reported in the literature. In [5], Schoof describes three algorithms for counting the points on an elliptic curve over a finite field.

4 Examples

We used Mathematica 10 as an aid in working out the following examples.

Example 1 This example concerns (8) above. Let $p = 7,\ f(x) = x^2 + 5x + 3$, and $g(x) = x^2 + 6x + 3$. Then $k_1 = 6,\ k_2 = 3,\ k_3 = 2$. We solve the system $x^2 + 5x + 3 = \gamma^2,\ x^2 + 6x + 3 = \phi^2$ for (x, γ, ϕ) over $F_7 \times F_7^* \times F_7^*$. The corresponding uv- equation is $u^2v - uv^2 + 2uv - 3u + 6v = 0$, which we solve over $F_7^* \times F_7^*$. The following table shows the one-to-one correspondence between the solutions of the system and the solutions of the uv-equation.

(u, v)	(x, γ, ϕ)
(3, 5)	(4, 5, 1)
(3, 2)	(4, 5, 6)
(2, 2)	(4, 2, 6)
(2, 5)	(4, 2, 1)
(5, 1)	(5, 5, 4)
(5, 3)	(5, 5, 3)
(4, 3)	(5, 2, 3)
(4, 1)	(5, 2, 4)

So, $\alpha = \frac{1}{4} \times 8 = 2$, and $W_7(f, g) = 4\alpha - p + 2 = 3$.

Example 2 This example is about (9). As in Example 1, we take $p = 7$, $f(x) = x^2 + 5x + 3$, and $g(x) = x^2 + 6x + 3$. Then $k_1 = 6$, $k_2 = 3$, $k_3 = 2(\widetilde{b} - b) = 2$, $\mathrm{Re}\,s(f(x), g(x)) = (c - \widetilde{c})^2 + \left(b - \widetilde{b}\right)\left(b\widetilde{c} - \widetilde{b}c\right) = 3 \bmod 7 \neq 0$ $t_0 = k_2/k_1{=}4$, $t_1 = (k_2 - AB)/(k_1 - AB) = 6$, $c = 4t_0(k_2 - k_1) = 1$, $A{=}k_2k_3/(k_1 - k_2){=}2$, $B{=}k_1 k_3/(k_1 - k_2) = 4$, $\phi_1(t) = AB - k_2 + (k_1 - AB)t = 5t - 2$, $\phi_2(t){=}Bt^2 - k_3t - A = 4t^2 - 2t - 2$, $\phi_3(t) = t - t^2$, $\delta(t){=}\phi_2^2 - 4\phi_1\phi_3 = 2t^4 - 3t^3 + 2t^2 + 2t + 4$. $(**)$, the uv-equation is

$$u^2v - uv^2 + k_3uv - k_2u + k_1v = 0 \Leftrightarrow$$
$$u^2v - uv^2 + 2uv - 3u + 6v = 0 \qquad (**)$$

$(***)$, the xy-equation is

$$y^2 = x^3 + \left(k_3^2 + 8k_2 - 4k_1\right)x^2 + 8k_2\left(k_3^2 + 2k_2 - 2k_1\right)x + 16k_2^2k_3^2 \Leftrightarrow$$
$$y^2 = x^3 + 4x^2 + x + 2 \qquad (***)$$

The 9 solutions of $(**)$ and the 11 solutions of $(***)$ are contained in the following table.

x	y	u	v	t	δ	ϕ_1	ϕ_2	ϕ_3	$\sqrt{\delta}$
0	4			ND					
0	3			ND					
1	1	5	1	5	1	2	4	1	1
1	6	4	3	5	1	2	4	1	1
2	0			1	0	3	0	0	0
4	1	3	5	6	2	0	4	5	3
4	6	5	3	6	2	0	4	5	3
5	1			0	4	5	5	0	2
5	6			0	4	5	5	0	2
6	2	4	1	3	4	6	0	1	2
6	5	2	2	3	4	6	0	1	2
		0	0	$t_0 = 4$	0	4	5	2	0
		3	2	6	2	0	4	5	3
		2	5	0	4	5	5	0	2

$$\Lambda_1 = \left\{(u, v) \in \Lambda : u \neq 0,\ (u, v) \neq \left(-B, \frac{-k_1}{B}\right),\ t \neq 0,\ t \neq 1\right\}$$
$$= \{(4, 3), (5, 1), (3, 5), (5, 3), (4, 1), (2, 2)\}$$

,

$$\Lambda_2 = \left\{(u, v) \in \Lambda : u = 0, (u, v) = \left(-B, \frac{-k_1}{B}\right), \ t = 0 \text{ (that is } (u, v) = \left(-\frac{k_2}{A}, -A\right)), \ t = 1\right\}$$
$$= \{(0, 0), (3, 2), (2, 5)\},$$
$$\Omega_1 = \{(x, y) \in \Omega : x \neq 0, \ t \neq 0, \ t \neq 1\} = \{(1, 1), (1, 6), (4, 1), (4, 6), (6, 2), (6, 5)\}$$
$$\Omega_2 = \{(0, 4k_2k_3), (0, -4k_2k_3), (4(k_1 - k_2), 4k_1k_3), (4(k_1 - k_2), -4k_1k_3), (-4k_2, 0)\}$$
$$= \{(x, y) \in \Omega : x = 0, \ t = 0, \ t = 1\} = \{(0, 3), (0, 4), (5, 1), (5, 6), (2, 0)\}$$

Example 3 This example is about (9) with $k_3 = 0$. Here, we take $p = 13$, $f(x) = x^2 + x + 12$, and $g(x) = x^2 + x + 4$. Then $k_1 = 5$, $k_2 = 11$, $k_3 = 2(\widetilde{b} - b) = 0$, $\operatorname{Re}s(f(x), g(x)) = (c - \widetilde{c})^2 + (b - \widetilde{b})(b\widetilde{c} - \widetilde{b}c) = 64 \bmod 13 \neq 0$, $t_0 = k_2/k_1 = 10$, $t_1 = (k_2 - AB)/(k_1 - AB) = 10$, $c = 4t_0(k_2 - k_1) = 6$, $A = k_2k_3/(k_1 - k_2) = 0$, $B = k_1k_3/(k_1 - k_2) = 0$, $\phi_1(t) = AB - k_2 + (k_1 - AB)t = 5t + 2$, $\phi_2(t) = Bt^2 - k_3t - A = 0$, $\phi_3(t) = t - t^2$, $\delta(t) = \phi_2^2 - 4\phi_1\phi_3 = 9(5t + 2)(t - t^2)$. (**), the uv-equation is

$$u^2v - uv^2 + k_3uv - k_2u + k_1v = 0 \Leftrightarrow$$
$$u^2v - uv^2 + 2u + 5v = 0 \qquad (**)$$

$(* * *)$, the xy-equation is

$$y^2 = x^3 + \left(k_3^2 + 8k_2 - 4k_1\right)x^2 + 8k_2\left(k_3^2 + 2k_2 - 2k_1\right)x + 16k_2^2k_3^2 \Leftrightarrow$$
$$y^2 = x^3 + 3x^2 + 3x = x(x + 5)(x - 2) \qquad (* * *)$$

The 9 solutions of (**) and the 11 solutions of $(* * *)$ are contained in the following table.

x	y	u	v	t	δ	ϕ_1	ϕ_2	ϕ_3	$\sqrt{\delta}$
0	0			ND					
2	0			0	0	2	0	0	0
6	2	9	8	11	3	5	0	7	9
6	11	4	5	11	3	5	0	7	9
7	2	4	10	9	3	8	0	6	9
7	11	9	3	9	3	8	0	6	9
8	0			1	0	7	0	0	0
10	2	2	3	8	9	3	0	9	3
10	11	11	10	8	9	3	0	9	3
12	8	11	5	4	3	9	0	1	9
12	5	2	8	4	3	9	0	1	9
		0	0	$t_0 = 10$	0	0	0	1	0

$$\begin{aligned}\Lambda_1 &= \{(u, v) \in \Lambda : u \neq 0,\ t \neq 0,\ t \neq 1\}\\ &= \{(9, 8), (4, 5), (4, 10), (9, 3), (2, 3), (11, 10), (11, 5), (2, 8)\}\end{aligned}$$

$$\begin{aligned}\Lambda_2 &= \{(u, v) \in \Lambda : u = 0, t = 0, t = 1\} = \{(0, 0)\},\\ \Omega_1 &= \{(x, y) \in \Omega : x \neq 0,\ t \neq 0,\ t \neq 1\}\\ &= \{(6, 2), (6, 11), (7, 2), (7, 11), (10, 2), (10, 11), (12, 8), (12, 5)\}\\ \Omega_2 &= \{(x, y) \in \Omega : x = 0,\ t = 0,\ t = 1\} = \{(0, 0), (4(k_1 - k_2), 0), (-4k_2, 0)\}\\ &= = \{(0, 0), (2, 0), (8, 0)\}\end{aligned}$$

Example 4 The smallest positive prime for which the conditions in (12) are satisfied is $p = 17 = (\pm 4)^2 + (\pm 1)^2$. We can take $f(x) = x^2 + 5x + c$, and $g(x) = x^2 + 5x + \widetilde{c}$. 6 and 3 are two non-squares in F_{17}. So, taking $k_1 = 6$ and $k_2 = 3$, we get, $5^2 - 4c = 6$, $5^2 - 4\widetilde{c} = 3$, $c = 9$, $\widetilde{c} = 14$. So, with $k_3 = 0$, the elliptic curve becomes $y^2 = x^3 - 16k_2^2 x \Leftrightarrow y^2 = x^3 + 9x$. Now, 3 is a generator for F_{17}^*. $l(9)$ is given by $9 = 3^{l(9)} \bmod 17$. So, $l(9) = 2$. Now, $\chi(2) = 1$, since 2 is a square in F_{17}. $p = m^2 + n^2$ with $m \equiv -\chi(2) \bmod 4$ and $n \equiv m3^{(p-1)/4} \bmod 17$, we get $m = -1$ and $n = 4$, $N_p = p + 2m(-1)^{(p+3)/4} = 17 + 2 = 19$, $W_p = N_p - p - 1 = 1$.

Example 5 Take $p = 1217 = (\pm 31)^2 + (\pm 16)^2$. Then $\chi(2) = 1$, 3 is a generator for F_p^*, $m = 31$, $n \equiv m3^{(p-1)/4} \bmod 1217 = 1201$, so, $n = -16$. We take $k_1 = 6$ and $k_2 = 3$. We get $f(x) = x^2 + 607$ and $g(x) = x^2 + 912$. The elliptic curve is $y^2 = x^3 + 1073x$. $l(1073)$ is given by $1073 = 3^{l(1073)} \bmod 1217$. This gives $l(1073) = 258 \equiv 2 \bmod 4$. So, $N_p = p + 2m(-1)^{(p+3)/4} = p - 62 = 1155$, and $W_p = N_p - p - 1 = 1155 - 1217 - 1 = -63$.

Example 6 Let $p = 1299721 = (\pm 1140)^2 + (\pm 11)^2$. Then 7 is a generator of F_p^*, $m \equiv -\chi(2) \bmod 4 = -1 \bmod 4$. So, m=11, and $n \equiv m7^{(p-1)/4} \bmod p$=1140. We take $f(x) = x^2 + c$, $g(x) = x^2 + \widetilde{c}$, $k_1 = 14$, $k_2 = 7$. So, $c = 649857$, $\widetilde{c} = 974789$. The elliptic curve is $y^2 = x^3 - 16k_2^2 x \Leftrightarrow y^2 = x^3 + 1298937x$. $1298937 = 7^{l(1298937)} \bmod p$ gives $l(1298937)$=873558 $\equiv 2 \bmod 4$. So, N_p=$p + 2m(-1)^{(p+3)/4}$ $= p - 22 = 1299699$, $W_p = N_p - p - 1 = -23$.

5 Conclusion

We mention here that there are papers (see for example [7]) that are concerned with evaluating the character sum W_p. In [7], Williams gives a different elliptic curve with N_p rational points. However, his approach does not reveal the fact that $N_p - 3$

is divisible by 4. This fact can be used to significantly increase the efficiency of computing N_p. Also, we are hopeful that we may find a closed form expression for α and therefore a closed form expression for W_p.

References

1. Berndt, B, Evans, R and Williams, K. : Gauss and Jacobi Sums, 1998, Wiley.
2. Connell, I. : Elliptic Curve Handbook, 1999. Available at https://webs.ucm.es/BUCM/mat/doc8354.pdf.
3. Ireland, K and Rosen, M. :A classical introduction to modern number theory, 2nd edition, Graduate Texts in Mathematics 84, Springer-Verlag.
4. Perel'muter, G.I. : On certain sums of characters, uspehi Mat. 18, 1963.
5. Schoof, R.: Counting points on elliptic curves over finite fields, Journal de Theorie des Nombres de Bordeaux, tome 7, no:1 (1995), p. 219–254.
6. Silverman, J. : The Arithmetic of Elliptic Curves, Graduate Texts in Mathematics, 2nd Edition, Springer.
7. Williams, K. S. : Evaluation of character sums connected with elliptic curves, Proceedings of the American Mathematical Society, Volume 73, Number 3, March 1979.

On Minimum Index Stanton 4-Cycle Designs

R. C. Bunge, A. Hakes, J. Jeffries, E. Mastalio, and J. Torf

Abstract Let G be a multigraph with the underlying structure of a 4-cycle where each edge multiplicity in the set $\{1, 2, 3, 4\}$ is represented. There are three such multigraphs, and we call each of these a Stanton 4-cycle. For each such multigraph G and integer $n \geq 4$, we consider the minimum λ such that there exists a G-decomposition of ${}^{\lambda}K_n$.

Keywords Graph design theory · Graph decomposition · Stanton graph

1 Introduction

Throughout this paper, we use the term graph to refer to both simple graphs and multigraphs, but always without loops. For a graph G, we use $V(G)$ and $E(G)$ to denote the vertex set and edge set (or multiset) of G, respectively; the order and size of G are $|V(G)|$ and $|E(G)|$, respectively. For a positive integer λ and a set A, we use ${}^{\lambda}A$ to refer to the multiset containing λ copies of each element of A. For a simple graph G, by ${}^{\lambda}G$ we mean the graph with vertex set $V(G)$ and edge multiset ${}^{\lambda}E(G)$. In particular, ${}^{\lambda}K_n$ is the λ-fold complete graph on n vertices. For positive integers x, r, and s, we use xG to denote the graph with x edge-disjoint copies of G, and we use $K_{r\times s}$ to denote the complete multipartite graph with r parts of size s.

For graphs G and H with G a subgraph of H, a *G-decomposition* of H (or *(H, G)-design*) is a set (or multiset) Δ of graphs isomorphic to G such that the edge

R. C. Bunge · J. Torf
Department of Mathematics, Illinois State University, Normal, IL 61790–4520, USA

A. Hakes
Department of Mathematics, Purdue University, West Lafayette, IN 47907–2067, USA

J. Jeffries (✉)
Department of Mathematics, Iowa State University, Ames, IA 50011, USA
e-mail: joel.a.jeffries@gmail.com

E. Mastalio
Mathematics Department, University of Northern Iowa, Cedar Falls, IA 50614–0506, USA

F. Hoffman (ed.), *Combinatorics, Graph Theory and Computing*,
Springer Proceedings in Mathematics & Statistics 388,
https://doi.org/10.1007/978-3-031-05375-7_6

sets (or multisets) of the graphs in Δ partition $E(H)$. The elements of Δ are called *G-blocks*. If there exists a G-decomposition of H, we say G divides H, or we may simply write $G \mid H$. In particular, a $({}^\lambda K_n, G)$-design is called a *G-design of order n and index* λ. For results of G-designs of index 1, see [2, 3].

More recently, G-designs of higher indices have been studied for multigraphs. For example, in [10] Carter determined the spectra for G-designs of any index λ for all connected cubic multigraphs G of order at most 6. The G-designs of any order n and index λ have been investigated for various multigraphs of small order. Some examples include multigraphs with 5 edges (see [7, 13, 16]), 6 edges (see [1, 9]), 7 edges (see [4]), and 8 edges (see [5]).

2 Stanton 4-Cycles

The concept of a Stanton graph was first introduced by Chan and Sarvate in [11] as a multigraph S_k with the complete graph K_k as its underlying simple graph, but where each edge of K_k is replaced by parallel edges such that each edge multiplicity from 1 to $\binom{k}{2}$ is represented. In [6], the authors of this paper generalized this concept of a Stanton graph. Given a simple graph G with edge set $\{e_1, e_2, \ldots, e_q\}$, a *Stanton graph SG* is formed by replacing edge e_i, for each $1 \le i \le q$, with i parallel edges. For example, the graph S_3 can be considered as SK_3. However, for larger k, S_k is not unique, nor is SG for most G of size larger than 3. For example, there are three non-isomorphic Stanton 4-cycles, as seen in Fig. 1. The latter two of these graphs, G_2 and G_3, are the focus of this paper.

Formally, we define $G_2[a, b, c, d]$ to be the multigraph with vertex set $\{a, b, c, d\}$ and edge multiset $\{\{a, b\}, \{a, d\}, \{a, d\}, \{a, d\}, \{c, d\}, \{c, d\}, \{b, c\}, \{b, c\}, \{b, c\}, \{b, c\}\}$, and we define $G_3[a, b, c, d]$ to be the multigraph with vertex set $\{a, b, c, d\}$ and edge multiset $\{\{a, b\}, \{a, d\}, \{a, d\}, \{c, d\}, \{c, d\}, \{c, d\}, \{c, d\}, \{b, c\}, \{b, c\}, \{b, c\}\}$.

In [11], Chan and Sarvate found S_3-designs for all order n and minimum index λ. In [12], El-Zanati et al. extended this result to find S_3-designs for all index λ. In [6],

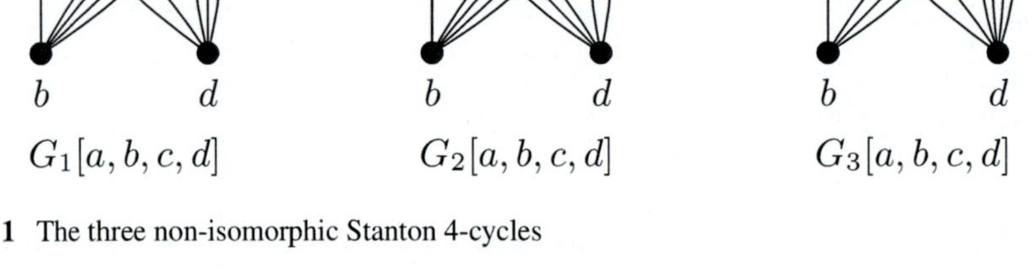

Fig. 1 The three non-isomorphic Stanton 4-cycles

the authors of this paper found G_1-designs for minimum index λ, where G_1 is the first Stanton 4-cycle shown in Fig. 1. In [14, 15], Hein and Sarvate study G-designs for minimum index λ for several graphs G of order 4 and size 3.

In this paper, we find G_2-designs and G_3-designs of minimum index λ. That is, we are interested in the following problems:

Problem 1 For each integer $n \geq 4$, find the minimum λ such that there exists a G_2-design of order n and index λ.

Problem 2 For each integer $n \geq 4$, find the minimum λ such that there exists a G_3-design of order n and index λ.

Some necessary conditions for a G-decomposition of ${}^{\lambda}K_n$ are that n must be at least the order of G, λ must be at least the largest edge multiplicity in G, and $|E(G)|$ must divide $|E({}^{\lambda}K_n)| = \lambda n(n-1)/2$. Since both G_2 and G_3 are of order 4 and size 10 and both have a maximum edge multiplicity of 4, we arrive at the following necessary conditions for G_2- and G_3-designs.

Lemma 1 *Let $n \geq 4$ and $G \in \{G_1, G_2\}$. The minimum λ for the existence of a G-decomposition of ${}^{\lambda}K_n$ is at least*

- $\lambda = 4$ *if* $n \equiv 0$ *or* $1 \pmod 5$,
- $\lambda = 5$ *if* $n \equiv 0$ *or* $1 \pmod 4$ *but* $n \not\equiv 0$ *or* $1 \pmod 5$,
- $\lambda = 10$ *otherwise.*

The task that remains is to provide sufficient conditions for G-designs for these indices, or argue the non-existence of such a design.

3 Small Decompositions

This section shows some decompositions of graphs that are used to construct larger ${}^{\lambda}K_n$ in Sect. 5.

Example 1 Let $V({}^4K_{5,5}) = \mathbb{Z}_5 \times \mathbb{Z}_2$ with the obvious bipartition and let $\Delta = \{G_2[(2+i,0),\ (4+i,1),\ (i,0),\ (i,1)] : i \in \mathbb{Z}_5\} \cup \{G_2[(3+i,0),\ (1+i,1),\ (i,0),\ (i,1)] : i \in \mathbb{Z}_5\}$. Then Δ is a G_2-decomposition of ${}^4K_{5,5}$.

Example 2 Let $V({}^6K_5) = \mathbb{Z}_5$ and let $\Delta = \{G_2[1,4,3,2],\ G_2[4,2,3,0],\ G_2[0,1,3,4],\ G_2[2,0,3,1],\ G_2[4,2,0,1],\ G_2[0,2,4,1]\}$. Then Δ is a G_2-decomposition of ${}^6K_5$.

Example 3 Let $V({}^5K_{2,2}) = \mathbb{Z}_4$ with bipartition $\{\{0,1\},\{2,3\}\}$ and let $\Delta = \{G_2[0,2,1,3],\ G_2[1,2,0,3]\}$. Then Δ is a G_2-decomposition of ${}^5K_{2,2}$.

Example 4 Let $V = V({}^5K_{2,3}) = \mathbb{Z}_5$ with bipartition $\{\{0,1,2\},\ \{3,4\}\}$ and let $\Delta = \{G_2[1,4,0,3],\ G_2[0,4,2,3],\ G_2[2,4,1,3]\}$. Then Δ is a G_2-decomposition of ${}^5K_{2,3}$.

Example 5 Let $V = V({}^5K_{2,2}) = \mathbb{Z}_4$ with bipartition $\{\{0, 1\}, \{2, 3\}\}$ and let $\Delta = \{G_3[0, 2, 1, 3], G_3[1, 3, 0, 2]\}$. Then Δ is a G_3-decomposition of ${}^5K_{2,2}$.

Example 6 Let $V({}^{10}K_{2,3}) = \mathbb{Z}_5$ with bipartition $\{\{0, 1\}, \{2, 3, 4\}\}$ and let $\Delta = \{G_3[0, 2, 1, 4], G_3[0, 2, 1, 3], G_3[1, 3, 0, 4], G_3[1, 3, 0, 2], G_3[0, 4, 1, 3], G_3[1, 4, 0, 2]\}$. Then Δ is a G_3-decomposition of ${}^{10}K_{2,3}$.

4 Decompositions via Graph Labellings

Let $V({}^\lambda K_n) = \mathbb{Z}_n$ and let G be a subgraph of ${}^\lambda K_n$. By *clicking* G, we mean applying the permutation $i \mapsto i + 1$ to $V(G)$. Moreover in this case, if $j \in \mathbb{N}$, then $G + j$ is the graph obtained from G by successively clicking G a total of j times. Also note that $G + j$ is isomorphic to G for every $j \in \mathbb{N}$.

The *length* of an edge $\{i, j\}$ in ${}^\lambda K_n$ is defined to be $\min\{|i - j|, n - |i - j|\}$. Note that if n is odd, then ${}^\lambda K_n$ consists of λn edges of length i for $i \in \{1, 2, \ldots, \frac{n-1}{2}\}$. If n is even, then ${}^\lambda K_n$ consists of λn edges of length i for $i \in \{1, 2, \ldots, \frac{n}{2} - 1\}$, and $\lambda \frac{n}{2}$ edges of length $\frac{n}{2}$.

Alternatively, we may let $V({}^\lambda K_n) = \mathbb{Z}_{n-1} \cup \{\infty\}$. Clicking a subgraph G of ${}^\lambda K_n$ in this case continues to mean applying the permutation $i \mapsto i + 1$ to $V(G)$, with the convention that $\infty + 1 = \infty$. If $i, j \in \mathbb{Z}_{n-1}$, then the length of the edge $\{i, j\}$ are defined as if $\{i, j\}$ were an edge in ${}^\lambda K_{n-1}$. The length of an edge $\{i, \infty\}$ is defined to be ∞. In this case, if n is even, there are λn edges of length i for $i \in \{1, 2, \ldots, \frac{n}{2} - 1, \infty\}$ in ${}^\lambda K_n$, and if n is odd, there are λn edges of length i for $i \in \{1, 2, \ldots, \frac{n-1}{2} - 1, \infty\}$ and $\lambda \frac{n}{2}$ edges of length $\frac{n-1}{2}$. As before, $G + j$ is defined as before, and clicking an edge does not change its length.

A G-decomposition Δ of ${}^\lambda K_n$ is said to be *cyclic* if clicking preserves the G-blocks in Δ. If $V({}^\lambda K_n) = \mathbb{Z}_{n-1} \cup \{\infty\}$, then a cyclic $({}^\lambda K_n, G)$-design is also called a 1-*rotational* $({}^\lambda K_n, G)$-design. The preservation of edge lengths in these types of designs lends itself to the idea of graph labellings.

Let n, k, and λ be positive integers such that $n = \lambda k$ or such that λ is even and $n = \lambda k + \frac{\lambda}{2}$. Let G be a multigraph of size n, order at most $\frac{2n}{\lambda} + 1$, and edge multiplicity at most λ. A λ-*fold* ρ-*labeling* of G is a one-to-one function $f\colon V(G) \to \{0, 1, \ldots, \frac{2n}{\lambda}\}$ such that the multiset

$$\left\{\min\left\{|f(u) - f(v)|, \tfrac{2n}{\lambda} + 1 - |f(u) - f(v)|\right\} : \{u, v\} \in E(G)\right\}$$
$$= \begin{cases} {}^\lambda[1, k] & \text{if } n = \lambda k, \\ {}^\lambda[1, k] \cup {}^{\frac{\lambda}{2}}\{k + 1\} & \text{if } n = \lambda k + \frac{\lambda}{2}. \end{cases}$$

Thus a λ-fold ρ-labeling of such a G induces an embedding of G in ${}^\lambda K_{\frac{2n}{\lambda}+1}$ so that G has either (1) λ edges of length i for each $i \in [1, k]$ when $n = \lambda k$ or (2) λ edges of length i for each $i \in [1, k]$ and $\frac{\lambda}{2}$ edges of length $k + 1$ when $n = \lambda k + \frac{\lambda}{2}$.

If f is a λ-fold ρ-labeling of a bipartite multigraph G with vertex bipartition $\{A, B\}$ and if for each $\{a, b\} \in E(G)$ with $a \in A$ and $b \in B$ we have $f(a) < f(b)$, then f is called an *ordered λ-fold ρ-labeling*, or λ-fold ρ^+-labeling.

Now, let G of size n be a subgraph of ${}^\lambda K_{\frac{2n}{\lambda}}$. Let w be a vertex in $V(G)$ of degree λ and let y and z be the neighbors of w (y and z need not be distinct). A *1-rotational λ-fold labeling* of G is a one-to-one function $f: V(G) \to \mathbb{Z}_{\frac{2n}{\lambda}-1} \cup \{\infty\}$ such that f restricted to $G - w$ is a λ-fold ρ-labeling, $f(w) = \infty$, $f(y) = 0$, and $f(z) \in \{0, 1\}$. If in addition G is bipartite and f restricted to $G - w$ is a λ-fold ρ^+-labeling, then f is *ordered*.

The next four theorems are proved in [8].

Theorem 1 *Let G be a subgraph of ${}^\lambda K_{\frac{2n}{\lambda}+1}$ such that $|E(G)| = n$. There exists a cyclic $({}^\lambda K_{\frac{2n}{\lambda}+1}, G)$-design if and only if G admits a λ-fold ρ-labeling.*

Theorem 2 *Let G be a bipartite subgraph of ${}^\lambda K_{\frac{2n}{\lambda}+1}$ such that $|E(G)| = n$. If G admits a 2-fold ρ^+-labeling, then there exists a cyclic $({}^\lambda K_{\frac{2n}{\lambda}x+1}, G)$-design for each positive integer x.*

Theorem 3 *Let G be a subgraph of ${}^\lambda K_{\frac{2n}{\lambda}}$ such that $|E(G)| = n$. There exists a 1-rotational G-decomposition of ${}^\lambda K_{\frac{2n}{\lambda}}$ if and only if G admits a 1-rotational λ-fold labeling.*

Theorem 4 *Let G be a bipartite subgraph of ${}^\lambda K_{\frac{2n}{\lambda}}$ such that $|E(G)| = n$. If G admits an ordered 1-rotational λ-fold labeling, then there exists a 1-rotational G-decomposition of ${}^\lambda K_{\frac{2n}{\lambda}x}$ for every positive integer x.*

The idea behind Theorems 1 and 3 can be explained easily: a ρ-labeling embeds G into ${}^\lambda K_{\frac{2n}{\lambda}+1}$ such that there are λ edges of each length in ${}^\lambda K_{\frac{2n}{\lambda}+1}$. Since edge lengths are preserved, clicking produces $\frac{2n}{\lambda}$ copies of G that provide a decomposition. The argument for 1-rotational labellings is similar.

We can also see how Theorems 2 and 4 work. Suppose we have a λ-fold ρ^+-labeling of G with bipartition $\{A, B\}$. By taking x copies of G and "stretching" the labels of the vertices in B by $\frac{2n}{\lambda}$ in each copy, we obtain a λ-fold ρ-labeling of xG. Then we get a cyclic (xG)-decomposition of ${}^\lambda K_{\frac{2n}{\lambda}x+1}$, where each copy of

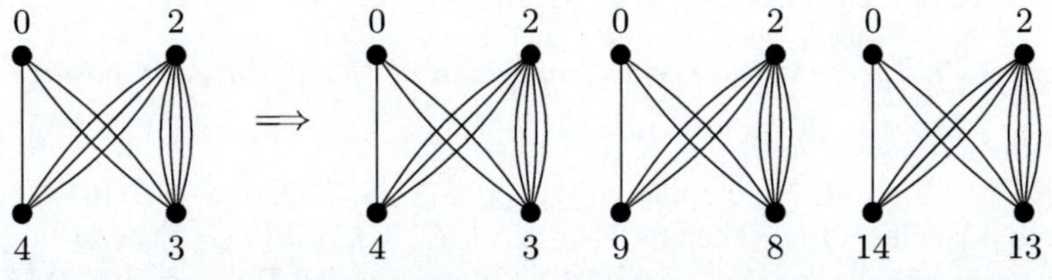

Fig. 2 A 4-fold ρ^+-labeling of G_3 and three G_3-blocks that can be used as starters for a cyclic G_3-decomposition of ${}^4K_{16}$

the x copies of G in xG are preserved by clicking. An example of this process is demonstrated in Fig. 2. A similar argument can be used for the ordered 1-rotational result. For a more formal explanation of this process, see [8] or [6].

5 Main Result

In this section, we aim to answer Problems 1 and 2. We do so by providing sufficient conditions for the existence of a G_2 or G_3-decomposition by construction. The constructions in this section rely heavily on four previous theorems. Another construction used throughout is to decompose ${}^{\lambda}K_n$ into multiple copies of smaller complete graphs connected by complete bipartite graphs. By decomposing these smaller graphs, we get a decomposition of the larger graph.

Theorem 5 *There exists a G_2-decomposition of ${}^4K_{5x}$ for every positive integer x.*

Proof Note that $G_2[\infty, 1, 0, 2]$ is a 1-rotational 4-fold labeling of G_2. Then by Theorem 3, $G_2 \mid {}^4K_5$. Consider ${}^4K_{5x}$ as $x{}^4K_5 \cup {}^4K_{x\times 5}$. By Example 1, $G_2 \mid {}^4K_{5,5}$. Then since ${}^4K_{5,5} \mid {}^4K_{x\times 5}$, we have $G_2 \mid {}^4K_{5x}$.

Theorem 6 *There exists a G_2-decomposition of ${}^4K_{5x+1}$ for every positive integer x.*

Proof Note that $G_2[1, 5, 0, 3]$ is a 4-fold ρ^+-labeling of G_2. Then by Theorem 2, $G_2 \mid {}^4K_{5x+1}$.

Theorem 7 *There exists a G_2-decomposition of ${}^5K_{4x}$ for every positive integer x.*

Proof Note that $G_2[0, \infty, 1, 2]$ is an ordered 1-rotational 5-fold labeling of G_2. Then by Theorem 4, $G_2 \mid {}^5K_{4x}$.

Note that $G_2 \nmid {}^5K_5$. This can be checked exhaustively.

Theorem 8 *There exists a G_2-decomposition of ${}^5K_{8x+1}$ for every positive integer x.*

Proof Let $G' = 2G_2$. Note that $G_2[1, 4, 0, 2] \cup G_2[0, 4, 1, 2]$ is a 5-fold ρ^+-labeling of G'. Then by Theorem 2, $G' \mid {}^5K_{8x+1}$. But $G_2 \mid G'$, so $G_2 \mid {}^5K_{8x+1}$.

Theorem 9 *There exists a G_2-decomposition of ${}^5K_{8x+5}$ for every positive integer x.*

Proof Let $G' = 3G_2$. Note that $G_2[1, 2, 0, 5] \cup G_2[1, 3, 0, 6] \cup G_2[0, 3, 2, 6]$ is a 5-fold ρ-labeling of G'. Then by Theorem 1, $G' \mid {}^5K_{13}$. But $G_2 \mid G'$, so $G_2 \mid {}^5K_{13}$.

Consider ${}^5K_{8x+5}$ as ${}^5K_{8(x-1)} \cup {}^5K_{13} \cup {}^5K_{8(x-1),13}$. By Theorem 7, $G_2 \mid {}^5K_{4x}$, and so $G_2 \mid {}^5K_{8(x-1)}$. By Examples 3 and 4, $G_2 \mid {}^5K_{2,2}$ and $G_2 \mid {}^5K_{2,3}$. Then since ${}^5K_{4x,13}$ can be decomposed into copies of ${}^5K_{2,2}$ and ${}^5K_{2,3}$, $G_2 \mid {}^5K_{8x+5}$.

Theorem 10 *There exists a G_2-decomposition of ${}^{10}K_{2x}$ for every integer $x \geq 2$.*

Proof Note that ${}^5K_{4x} \mid {}^{10}K_{4x}$. Then since $G_2 \mid {}^5K_{4x}$ by Theorem 7, we need only consider the ${}^{10}K_{4x+2}$ case.

Let $G' = 3G_2$. Note that $G_2[1, \infty, 0, 2] \cup G_2[0, 2, 1, \infty] \cup G_2[1, 3, 0, 2]$ is a 1-rotational 10-fold labeling of G'. Then by Theorem 3, $G' \mid {}^{10}K_6$. But $G_2 \mid G'$, so $G_2 \mid {}^{10}K_6$.

Consider ${}^{10}K_{4x+2}$ as ${}^{10}K_{4(x-1)} \cup {}^{10}K_6 \cup {}^{10}K_{4(x-1),6}$. By Example 4, $G_2 \mid {}^5K_{2,3}$. Then since ${}^5K_{2,3} \mid {}^{10}K_{2,3} \mid {}^{10}K_{4(x-1),6}$, $G_2 \mid {}^{10}K_{4x+2}$.

Theorem 11 *There exists a G_2-decomposition of ${}^{10}K_{2x+1}$ for every integer $x \geq 2$.*

Proof Note that ${}^5K_{4x+1} \mid {}^{10}K_{4x+1}$. Then since $G_2 \mid {}^5K_{4x+1}$ with one exception by Theorems 8 and 9, we need only consider the ${}^{10}K_{4x+3}$ case. The exception is that $G_2 \nmid {}^5K_5$. In this case, by Theorem 5, $G_2 \mid {}^4K_5$ and by Example 2, $G_2 \mid {}^6K_5$, and so indeed $G_2 \mid {}^{10}K_5$.

Let $G' = 3G_2$. Note that $G_2[0, 4, 1, 2] \cup G_2[0, 2, 1, 3] \cup G_2[3, 2, 0, 4]$ is a 10-fold ρ-labeling of G'. Then by Theorem 1, $G' \mid {}^{10}K_7$.But $G_2 \mid G'$, so $G_2 \mid {}^{10}K_7$.

Consider ${}^{10}K_{4x+3}$ as ${}^{10}K_{4(x-1)} \cup {}^{10}K_7 \cup {}^{10}K_{4(x-1),7}$. By Theorem 10, $G_2 \mid {}^{10}K_{4x}$. By Examples 3 and 4, $G_2 \mid {}^5K_{2,2}$ and $G_2 \mid {}^5K_{2,3}$. Then since ${}^{10}K_{4(x-1),7}$ can be decomposed into copies of ${}^5K_{2,2}$ and ${}^5K_{2,3}$, $G_2 \mid {}^{10}K_{4(x-1),7}$. Thus $G_2 \mid {}^{10}K_{4x+3}$.

Theorem 12 *There exists a G_3-decomposition of ${}^4K_{5x}$ for every positive integer x.*

Proof Note that $G_3[0, \infty, 1, 2]$ is an ordered 1-rotational 4-fold labeling of G_3. Then by Theorem 4, $G_3 \mid {}^4K_{5x}$.

Theorem 13 *There exists a G_3-decomposition of ${}^4K_{5x+1}$ for every positive integer x.*

Proof Note that $G_3[0, 4, 2, 3]$ is a 4-fold ρ^+-labeling of G_3. Then by Theorem 2, $G_3 \mid {}^4K_{5x+1}$.

Theorem 14 *There does not exists a G_3-decomposition of ${}^5K_{4x}$ for any positive integer x.*

Proof Assume there exists a G_3-decomposition of ${}^5K_{4x}$. Let Δ be such a set of G_3-blocks.

Consider $u, v \in V({}^5K_{4x})$ and the 5 parallel edges incident with both u and v, each of which must appear in some G_3-block of Δ. If four copies of edge $\{u, v\}$ appear in one G_3-block, then the remaining copy must appear in a G_3-block as an edge with multiplicity 1. Thus every edge of multiplicity 4 in a G_3-block of Δ must pair with an edge of multiplicity 1 in separate G_3-block. It similarly follows for the remaining edges of ${}^5K_{4x}$ that every edge of multiplicity 3 in a G_3-block must pair with an edge of multiplicity 2 in another G_3-block.

Now take a vertex v in ${}^5K_{4x}$. Let m be the number of G_3-blocks that contain v. By the structure of G_3, in each G_3-block v is incident with an edge with multiplicity 1 or multiplicity 4 and an edge with multiplicity 2 or multiplicity 3. This means there are m edges with multiplicity 1 or multiplicity 4 and m edges with multiplicity 2 or multiplicity 3.

Furthermore, since in the edges with multiplicity 1 must pair with edges with multiplicity 4 and edges with multiplicity 2 must pair with edges with multiplicity 3, there must be $2m$ sets of 5 parallel edges in ${}^5K_{4x}$ incident with v. However, v is adjacent to $4x - 1$ other vertices, a contradiction in parity. Thus, there cannot exist a G_3-decomposition of ${}^5K_{4x}$.

Theorem 15 *There exists a G_3-decomposition of ${}^5K_{4x+1}$ for every positive integer x.*

Proof Note that $G_3[0, 4, 1, 2]$ is a 4-fold ρ^+-labeling of G_3. Then by Theorem 2, $G_3 \mid {}^5K_{4x+1}$.

Theorem 16 *There exists a G_3-decomposition of ${}^{10}K_{2x}$ for every integer $x \geq 2$.*

Proof Let $G' = 2G_3$. Then $G_3[0, 2, 1, \infty] \cup G_3[0, \infty, 1, 2]$ is an ordered 1-rotational 10-fold labeling of G'. Then by Theorem 4, $G' \mid {}^{10}K_{4x}$. But $G_3 \mid G'$, so $G_3 \mid {}^{10}K_{4x}$. With this, we need only consider the ${}^{10}K_{4x+2}$ case.

Let $G'' = 3G_3$. Then $G_3[3, \infty, 0, 1] \cup G_3[0, 1, 3, \infty] \cup G_3[0, 2, 1, 4]$ is a 1-rotational 10-fold labeling of G'. Then by Theorem 3, $G'' \mid {}^{10}K_6$. But $G_3 \mid G''$, so $G_3 \mid {}^{10}K_6$.

Consider ${}^{10}K_{4x+2}$ as ${}^{10}K_{4(x-1)} \cup {}^{10}K_6 \cup {}^{10}K_{4(x-1),6}$. By Example 5, $G_3 \mid {}^5K_{2,2}$. But ${}^5K_{2,2} \mid {}^{10}K_{4(x-1),6}$, so $G_3 \mid {}^{10}K_{4(x-1),6}$. Thus $G_3 \mid {}^{10}K_{4x+2}$.

Theorem 17 *There exists a G_3-decomposition of ${}^{10}K_{2x+1}$ for every integer $x \geq 2$.*

Proof Note that ${}^5K_{4x+1} \mid {}^{10}K_{4x+1}$. Then since $G_3 \mid {}^5K_{4x+1}$ by Theorem 15, we need only consider the ${}^{10}K_{4x+3}$ case.

Let $G' = 3G_3$. Note that $G_3[1, 3, 0, 2] \cup G_3[1, 4, 0, 6] \cup G_3[0, 4, 2, 3]$ is a 10-fold ρ-labeling of G'. Then by Theorem 1, $G_3 \mid {}^{10}K_7$.But $G_3 \mid G'$, so $G_3 \mid {}^{10}K_7$.

Consider ${}^{10}K_{4x+3}$ as ${}^{10}K_{4(x-1)} \cup {}^{10}K_7 \cup {}^{10}K_{4(x-1),7}$. By Theorem 16, $G_3 \mid {}^{10}K_{4(x-1)}$. By Examples 5 and 6, $G_3 \mid {}^5K_{2,2}$ and $G_3 \mid {}^{10}K_{2,3}$. Then since ${}^{10}K_{4(x-1),7}$ can be decomposed into copies of ${}^5K_{2,2}$ and ${}^{10}K_{2,3}$, $G_3 \mid {}^{10}K_{4(x-1),7}$. Thus $G_3 \mid {}^{10}K_{4x+3}$.

We can combine these results to answer the main problems of this paper, Problems 1 and 2.

Theorem 18 *Given an integer $n \geq 4$, the minimum λ for which there is a G_2-decomposition of ${}^{\lambda}K_n$ are as follows:*

- $\lambda = 4$ *for* $n \equiv 0, 1, 5, 6, 10, 11, 15, 16 \pmod{20}$,
- $\lambda = 5$ *for* $n \equiv 4, 8, 9, 12, 13, 17 \pmod{20}$,
- $\lambda = 10$ *for* $n \equiv 2, 3, 7, 14, 18, 19 \pmod{20}$,

with the exception that the minimum λ for $n = 5$ is $\lambda = 6$.

Proof The necessity of these conditions are established in Lemma 1. All of these conditions are shown to be sufficient in the above theorems, with the exception for $n = 5$, where there is no G_2-decomposition of ${}^5K_5$. Instead, Example 2 provides sufficient conditions when $n = 5$.

Theorem 19 *Given an integer $n \geq 4$, the minimum λ for which there is a G_3-decomposition of ${}^{\lambda}K_n$ are as follows:*

- $\lambda = 4$ *for* $n \equiv 0, 1, 5, 6, 10, 11, 15, 16 \pmod{20}$,
- $\lambda = 5$ *for* $n \equiv 9, 13, 17 \pmod{20}$,
- $\lambda = 10$ *for* $n \equiv 2, 3, 4, 7, 8, 12, 14, 18, 19 \pmod{20}$.

Proof The necessity of these conditions are established in Lemma 1. All of these conditions are shown to be sufficient in the above theorems, with the exception for $n \equiv 4, 8, 12 \pmod{20}$, where there is no G_3-decomposition of ${}^5K_{4x}$ for any x as shown in Theorem 14. Then for $n \equiv 4, 8, 12 \pmod{20}$, the next value of λ such that $|E(G)|$ divides $\lambda\frac{n(n-1)}{2}$ is $\lambda = 10$. The sufficiency for a G_3-decomposition of ${}^{10}K_n$ for these value of n is shown in Theorem 16.

Acknowledgements This research is partially supported by grant number A1063038 from the Division of Mathematical Sciences at the National Science Foundation. This work was done in part while the last four authors were participants in *REU Site: Mathematics Research Experience for Pre-service and for In-service Teachers* at Illinois State University. In a forthcoming article, we plan to settle the spectrum for all n and λ for each of G_1, G_2, and G_3.

References

1. N. A. Avila Cervantes, R. C. Bunge, H. R. Bush, D. P. Roberts, A. E. Rocha, and K. M. Stilson, Spectrum for multigraph designs on four vertices and six edges, *J. Combin. Math. Combin. Comput.* **106** (2018), 69–81.
2. D. Bryant and S. El-Zanati, "Graph decompositions," *in* Handbook of Combinatorial Designs, C. J. Colbourn and J. H. Dinitz (Editors), 2nd ed., Chapman & Hall/CRC, Boca Raton, 2007, 477–485.
3. D. E. Bryant and T. A. McCourt, Existence results for G-designs, http://wiki.smp.uq.edu.au/G-designs/.
4. R. C. Bunge, M. K. Chwee, A. M. Cooper, S. I. El-Zanati, K. L. Kennedy, D. P. Roberts, and C. C. Wilson, The spectrum problem for a multigraph on 4 vertices and 7 edges, *Congr. Numer.*, **231** (2018), 239–248.
5. R. C. Bunge, S. I. El-Zanati, D. J. Gibson, D. P. Roberts, A. L. Sickman, L. A. States, and J. T. Ward, The lambda-fold spectrum problem for a multigraph on four vertices and eight edges, *Congr. Numer.*, **231** (2018), 249–257.
6. R. C. Bunge, S. I. El-Zanati, A. Hakes, J. Jeffries, E. Mastalio, and J. Torf, On decomposing lambda-fold complete graphs into the standard Stanton 4-cycle, *Congr. Numer.*, **215** (2013), 7–16.
7. R. C. Bunge, S. I. El-Zanati, L. Febles Miranda, J. Guadarrama, D. Roberts, E. Song, and A. Zale, On the lambda-fold spectra of tripartite multigraphs of order 4 and size 5, *Ars Combin.* **144** (2019), 323–343.

8. R. C. Bunge, S. I. El-Zanati, J. Mudrock, C. Vanden Eynden, and W. Wannasit, On lambda-fold Rosa-type labelings of bipartite multigraphs, *Electron. Notes Discrete Math.* **60** (2017), 11–23.
9. R. C. Bunge, J. Jeffries, J. Kirkpatrick, D. P. Roberts, and A. Sickman, Spectrum for multigraph designs on four vertices and six edges, *Congr. Numer.* **228** (2017), 29–49.
10. J. E. Carter, Designs on cubic multigraphs, Ph.D. Thesis, Department of Mathematics and Statistics, McMaster University, Canada, (1989).
11. H. Chan and D. G. Sarvate, Stanton graph decompositions, *Bull. Inst. Combin. Appl.* **64** (2012), 21–29.
12. S. I. El-Zanati, W. Lapchinda, P. Tangsupphathawat, and W. Wannasit, The spectrum for the Stanton 3-cycle, *Bull. Inst. Combin. Appl.* **69** (2013), 79–88.
13. S. Faruqi, S. A. Katre, and D. G. Sarvate, Decomposition of λK_v into Multigraphs with Four Vertices and Five Edges, *Ars Combin.* **136**, 287–316.
14. D. W. Hein, and D. Sarvate, Decompositions of λK_n using Stanton-type graphs. *J. Combin. Math. Combin. Comput.* **90** (2014), 185–195.
15. D. W. Hein, and D. Sarvate, Decompositions of λK_n into $S(4,3)$'s. *J. Combin. Math. Combin. Comput.* **94** (2015) 3–14.
16. S. Malick and D. G. Sarvate, Decomposition of λK_v into Multigraphs with Four Vertices and Five Edges, *J. Combin. Math. Combin. Comput.* **86** (2013), 221–237.

k-Plane Matroids and Whiteley's Flattening Conjectures

Brigitte Servatius

Abstract In this short note we consider the k-plane matroid, which is a matroid on the edge set, I, of a bipartite graph, $H = (A, B; I)$, defined by a counting condition. We show that $2k$-connectivity of H implies that I is a spanning set for the k-plane matroid on the edge set of the complete bipartite graph on (A, B). For $k = 2$ we explain the connections to rigidity in the plane.

1 k-Plane Matroids

Given a bipartite graph $H = (A, B; I)$, also called an incidence structure, we define the *generic k-plane matroid* $\mathbf{M}_k(H)$ on I by setting subsets $I' \subseteq I$ independent in $\mathbf{M}_k(H)$ if

$$|I''| \leq |A(I'')| + k|B(I'')| - k$$

holds for all subsets $I'' \subseteq I'$, where $A(I'')$ and $B(I'')$ are the *supports* of I'' in A and B respectively. An independent set I is *k-tight* if $|I'| = |A(I')| + k|B(I')| - k$.

Given an incidence structure $H = (A, B, I)$, we define its associated *butterfly graph* as follows. The vertex set consists of the *spine vertex set* A together with the set of *wing vertices* $B \times \{1, \ldots, k\}$. For each $(a, b) \in I$ there are k edges $(a, (b, i))$ in the butterfly graph.

Examples of butterfly graphs of 2-tight graphs, with vertices in A colored black, vertices in B colored white, are given in Figs. 1 and 2, where, after doubling one edge, a 2-tree decomposition of the butterfly graph is indicated by the edge colors to illustrate Theorem 1.

B. Servatius (✉)
Worcester Polytechnic Institute, Worcester, MA, USA
e-mail: bservat@wpi.edu

F. Hoffman (ed.), *Combinatorics, Graph Theory and Computing*,
Springer Proceedings in Mathematics & Statistics 388,
https://doi.org/10.1007/978-3-031-05375-7_7

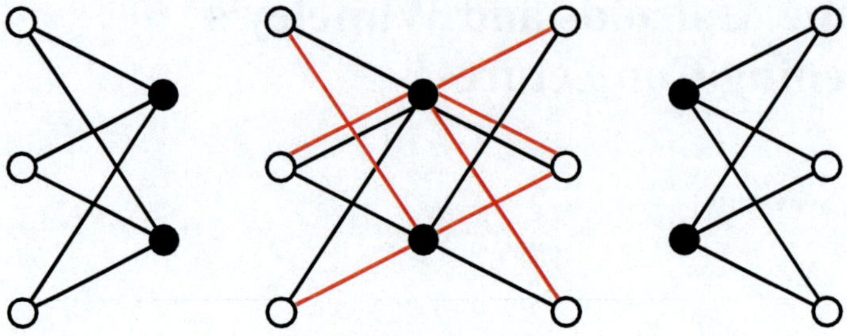

Fig. 1 $|A| = 2, |B| = 3, k = 2$; a butterfly graph and its wings

BRIGITTE SERVATIUS

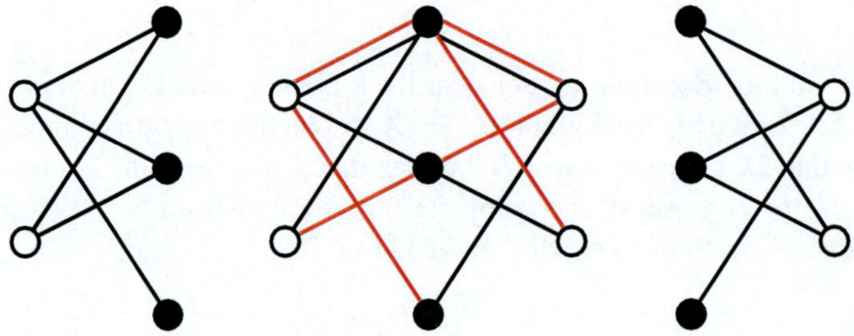

Fig. 2 $|A| = 3, |B| = 2, k = 2$, 2-tight

Theorem 1 *I is k-tight in the k-plane matroid on* $(A, B; I)$ *if and only if adding any* $k - 1$ *incidences yields an incidence structure whose associated butterfly graph decomposes into k spanning trees.*

Proof After adding $k - 1$ incidences, we have $k(|A| + k|B| - 1)$ edges in the butterfly graph. Consider any subset $A' \bigcup B'$ of its vertex set. By summing over the wings and using the fact that I is independent, we get, writing $B' = \bigcup B_i$ that the number of edges induced on $A' \bigcup B'$ is at most $\sum(a' + kb_i - 1) \leq k(a' + kb' - 1)$. □

Components of k-plane matroids. A *component* of the k-plane matroid is a maximal subincidence structure $(A', B'; I')$ for which there is an independent subset $I'' \subseteq I'$ with

$$|I''| = |A'| + k|B'| - k.$$

Theorem 2 *Two distinct components intersect in at most* $k - 1$ *a-vertices.*

Proof Independent sets for which equality holds are in one to one correspondence with subsets of the butterfly graph for which adding any $(k - 1)$ edges yields the union of k spanning trees. If two trees intersect in more than one vertex, the union is not a tree. □

Consider the k-plane matroid as a matroid on the edges of the complete bipartite graph, $\mathbf{M}_k(K_{|A|,|B|}) = \mathbf{M}_{k,(|A|,|B|)}$. We may characterize the bases as follows:

Theorem 3 *A subset $I \leq E(K_{|A|,|B|})$ is a basis of $\mathbf{M}_{k,(|A|,|B|)}$ if and only if doubling any edge of I yields the union of k spanning trees in the k-fold butterfly graph on I with A as the body, and B as the wings.*

Proof The k-fold butterfly graph has $k|I|$ edges and vertex set of size $|A| + k|B|$.

Since $|I| = |A| + k|B| - k$ for a basis, doubling an edge gives and edge set of size $|I| = |A| = k|B| - (k - 1)$ so in the k-fold butterfly graph we get $k(|A| + k|B| - 1)$ edges, just enough for k spanning trees. Since the inequalities have to be met for all subsets, we know by Nash-Williams's [1] theorem that we have the edge disjoint union of k-spanning trees. □

2 The 2-Plane Matroid and the Connectivity of the Incidence Graph

Whiteley conjectured in "Matroids from Discrete Geometry" [2] that a set of incidences will be 2-tight if the bipartite incidence graph is 4-connected.

We first show that there are 3-connected incidence graphs whose incidences are not 2-tight.

Consider $2n$ copies of $K_{3,4}$ where the black vertices of each copy i are $\{a_i, b_i, c_i\}$, $i = 1, 2, \ldots, 2n$. Identify vertex a_{i+1} with c_i, and vertex b_i with b_{i+n} indices modulo $2n$.

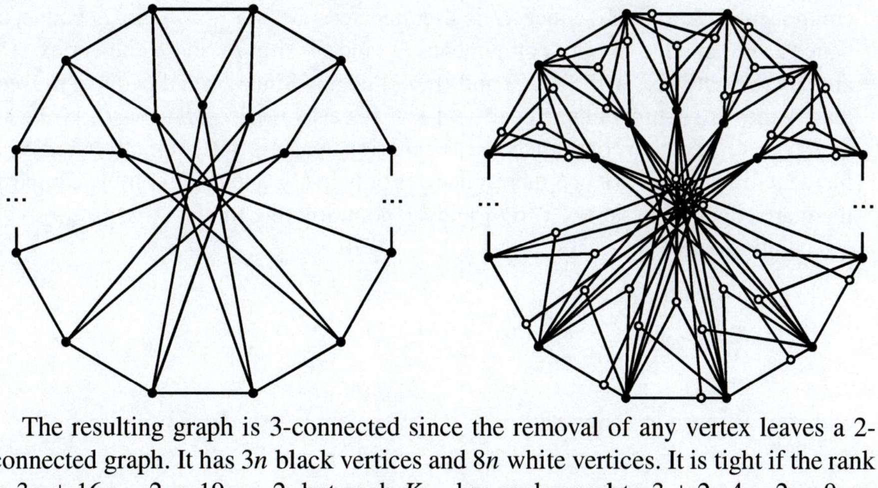

The resulting graph is 3-connected since the removal of any vertex leaves a 2-connected graph. It has $3n$ black vertices and $8n$ white vertices. It is tight if the rank is $3n + 16n - 2 = 19n - 2$, but each $K_{3,4}$ has rank equal to $3 + 2 \cdot 4 - 2 = 9$, so the rank of the whole graph can be at most $18n$, and $18n < 19n - 2$ for all $n > 2$.

For $n = 3$ we have

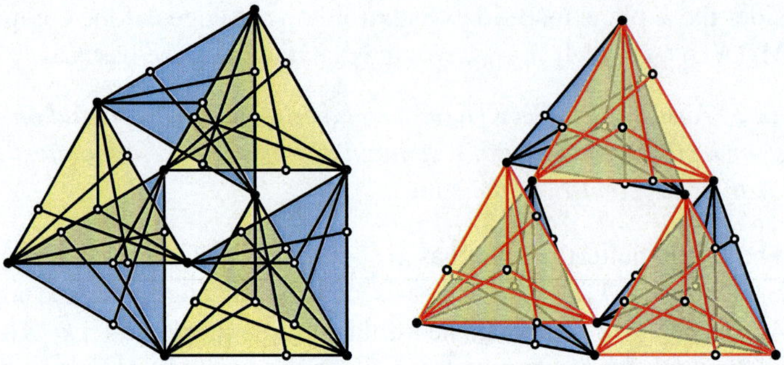

In [3] Lovász and Yemini showed that 6-connectivity of a graph G implies that G is generically rigid in the plane. We adapt their proof to bipartite graphs in the proof of the following theorem.

Theorem 4 *Let $G = (A, B; I)$ be an incidence graph. If G is vertex 4-connected then I is 2-tight.*

Proof Assume that there is a 4-connected graph which is not 2-tight. Among all counterexamples, choose one with $|A|$ minimal and among those one with $|I|$ maximal. Since I has rank less than $a + 2b - 2$, we may decompose G into 2-tight components $G = G_1 \cup G_2 \cup \cdots \cup G_d$, $d \geq 2$, where $G_i = K_{a_i,b_i}$. A pair of components can intersect in at most one a vertex. 2-tight components never intersect in a b vertex. Moreover, each a vertex is contained in at least two of the G_i's:

Assume for contradiction that there is a vertex $a \in G_1$ and $a \notin G_2, G_3, \ldots G_d$. Then $G - a$ is only 3-connected, since G is vertex minimal, and so there exist vertices v_1, v_2 and v_3 such that $G - a - v_1 - v_2 - v_3$ is disconnected, with connected components H_1 and H_2. Since G is 4-connected, $G - v_1 - v_2 - v_3$ is connected, hence a has an edge to both components H_1 and H_2. Let b_1 and b_2 be vertex of H_1 and H_2 respectively, with (a, b_1) and (a, b_2) edges. Since b_1 and b_2 have degree at least 4, they are in turn connected to a set S of at least 4 vertices. If $S = \{a, v_1, v_2, v_3\}$, then G is $K_{4,m}$, which contradicts the assumption that G is not 2-tight. If b_1 and b_2 are incident to a vertex x not in S then, since a is only in G_1, and G_1 is complete bipartite, there are edges from vertex x to b_1 and b_2, contradicting the fact that $\{v_1, v_2, v_3\}$ is a separating set of $G - a$.

We have

$$\operatorname{rank}(G) = \sum_{i=1}^{d}(a_i + 2b_i - 2) = \sum_{i=1}^{d}(a_i - 2) + 2b$$

and now we want to show that

$$\sum_{i=1}^{d}(a_i - 2) \geq a.$$

Since G is 4-connected, each component contains at least 4 a-vertices and every a-vertex is in at least two components, we have

$$\sum_{|A(G_i)|\ni a} \left(1 - \frac{2}{|A(G_i)|}\right) \geq 1,$$

which gives

$$\sum_{i=1}^{i=d} |A(G_i)| \left[1 - \frac{2}{|A(G_i)|}\right] = \sum_{i=1}^{i=d} (a_i - 2) \geq a.$$

We conclude that the rank of G,

$$\text{rank}(G) = \sum_{i=1}^{d} (a_i + 2b_i - 2) \geq a + 2b$$

which is impossible. □

Note that the proof may easily be adapted to show that if $G = (A, B; I)$ is vertex $2k$-connected then I is k-tight.

3 Connection to 2-d Rigidity

A bar-and-joint framework *realizes* an incidence structure $H = (A, B; I)$ if there is a joint for each vertex in B and each vertex $a \in A$ is replaced by a tree of collinear bars on the joints incident with a. Whiteley proved in [4] that an incidence graph $G(A, B; I)$ has a realization as an isostatic (minimally infinitesimally) rigid bar-and-joint framework in the plane if and only if $|I| = a + 2b - 3$ and for any proper subset $I' \subseteq I$, $|I'| \leq a' + 2b' - 3$. Note that if every b vertex of G has degree 2, this is Laman's theorem.

In Fig. 3, we give an incidence structure and two realizations. For the realization on four vertices, the black vertices of the bipartite graph represent the set B, while the white vertices represent lines. Note that by cutting off the rays and considering the line segments between vertices as bars, we find that the graph is overbraced, in fact a circuit in the 2 dimensional generic rigidity matroid, while the 12 incidences are 2-tight, because $|A| = 6$ and $|B| = 4$ yields $6 + 2 \cdot 4 - 2 = 12$.

In the second representation, the black vertices are realized as the four lines intersecting in six vertices. We now have $|A| = 4$ and $|B| = 6$, so $4 + 2 \cdot 6 - 2 = 14$, so the 12 incidences are not enough for 2-tightness, in fact the second realization is not rigid as a bar and joint framework (Fig. 4).

An older result of Whiteley, [4], showed that:

Theorem 5 *A generically independent body and pin framework in the plane remains independent for realizations generic under the condition that all pins of each body are collinear.*

Jackson and Jordán [5] have confirmed that the analog in the plane also holds:

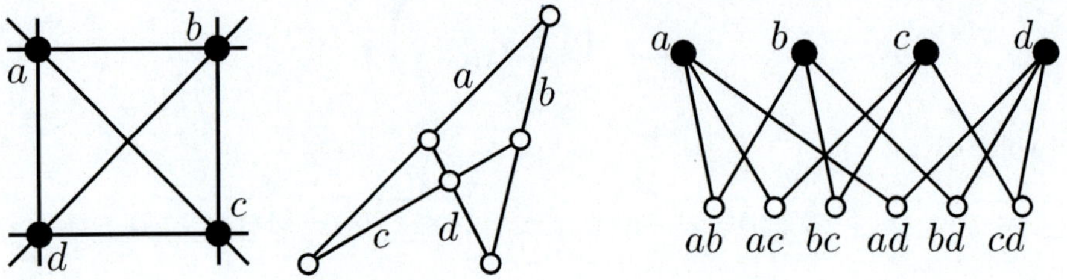

Fig. 3 Two realizations of a bipartite graph

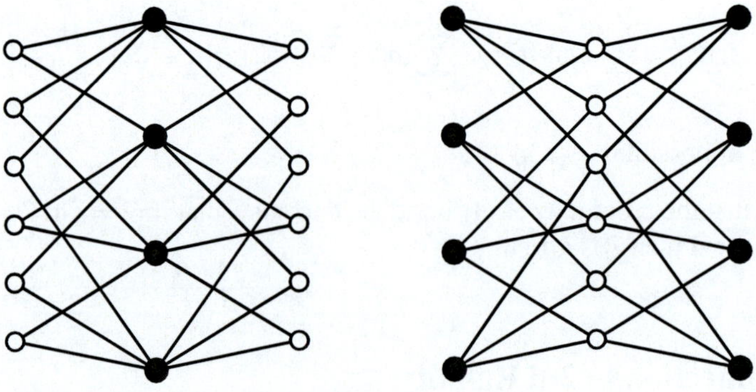

Fig. 4 Butterfly graphs for (A, B;I) and (B, A;I)

Theorem 6 (Jackson and Jordan) *A generically rigid body pin framework, with two bodies at each pin, remains first-order rigid for realizations generic under the condition that all pins of each body are collinear.*

Katoh and Tanigawa [6] proved that a graph can be realized as an infinitesimally rigid body-hinge framework in $\mathbb{R}^d$ if and only if it can be realized as an infinitesimal panel-hinge framework in $\mathbb{R}^d$. For $d = 2$ this is equivalent to the Jackson-Jordán result.

Whiteley's result is not restricted to two bodies at each pin, so a generalized conjecture remains open, even in the plane:

Conjecture 1 (Whiteley $<$ 2007) *A generically rigid body pin framework, remains first-order rigid for realizations generic under the condition that all pins of each body are collinear, without restriction of how many bodies a pin is incident to.*

Acknowledgements Thanks go to Walter Whiteley for discussions at York University back in 2007 and to the AIM workshop on Rigidity and Flexibility of Micro-structures in 2019, during which the proofs were finally written down.

References

1. C. St. J. A. Nash-Williams. Decomposition of finite graphs into forests. *J. London Math. Soc.*, 39:12, 1964.
2. Walter Whiteley. Some matroids from discrete applied geometry. In *Matroid theory (Seattle, WA, 1995)*, volume 197 of *Contemp. Math.*, pages 171–311. Amer. Math. Soc., Providence, RI, 1996.
3. L. Lovász and Y. Yemini. On generic rigidity in the plane. SIAM J. Algebraic Discrete Methods, 3(1):91–98, 0196-5212, 52A37 (51N99 73K99), 644960, Paul Bankston 1982. https://doi.org/10.1137/0603009.
4. Walter Whiteley. A matroid on hypergraphs, with applications in scene analysis and geometry. *Discrete Comput. Geom.*, 4(1):75–95, 1989.
5. Bill Jackson and Tibor Jordán. Pin-collinear body-and-pin frameworks and the molecular conjecture. *Discrete Comput. Geom.*, 40(2):258–278, 2008.
6. Naoki Katoh and Shin-ichi Tanigawa. A proof of the molecular conjecture. *Discrete Comput. Geom.*, 45(4):647–700, 2011.

Bounding the Trace Function of a Hypergraph with Applications

Farhad Shahrokhi

Abstract An upper bound on the trace function of a hypergraph H is derived and its applications are demonstrated. For instance, a new upper bound for the VC dimension of H, or $vc(H)$, follows as a consequence and can be used to compute $vc(H)$ in polynomial time provided that H has bounded degeneracy. This was not previously known. Particularly, when H is a hypergraph arising from closed neighborhoods of a graph, this approach asymptotically improves the time complexity of the previous result for computing $vc(H)$. Another consequence is a general lower bound on the *distinguishing transversal number* of H that gives rise to applications in domination theory of graphs. To effectively apply the methods developed here, one needs to have good estimates of the degeneracy of a hypergraph and its variation the reduced degeneracy which is introduced here.

1 Introduction and Summary

Many important combinatorial problems in computer science, mathematics, and operations research arise from the set systems or *hypergraphs*. We recommend [3] and thesis [4] as references on hypergraphs. Formally, a hypergraph $H = (V, E)$ has the vertex set V and the edge set E, where each $e \in E$ is a subset of V. We do not allow multiple edges in our definition of a hypergraph, unless explicitly stated. When multiple edges exist, we slightly modify the concept. Let $S \subseteq V$ and $e \in E$. The *trace* of e on S is $e \cap S$. The *restriction* of H to S, denoted by $H[S]$, is the hypergraph on vertex set S whose edges are the set of all *distinct* traces of edges in E on S. $H[S]$ is also referred to as the *induced subhypergraph* of H on S. A *Pseudo induced subhypergraph* on the vertex set S is obtained from H by removing the set

F. Shahrokhi (✉)
UNT, P.O.Box 13886, Denton, TX 76203-3886, USA
e-mail: Farhad.Shahrokhi@unt.edu

F. Hoffman (ed.), *Combinatorics, Graph Theory and Computing*,
Springer Proceedings in Mathematics & Statistics 388,
https://doi.org/10.1007/978-3-031-05375-7_8

$V - S$ and the set of all edges of H that have non-empty intersection with $V - S$. Note that any edge of such hypergraph is an edge e of H if $e \subseteq S$. S is *shattered* in H, if any $X \subseteq S$ is a trace. Thus if S is shattered, then it has $2^{|S|}$ traces, that is, $H[S]$ has $2^{|S|}$ edges. The Vapnik–Chervonenkis (VC) dimension of a hypergraph H, denoted by $vc(H)$, is the cardinality of the largest subset of V which is shattered in H. It was originally introduced for its applications in statistical learning theory [26] but has shown to be of crucial importance in combinatorics and discrete geometry [11]. Let $S \subseteq V$, then, S is a *transversal*, or a *hitting set*, if $e \cap S \neq \emptyset$, for all $e \in E$. A set S is a *distinguishing* set if any two distinct edges of H have different traces on (intersections with) S. Let $dt(H)$ denote the size of a smallest distinguishing transversal set in H. Note that if S is a smallest distinguishing transversal set, then it can not have an empty trace on it.

For any $x \in V$, let *degree* of x, denoted by $d_H(x)$, denote the number of edges that contain x. We denote by $\delta(H)$, the smallest degree of any vertex in H.

Any definition for a hypergraph readily extends to a subhypergraph. A hypergraph I is a subhypergraph of H if it can be obtained by deleting some edges in $H[S]$ for some $S \subseteq V$. (Note that there are subhypergraphs of H that may not be induced.) Particularly, for any $x \in S$, the degree of x in I is denoted by $d_I(x)$. Furthermore $\delta(I)$ denotes the minimum degree of I. The degeneracy of H, denoted by $\hat{\delta}(H)$, is the largest minimum degree of any subhypergraph of H. Observe that one can define $\hat{\delta}(H)$ as the largest minimum degree of any induced subhypergraph of H, since the addition of new edges to a hypergraph does not decrease the degrees of vertices. The *pseudo degeneracy* of H, denoted by $\delta^*(H)$, is the largest minimum degree of any pseudo induced subhypergraph of H. Finally, the *reduced degeneracy* of H, denoted by $\tilde{\delta}(H)$ is the largest pseudo degeneracy of any induced subhypergraph of H.

Proposition 1 *For any induced subhypergraph I of H, one has $\delta^*(I) \leq \tilde{\delta}(I) \leq \hat{\delta}(I)$, consequently, $\delta^*(H) \leq \tilde{\delta}(H) \leq \hat{\delta}(H)$.*

The *trace function* of H denoted by $T[H, k]$, is the largest number of traces of H on a set S, $|S| = k$. Unless otherwise stated, we assume that $T[H, k]$ counts the number of non empty traces only.

A powerful tool in studying hypergraph problems with a very broad range of applications is the Sauer Shelah Lemma [20, 23]. The Lemma asserts for any hypergraph H with $vc(H) = d$ and any $k \geq 0$, one has:

$$T[H, k] \leq \sum_{i=0}^{d} \binom{k}{i} = O(k^d) \tag{1}$$

The concept of a trace function is also studied as the Max Partial VC Dimension [2]. Particularly, it was shown in [2] that

$$T[H, k] \leq k(\Delta(H) + 1)/2 + 1 \tag{2}$$

Our main result in this paper is Lemma 1, which is an upper bound on $T[H, k]$. A simple consequence of this upper bound is

$$T[H, k] \leq k\widetilde{\delta}(H) \tag{3}$$

This upper bound is within a multiplicative factor of $\widetilde{\delta}(H)$ from the lower bound of $L(H, k) = \min\{|E|, k+1\}$ (when H does not have multiple edges) that has also been recently constructed in [2]; Thereby, $T(H, k)$ is proportional to k, provided that reduced degeneracy of H is "small", and hence in light of our upper bound for $T(H, k)$, the lower bound $L(H, k)$ (constructed in [2]), actually approximates $T(H, k)$ (for any k) to within a factor of $\widetilde{\delta}(H)$ which is an improvement of the factor $(\Delta(H) + 1)/2 + 1$ as authors stated in [2].

This paper is organized as follows. Section two contains our main lemma as well as the lower bound on distinguishing transversal number. Section three contains the applications to VC dimension. Section four contains the applications to domination theory by deriving general lower bounds for several variations of *domination number* of a graph [12]. The bounds are derived using the lower bound on distinguishing transversal number. For trees, our bounds are shown to match some of the best known results, or come close to them.

We finish this section by stating two folklore results for computing degeneracy and pseudo degeneracy of a hypergraph. The properties of the output of algorithm will help to establish some of our claims more easily.

Theorem 1 *Let $H = (V, E)$ be a hypergraph, then $\hat{\delta}(H)$ can be computed in $O(|V| + \sum_{e \in E} |e|)$ time.*

Proof For $i = 1, 2, \ldots n$, let x_i be a vertex of degree $d_i = d_{H_i}(x_i) = \delta(H_i)$ in the induced subhypergraph $H_i = H[V_i]$ on the vertex set $V_i = V - \{x_1, x_2, \ldots, x_{i-1}\}$. Let $d = \max\{d_i, i = 1, 2, \ldots, n\}$. We claim that $\hat{\delta}(H) = d$. Clearly, $\hat{\delta}(H) \geq d$, and it suffices to show that $\hat{\delta}(H) \leq d$. Now let I be any (induced) subhypergraph of H, and let j be the smallest integer so that x_j is a vertex of I. Then $d_I(x_j) \leq d_j = \delta(H_j) \leq d$. Thus, $\delta(I) \leq d$, and consequently, $\hat{\delta}(H) \leq d$ as stated. Details of deriving time complexity that include representation of H as a bipartite graph and utilization of elementary data structures are omitted. □

For a subhypergraph $I = (U, F)$ of H, and any $x \in U$, let F_x denote the set of edges in F containing x. The next result almost copies Theorem 1.

Theorem 2 *Let $H = (V, E)$, be a hypergraph, then, $\delta^*(H)$ can be computed in $O(|V| + \sum_{e \in E} |e|)$ time.*

Proof For $i = 1, .2, \ldots n$, let x_i be a vertex of degree $d_i = d_{H_i}(x_i) = \delta(H_i)$ in the subhypergraph H_i on the vertex set $V_i = V - \{x_1, x_2, \ldots, x_{i-1}\}$ and edge set $E_i = E - \{E_{x_1}, E_{x_2}, \ldots, E_{x_{i-1}}\}$. Let $d = \max\{d_i, i = 1, 2, \ldots, n\}$. Clearly, $\delta^*(H) \geq d$. Now let I be any pseudo induced subhypergraph of H, and let j be the smallest integer so that x_j is a vertex of I. Then, vertex set of I does note contain $x_i, i =$

$1, 2, \ldots, j-1$; Consequently, the edge set of I is a subset of E_j. Then $d_I(x_j) \leq d_j = \delta(H_j) \leq d$ proving the claim. Details of deriving time complexity that include representation of H as a bipartite graph and utilization of elementary data structures are omitted. □

Remark 1 The sequences $d_1, d_2, \ldots, d_n$ generated in Theorems 1 and 2 are called the *degeneracy sequence*, and *pseudo degeneracy sequence*, respectively.

2 Main Results

For a subhypergraph $I = (U, F)$ of H, and any $x \in U$, let F_x denote the set of edges in F containing x.

Lemma 1 *Let $H = (V, E)$, let $S \subseteq V$, $|S| = k$, and let $I = H[S] = (S, F)$ be the restriction of H to S. For $i = 1, \ldots, k$, let x_i be a vertex in subhypergraph I_i on the vertex set $S_i = S - \{x_1, x_2, \ldots, x_{i-1}\}$ and edge set $F_i = F - \{F_{x_1}, F_{x_2}, \ldots, F_{x_{i-1}}\}$. and let $k, j, l \geq 0$ be integers with $k = l + j$. Then,*

$$|F| = \sum_{i=1}^{k} |F_{x_i}| = \sum_{i=1}^{k} d_{I_i}(x_i) \tag{4}$$

$$= \sum_{i=1}^{l} d_{I_i}(x_i) + |F_{l+1}| \tag{5}$$

$$\leq \sum_{i=1}^{l} d_{I_i}(x_i) + T[H, j] \tag{6}$$

Consequently,

$$T[H, k] \leq \delta^*(I) \times l + T[H, j] \tag{7}$$

$$\leq \delta^*(I) \times k \tag{8}$$

$$\leq \widetilde{\delta}(H) \times k \tag{9}$$

Proof For (4) observe that $F = \cup_{i=1}^{k} F_{x_i}$, that for $i = 1, 2 \ldots, k$, F_{x_i}'s are disjoint and $|F_{x_i}| = d_{I_i}(x_i)$. For (5) note that $F_{l+1} = \cup_{i=l+1}^{k} F_{x_i}$. Next, note that the hypergraph I_{l+1} has the vertex set $S_{l+1} = \{x_l, x_{l+1}, \ldots, x_k\}$, thus, $|S_{l+1}| = k - l = j$. Consequently, (6) follows, since $|F_{l+1}| \leq T[H, j]$. For (7), for $i = 1, 2, \ldots, k$, let x_i to be a vertex of minimum degree in I_i, that is $d_{I_i}(x_i) = \delta(I_i)$, note that $\delta(I_i) \leq \delta^*(I) = \max\{\delta(I_i), i = 1, 2, \ldots, k\}$ (by Theorem 2) and use (6); Now set $j = 0$ to obtain (8) and note that $\delta^*(I) \leq \widetilde{\delta}(H)$ to obtain (9). □

Remark 2 Note that $S_1 = S - \{x_0\} = S - \emptyset = S$, and similarly $F_1 = F$, in the above Lemma.

Theorem 3 *For any hypergraph $H = (V, E)$, and any integer $0 \le j \le dt(H)$, one has*

$$dt(H) \ge \frac{|E| - T[H, j]}{\widetilde{\delta}(H)} + j.$$

Consequently,

$$dt(H) \ge \frac{|E| - 2^j + 1}{\widetilde{\delta}(H)} + j.$$

Proof Let S with $|S| = dt(H)$ be the smallest cardinality distinguishing transversal set; Thus S must have exactly $|E|$ non empty distinct traces, that is, $T(H, d(H)) = |E|$. Now applying Lemma 1, we have $|E| \le \delta^*(H[S])(dt(H) - j) + T[H, j]$ which proves the main claim, since $\delta^*(H[S]) \le \widetilde{\delta}(H)$. To verify the second claim note that $T[H, j] \le 2^j - 1$. □

3 Applications to VC Dimension

It is easy to verify that $vc(H) \le \log(|E|)$ for any hypergraph H. It was previously known that when H has an explicit representation by an $m \times n$ incident matrix, $vc(H)$ can be computed in $n^{O(\log(n))}$ [16]. Also, the decision version of the problem is LOGNP-complete [17] and remains in this complexity class for neighborhood hypergraphs of graphs [15]. A simple and immediate consequence of our work is that $vc(H) \le \log(\hat{\delta}(H)) + 1$ (which was not known before) and hence $vc(H)$ can be computed in $n^{O(\log(\hat{\delta}(H))}$. Consequently, $vc(H)$ can be computed in polynomial time for hypergraphs of bounded degeneracy, which had not been known. Moreover, these results give rise to an algorithm for computing $vc(H)$ in $n2^{O(\log^2(\Delta(G)))}$ time, when H is the set of all closed neighborhoods of vertices of a graph G with maximum degree $\Delta(G)$. This is an asymptotic improvement of the best known time complexity of $O(n2^{\Delta(G)})$ for solving the problem which was derived in [15].

Theorem 4 *Let $H{=}(V, E)$, $|V| = n$, then, $vc(H) \le \log(\hat{\delta}(H)) + 1$. Consequently, for any n vertex hypergraph H, $vc(H)$ can be computed in $n^{O(\log(\hat{\delta}(H)))}$ time. Particularly, if H is the closed neighborhood hypergraph of an n vertex graph with maximum degree Δ, then $vc(H)$ can be computed in $n2^{O(\log^2(\Delta))}$ time.*

Proof Let S with $|S| = d$ be a largest shattered set in H. We apply Lemma 1 with $j = d - 1$. Thus, $2^d - 1 = T(H, d) \le \hat{\delta}(H)(d - d + 1) + 2^{d-1} - 1$, which gives $d \le \log(\hat{\delta}(H)) + 1$ as claimed.

To compute $vc(H)$, one can represent H in its incidence matrix form, requiring $O(nm)$ space, or in $O(n^2\hat{\delta}(H))$ space, where m is the number of edges of H, since by Lemma 1 with $k = n$ one has $m \le n\hat{\delta}(H)$. Now one can find $vc(H)$ by exhaustive enumeration. Note that the largest shattered subset has size $O(\log(\hat{\delta}))$; Hence in $n^{O(\log(\hat{\delta}(H)))}$ time, one can compute $vc(H)$. To prove the claim when H is the closed neighborhood hypergraph, note that $\hat{\delta}(H) \leq \Delta(G) + 1$, and hence $vc(H) = O(\log(\Delta(G)))$. Since the largest shattered set must be contained in the closed neighborhood of one vertex of G, the enumeration algorithm takes $n\Delta(G)^{O(\log(\Delta(G)))}$ or in $n2^{O(\log^2(\Delta(G)))}$ time. □

Remark 3 Note that the enumeration algorithm in Theorem 4 does not require knowing $\hat{\delta}(H)$, although $\hat{\delta}(H)$ can be computed in polynomial time. Also note that the run time of $n2^{O(\log^2(\Delta(G)))}$ for computing VC dimension of neighborhood system of graphs compares favorable with the time complexity of $O(n2^{\Delta(G)})$ derived in [15].

4 Applications to Domination Theory

We recommend [12] as a reference on domination theory. For a graph $G = (V, E)$ and a vertex x, $N(x)$ denotes the *open neighborhood* of x, that is the set of all vertices adjacent to x, not including x. The *closed neighborhood* of x is $N[x] = N(x) \cup \{x\}$. The closed (open) neighborhood hypergraph of an n vertex graph G is a hypergraph on the same vertices as G whose edges are all n closed (open) neighborhoods of G. A subset of vertices S in G is a *dominating set* [12], if for every vertex x in G, $N[x] \cap S \neq \emptyset$. S is a *total or open domination set* [6] if, $N(x) \cap S \neq \emptyset$. A subset of vertices S is *locative* in G, if for every two distinct vertices $x, y \in V - S$, one has $N(x) \cap S \neq N(y) \cap S$. S is *totally locative* in G, if for every two distinct vertices $x, y \in V$, one has $N(x) \cap S \neq N(y) \cap S$. A subset S of vertices in G is a *locating dominative* (*locating total dominative*) if it is a dominating (total dominating) set and it is also a locative set [24, 25]. S is an *identifying code* if it is a dominating set and for every two distinct vertices $x, y \in V$, one has $N[x] \cap S \neq N[y] \cap S$ [14]. S is an *open locating dominative* set, if S is a totally domination set and also totally locative in G [22]. Let $\gamma^{LD}(G)$ and $\gamma^{ID}(G)$ denote the sizes of a smallest location domination and identifying code sets in G, respectively. Let $\gamma^{OLD}(G)$ denote the size of a smallest open location domination in G. Computing $\gamma^{LD}(G)$, $\gamma^{ID}(G)$ and $\gamma^{OLD}(G)$ are known to be NP-hard problems and hence estimations of these parameters or their computational complexities have been an active area of research [1, 2, 5, 7–10, 18, 19, 21, 22]. Recall that the *distinguishing transversal number* of H, denote by $dt(H)$, is the minimum size of any distinguishing transversal set [13]. A consequence of our upper bound for $T[H, k]$, is that for any hypergraph $H = (V, E)$ and any integer $0 \leq j \leq dt(H)$ one has $dt(H) \geq \frac{|E|-T[H,j]}{\tilde{\delta}(H)} + j$. By properly applying this result to suitable neighborhood hypergraphs of a graph, one obtains some general lower bounds on $\gamma^{LD}(G)$, $\gamma^{ID}(G)$ and $\gamma^{OLD}(G)$. For a specific application, one needs to determine the exact value or a good estimate for $\tilde{\delta}(H)$ or $\hat{\delta}(H)$, and this can become a challenging task.

Theorem 5 *Let G be an n vertex graph with closed and open neighborhood hypergraphs H and H^o, respectively, let $\delta^{**}(H) = \min\{\tilde{\delta}(H), \tilde{\delta}(H^o)\}$. Then the following hold for any $0 \leq j \leq \gamma^{ID}$ in (i), $0 \leq j \leq \gamma^{OLD}$ in (ii) and $0 \leq j \leq \gamma^{LD}$ in (iii), where H and H^o do not have multiple edges in (ii) and (iii), respectively.*

(i) $\gamma^{LD}(G) \geq \frac{n+\delta^{**}(H).j-T[H,j]}{\delta^{**}(H)+1}$.
(ii) $\gamma^{ID}(G) \geq \text{Max}\{\frac{n-T[H,j]}{\widetilde{\delta}(H)} + j, \frac{n+\delta^{**}(H).j-T[H,j]}{\delta^{**}(H)+1}\}$.
(iii) $\gamma^{OLD}(G) \geq \text{Max}\{\frac{n-T[H,j]}{\widetilde{\delta}(H^o)} + j, \frac{n+\delta^{**}(H).j-T[H,j]}{\delta^{**}(H)+1}\}$.

Proof For (i), let S be the smallest cardinality locative dominative set in G. Now, let $H^1 = (V, E^1)$, where $E^1 = \{N(x)|x \in V - S\}$ and $H^2 = (V, E^2)$ where $E^2 = \{N[x]|x \in V - S\}$. Note that for $i = 1, 2$, $T(H^i, |S|) = n - |S| \leq \widetilde{\delta}(H^i)(|S| - j) + T[H^i, j]$ where last inequality is obtained by the application of Lemma 1. Furthermore, H^1 is a subhypergraph of H', and H^2 is a subhypergraph of H. Consequently, $\widetilde{\delta}(H^1) \leq \widetilde{\delta}(H')$ and $\widetilde{\delta}(H^2) \leq \widetilde{\delta}(H)$. It follows that $n - |S| \leq \delta^{**}(H)(|S| - j) + T[H, j]$. To finish the proof note that $LD(G) = |S|$, and do the algebra.

For (ii), note that $\gamma^{ID}(G) \geq \gamma^{LD}(G)$ and hence the lower bond in (i) is also a lower bound for $\gamma^{ID}(G)$. To complete the proof, observe that S is an identifying code set in G if and only if S is a distinguishing transversal in H. Thus, $dt(H) = \gamma^{ID}(G)$. Now apply Theorem 3.

Similarly for (iii) note that $\gamma^{OLD}(G) \geq \gamma^{LD}(G)$, and that, S is an totally dominative and totally locative set in G, if and only if, S is a distinguishing transversal set in H' and thus $dt(H') = \gamma^{OLD}(G)$. Now apply Theorem 3. □

Remark 4 Let G be an n vertex graph of maximum degree $\Delta(G)$ with closed and open neighborhood hypergraphs H and H^o, respectively. Then clearly $\hat{\delta}(H) \leq \Delta(G) + 1$ and $\hat{\delta}(H^o) \leq \Delta(G)$, since the largest sets in H and H^0 are of cardinalities $\Delta(G) + 1$ and $\Delta(G)$, respectively. As we will see, one can get much stronger results in trees.

Remark 5 Let L denote the set of leaves in a tree T, and note that after removal of all vertices in L from T we obtain another tree T'. Let S denote the set of all leaves of the tree T'. Then each vertex in S is a support vertex in T and is called a *canonical support vertex* in T.

Theorem 6 *If T is a $n \geq 2$ vertex tree with closed and open neighborhood hypergraphs H and H^o, respectively, then the following hold.*

(i) $\hat{\delta}(H) \leq 3$.
(ii) $\hat{\delta}(H^o) \leq 2$.
(iii) $\widetilde{\delta}(H^o) \leq 2$.
(iv) $\delta^*(H) \leq 2$.
(v) $\delta^*(H^o) \leq 2$.

Proof For $n \leq 2$ the claims are valid so let $n \geq 3$. For (i) note that for any vertex x, $d_H(x)$ equals degree of x in T plus one, and hence for any leaf x, one has $d_H(x) = \delta(H) = 2$. Now apply Theorem 1, and let $d_1, d_2, \ldots d_n$, be the sequence or minimum degrees generated by the algorithm associated with vertices $x_1, x_2, \ldots x_n$, in the subhypergraph $H_1, H_2, \ldots H_n$. Note that for any leaf of $x = x_i$ of T, we have $d_{H_i}(x_i) = d_i \leq 2$, where $1 \leq i \leq n$. Note further that by the previous remark any leaf in the new tree T' is a canonical support vertex of T and will of degree at most 3 in the hypergraph obtained after removing all leaves attached to it. Thus after removal of all leaves of T, we obtain a tree T' whose leaves have degree at most three in the associated hypergraph. Now iterate on this process by removing all leaves of T' to obtain a tree T'', and note that the degree of any leaf of T'' in the associate hypergraph is at most three. Consequently for $i = 1, 2, \ldots, n$ we have $d_i \leq 3$. For (ii), a similar argument is carried out, but we need to observe that initially $d_{H^o}(x) = \delta(H^o) = 1$ and that after removal of leaves in T, any leaf of the resulting tree T' has degree at most two in the corresponding hypergraph. (iii) follows from (ii). For (iv), we follow the arguments in (i), and note that degree of any leaf x of T is initially two in H. Now apply Theorem 2 and note that after removing any leaf x, the degree of all leaves with the same support vertex becomes one in the corresponding hypergraph, and after removing all leaves joined to a canonical support vertex s, the degree of s becomes one in the resulting hypergraph.

Finally, (iv) follows from (iii). □

Remark 6 Corollary 1 summarizes some specific applications of our results in this section by eliminating past ad hoc approaches. Particularly, the lower bound in part (i) matches the best previously known lower bound of $\frac{n+1+2(L-S)}{3}$ in [22], however, is weaker (by a multiplicative factor of 3/2) in part (ii) than a recent result in [18], and in part (iii) is weaker only by an additive factor of 1 when n is odd compared to the result in [22].

Corollary 1 *Let T be an $n \geq 4$ vertex tree, with L leaves and S support vertices. Then the following hold. For (ii) assume that every support vertex is adjacent to only one leaf.*

(i) $\gamma^{LD}(T) \geq \frac{n+1+2(L-S)}{3}$.
(ii) $\gamma^{ID}(T) \geq \frac{n+3}{3}$.
(iii) $\gamma^{OLD}(T) \geq \frac{n+1}{2}$.

Proof For (i) let D be an LD set and let s be a support vertex. We assume WLOG that $s \in D$, otherwise by placing s and all but one leaf attached to s in D, we obtain another LD set of the same size. Now follow Theorem 5 and Lemma 1 and note that a total of $L - S$ leaves have degree zero in hypergraph H^1 (defined in Theorem 5). Thus, we have

$$\begin{aligned} n - |D| &\leq L^* + T[H^1, D - (L - S)] && (10)\\ &\leq T[H^1, D - (L + L^* - S) - 1] + 1 && (11)\\ &\leq \widetilde{\delta}(H^1)(|D| - (L - S) - 1) + 1) && (12)\\ &\leq 2((|D| - (L - S) - 1) + 1 && (13)\end{aligned}$$

where the last three inequalities are obtained by the application of Lemma 1, Theorem 6 and noting that $T[H^1, 1] = 1$. Now (i) follows.

For (ii) use $j = 2$, and $\delta^*(H) \leq 3$ from Theorem 6 and use Theorem 4.1. For (iii) use Theorem 5 with $j = 1$ and $\widetilde{\delta}(H^0) \leq 2$ from Theorem 6. □

Acknowledgements We thank N. Bousquet for pointing out an error on an earlier version of this paper and email discussions.

References

1. Auger D., Charon I., Hudry O., Lobstein A.: Complexity results for identifying codes in planar graphs. International Transactions in Operational Research 17(6), pp. 691–10 (2010).
2. Bazgan C., Foucaud F., Sikora F.: Parameterized and Approximation Complexity of Partial VC Dimension, Theoretical Computer Science 766(25), pp. 1–15 (2019).
3. Berge C..:Theory of Graphs and its Applications. Methuen, London (1962)). RI, 1962.
4. Bousquet N.: Hitting sets: VC-dimension and Multicuts. Université Montpellier II-Sciences et Techniques du Languedoc (2013).
5. Bousquet N., Lagoutte A., Li Z., Parreau A., Thomasse A.: Identifying codes in hereditary classes of graphs and VC-dimension. SIAM Journal on Discrete Mathematics 29(4), pp. 2047–2064 (2015).
6. Cockayne E.J., Dawes R.M., Hedetniemi S.T.: Total domination in graphs, Networks 10, pp. 211–219 (1980).
7. Foucaud F., Gravier S., Naserasr R.: A. Parreau and P. Valicov. Identifying codes in line graphs. Journal of Graph Theory 73(4), pp. 425–448 (2013).
8. Foucaud F., Mertzios G., Naserasr R., Parreau A., Valicov P.:Algorithms and complexity for metric dimension and location-domination on interval and permutation graphs. Proceedings of the 41st International Workshop on Graph-Theoretic Concepts in Computer Science, WG 2015, 456–471 (2015).
9. Foucaud F.: Decision and approximation complexity for identifying codes and locating-dominating sets in restricted graph classes. Journal of Discrete Algorithms 31, pp. 48–68 (2015).
10. Foucaud F., Mertzios G., Naserasr R., Parreau A., Valicov P.: Identification, location-domination and metric dimension on interval and permutation graphs. I. Bounds. Algorithmica 78, pp. 914–944 (2017).
11. Haussler D., and Welzl E.: $\epsilon-$nets and simplex range queries. Discrete and Computational Geometry 2, pp. 127–151 (1987).
12. Haynes T.W., Hedetniemi S. T., Slater S.: Fundamentals of Domination in Graphs. CRC press (1988).
13. Henning M., Yeo A.: Distinguishing-transversal in hypergraphs and identifying open codes in cubic graphs. Graphs and Combinatorics 30(4), pp. 909–932 (2014).
14. Karpovsky M.G., Chakrabarty K., Levitin L.B.: On a new class of codes for identifying vertices in graphs. IEEE Transactions on Information Theory 44, pp. 599–611 (1998).
15. Kranakis E., Krizanc D., Ruf B., Urrutia J., Woeginger G., The VC-dimension of set systems defined by graphs, Discrete Appl. Math. 77, pp. 237–257 (1997).
16. Linial N., Mansour Y., Rivest R. L.: Results on learnability and the Vapnik-Cervonenkis dimension. In Proceedings of FOCS, pp. 120–129 (1988).
17. Papadimitriou C.H., Yannakakis M.: On limited nondeterminism and the complex- ity of the VC dimension. Journal of Computer and System Sciences 53(2), pp. 161–70 (1996).
18. Rahbani H., Jafari Rad N., MirRezaei S.M.: Bounds on the Identifying Codes in Trees, Graphs and Combinatorics 35, pp. 599–609 (2019).
19. Rall D.and Slater P.: On location-domination numbers for certain classes of graphs. Congressus Numerantium 45, pp. 97–106 (1984).
20. Sauer, N. On the density of families of sets, Journal of Combinatorial Theory, Series A 13, pp. 145–147 (1972).

21. S. J. Seo, Slater P.J.: Open neighborhood locating-dominating sets. The Australasian Journal of Combinatorics 46, pp. 109–120 (2010).
22. Seo S.J., Slater P.J.: Open neighborhood locating-dominating in trees. Disc. App. Math. 159, pp. 484–489 (2011).
23. Shelah, S. A combinatorial problem;.:stability and order for models and theories in infinitary languages, Pacific Journal of Mathematics 41, pp. 247–261 (1972).
24. Slater P.J.:Domination and location in acyclic graphs. Networks 17(1), pp. 55–64 (1987).
25. Slater P. J.: Dominating and reference sets in a graph. Journal of Mathematical and Physical Sciences 22(4), pp. 445–455 (1988).
26. Vapnik, V. N., Chervonenkis A.: On the Uniform Convergence of Relative Frequencies of Events to Their Probabilities. Theory of Probability and Its Applications. 16(2), pp. 264–279, Springer (1971).

A Generalization on Neighborhood Representatives

Sarah Heuss Holliday

Abstract In 2017, Hedetniemi asked the question "for which graphs G does the indexed family $\{N_G(v)|V \in V(G)\}$ of open neighborhoods have a system of distinct representatives?" In [2, 3] we answered that question, and explored necessary conditions and associated parameters. In [1], Haenel and Johnson looked over longest paths and cycles. We are now generalizing and deepening our examination.

Keywords SDR · Neighborhoods · Matchings · Independent set

All graphs will be finite and simple. In [2], we called a graph G SDR-good if the indexed collection $\mathcal{N}(G) = \{N_G(v)|v \in V(G)\}$ of open neighborhoods has a system of distinct representatives (SDR). An SDR for $\mathcal{N}(G)$ is a one-to-one function $\phi : V(G) \to V(G)$ such that $\phi(v) \in N_G(v)$ for all $v \in V(G)$.

Theorem 1 ([2]) *A graph G is SDR-good if and only if G has a spanning subgraph the components of which are either single edges or cycles.*

In [3], we developed analogous results for sets of maximum matchings and maximum independent sets. Let $\mathcal{M}(G)$ be the set of all maximum matchings in G. We shall say that a graph is SDR-$\mathcal{M}(G)$-good if $\mathcal{M}(G)$ has a system of distinct representatives (SDR). An SDR for $\mathcal{M}(G)$ is a one-to-one function $\phi : \mathcal{M}(G) \to E(G)$ such that $\phi(M) \in M$ for all $M \in \mathcal{M}(G)$. Note that the existence of such a function requires $|E(G)| \geq |\mathcal{M}(G)|$. So, $|E(G)| < |\mathcal{M}(G)|$ implies that G is not SDR-$\mathcal{M}(G)$-good.

Theorem 2 ([3]) $|\mathcal{M}| \leq |E|$ *is necessary but not sufficient for G to be SDR-*$\mathcal{M}(G)$*-good.*

We developed the following result for maximum independent sets: Let $\mathcal{I}(G)$ be the set of maximum independent sets in G. We shall say that a graph is SDR-$\mathcal{I}(G)$-good if the set $\mathcal{I}(G)$ has a system of distinct representatives (SDR). An SDR for $\mathcal{I}(G)$ is a one-to-one function $\phi : \mathcal{I}(G) \to V(G)$ such that $\phi(I) \in I$ for all $I \in \mathcal{I}(G)$. Using

S. H. Holliday (✉)
Kennesaw State University, Kennesaw, GA, USA
e-mail: shollid4@kennesaw.edu

F. Hoffman (ed.), *Combinatorics, Graph Theory and Computing*,
Springer Proceedings in Mathematics & Statistics 388,
https://doi.org/10.1007/978-3-031-05375-7_9

properties of line graphs and the preceding result, we're able to show the following result.

Theorem 3 ([3]) *If G is SDR-$\mathcal{M}(G)$-good, then $L(G)$ is SDR-$\mathcal{I}(L(G))$-good.*

We were also able to individually sort the Beinecke graphs into those which are and aren't SDR-$\mathcal{I}(G)$-good.

For a graph G, denote the set of maximum paths in G by $\mathcal{P}(G)$. Denote by $V^*(\mathcal{P}(G))$ the union of the vertex sets of the maximum paths of G; in other words, $V^*(\mathcal{P}(G))$ is the set of all vertices that have the good fortune of lying on a maximum path in G. G is SDR-$\mathcal{P}$-good if $\mathcal{P}(G)$ has a system of distinct vertex representatives (SDR). An SDR for $\mathcal{P}(G)$ is a one-to-one function $\phi : \mathcal{P}(G) \to V(G)$ such that $\phi(P) \in V(P)$, for all $P \in \mathcal{P}(G)$.

Theorem 4 ([1]) $|\mathcal{P}| \leq |V^*(\mathcal{P}(G))|$ *is necessary but not sufficient for G to be SDR-$\mathcal{P}(G)$-good.*

Corollary 1 ([1]) *Suppose that G is a graph in which the maximum paths have order q. If every vertex of G lies on no more than q maximum paths then G is SDR-$\mathcal{P}(G)$-good.*

Corollary 2 ([1]) *Suppose that every maximum path in $\mathcal{P}(G)$ has order q and that every vertex in $V^*(\mathcal{P}(G))$ lies on at least q maximum paths in G. Suppose that at least one vertex of G lies on more than q maximum paths in G; then G is not SDR-$\mathcal{P}(G)$-good.*

For a graph G, denote the set of maximum cycles in G by $\mathcal{C}(G)$. Denote by $V^*(\mathcal{C}(G))$ the union of the vertex sets of the maximum cycles of G; in other words, $V^*(\mathcal{C}(G))$ is the set of all vertices that have the good fortune of lying on a maximum cycle in G. G is SDR-$\mathcal{C}$-good if $\mathcal{C}(G)$ has a system of distinct vertex representatives (SDR). An SDR for $\mathcal{C}(G)$ is a one-to-one function $\phi : \mathcal{C}(G) \to V(G)$ such that $\phi(C) \in V(C)$, for all $C \in \mathcal{C}(G)$.

Theorem 5 ([1]) $|\mathcal{C}| \leq |V^*(\mathcal{C}(G))|$ *is necessary but not sufficient for G to be SDR-$\mathcal{C}(G)$-good.*

Corollary 3 ([1]) *Suppose that G is a graph in which the maximum cycles have order q. If every vertex of G lies on no more than q maximum cycles then G is SDR-$\mathcal{C}(G)$-good.*

Corollary 4 ([1]) *Suppose that every maximum cycle in $\mathcal{C}(G)$ has order q and that every vertex in $V^*(\mathcal{C}(G))$ lies on at least q maximum cycles in G. Suppose that at least one vertex of G lies on more than q maximum cycles in G; then G is not SDR-$\mathcal{C}(G)$-good.*

G is a finite simple graph with no isolated vertices. $\mathbb{N} = \{0, 1, 2, \ldots\}$. Suppose $f : V(G) \to \mathbb{N}$ is a function $\mathcal{N} = [N_G(v); v \in V(G)]$, where ("[ ]" means it's an indexed collection, not a set).

An f-satisfying choice of subset representatives for $\mathcal{N}(G)$ is a function $U : V(G) \to 2^{V(G)}$ such that for all $v, w \in V(G)$:

1. $U(v) \subseteq N_G(v)$;
2. $|U(v)| = f(v)$; and
3. if $v \neq w$ then $U(v) \cap U(w) = \emptyset$.

Let $\mathcal{F}(G) = \{f : V(G) \to \mathbb{N}|$ there is a an f-satisfying choice of subset representatives for $\mathcal{N}(G)\}$.

Obviously, the constant function $f \equiv 0$ is in $\mathcal{F}(G)$. The assumption of no isolated vertices implies that, for each $v \in V(G)$, a function defined by $f(v) = k$, for any $1 \leq k \leq |N_G(v)|$, and $f = 0$ on $V(G)\backslash\{v\}$, is in $\mathcal{F}(G)$. Theorem 1 characterizes the graphs such that the constant function $f \equiv 1$ (on $V(G)$) is an element of $\mathcal{F}(G)$.

Problem 1 If G and H are graphs on the same set of vertices, $V = V(H) = V(G)$, and $\mathcal{F}(G) = \mathcal{F}(H)$, does it follow that $G = H$?

In order for $\mathcal{F}(G) = \mathcal{F}(H)$, it is necessary that the degree sequences of G and H are also the same, so that the $[0, 0, \ldots, 0, deg(v), 0, \ldots, 0, 0]$ element appears with the same largest possible value in the same position in both $\mathcal{F}(G)$ and $\mathcal{F}(H)$ for each $v \in V$.

Lemma 1 *If G and H are graphs on the same set of vertices, $V = V(H) = V(G)$, and $G \neq H$, then $\mathcal{F}(G) \neq \mathcal{F}(H)$.*

Proof Consider the following pair of graphs:

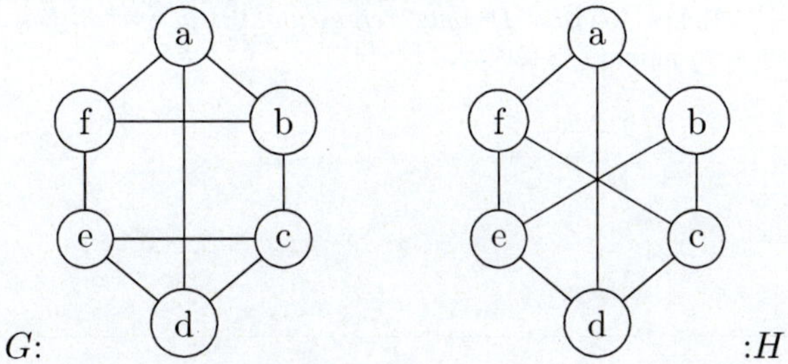

For the given graphs, they both include the $[0, 0, 0, 0, 0, 0]$ and $[1, 1, 1, 1, 1, 1]$ in their respective $\mathcal{F}$, as well as $[3, 0, 0, 0, 0, 0]$, but only $\mathcal{F}(G)$ has $[3, 0, 1, 0, 0, 0]$. This example can be quickly reproduced by any pair of graphs with the same vertex set and degree sequence by ensuring they have different shortest cycles. □

Theorem 6 *If G and H are graphs on the same set of vertices, $V = V(H) = V(G)$, and $\mathcal{F}(G) = \mathcal{F}(H)$, then $G = H$.*

Proof We know that the two graphs have the same degree sequence, and each graph has a shortest cycle.

Suppose the shortest cycle in G has a different number of vertices than the shortest cycle in H. As in [2], the representative from each neighborhood can be defined by the orbit of a vertex along its cycle memberships. This means a vertex belonging to a cycle of length x will have an orbit of that length, and a cycle of length y respectively. Thus, the $f \in \mathcal{F}$ will have different representations when the vertices on a cycle of length x and y are compared. Therefore, if the shortest cycle in G has length x and the shortest cycle in H has length y, $\mathcal{F}(G) \neq \mathcal{F}(H)$.

Having shown that the shortest cycles in G and H must be the same, it is now possible to induct on the remaining $v \in V$ and see $G = H$. □

The proposer of the original problem that inspired this paper, S. Hedetniemi, now proposes looking at a large class of similar problems, in which the roles of the indexed collections of open neighborhoods, or closed neighborhoods, are replaced by other iconic collections of graph elements. For instance, we could ask: for which finite simple graphs G does the list of maximal cliques in G have a system of distinct vertex representatives, or a system of distinct edge representatives? As with the cases dealt with in this paper, the answers could be interesting, or not. The authors may also investigate whether it is satisfying to redefine $\mathcal{F}$ using the matching or independent sets instead of neighborhoods.

References

1. Haenel, Johnson. *On a Problem of Hedetniemi*, Alabama Journal of Mathematics 43 (2019), 4.
2. Hedetniemi, Holliday, Johnson. *Neighborhood Representatives*, Congressus Numerantium 231 (2018), 117–119.
3. Hedetniemi, Holliday, Johnson. *Distinct Representatives in Special Set Families in Graphs*, Congressus Numerantium 233 (2019), 111–118.

Harmonious Labelings of Disconnected Graphs Involving Cycles and Multiple Components Consisting of Starlike Trees

Atif Abueida and Kenneth Roblee

Abstract A *harmonious labeling of a (simple) graph* $G = (V, E)$ on $m > 0$ edges is a one-to-one function $f : V \to \mathbf{Z}_m$ such that if $e_1, e_2 \in E$ with respective endpoints u_1, v_1 and u_2, v_2, then $f(u_1) + f(v_1) \not\equiv f(u_2) + f(v_2) (\text{mod } m)$. If such a function exists, then G is said to be *harmonious*. If G were a tree, then precisely one vertex label is allowed to be used twice. A starlike tree is a tree with a central vertex adjacent to one endpoint of some number of paths each with the same number of vertices. It has been shown using cyclic groups that the disjoint union of an odd cycle on s vertices and starlike trees with the central vertex adjacent to some even $t \geq 2$ many s-paths is harmonious. We now consider the disjoint union of an odd cycle with at least two starlike trees with new notions of harmonious labelings to accommodate the case where $|V| > |E|$, one of which is a basic generalization of harmonious labeling and the other of which is a stricter and more balanced harmonious labeling.

Keywords Cycle · Tree · Harmonious labelling

1 Introduction

Suppose that $G = (V, E)$ is a simple and non-edgeless graph with $|V| = n$ and $|E| = m$. A *harmonious labeling of* G is a one-to-one function $f : V \to \mathbf{Z}_m$ such that whenever $e_1 = u_1v_1$ and $e_2 = u_2v_2$ are distinct edges with their respective endpoints as indicated, we have that $f(u_1) + f(v_1) \not\equiv f(u_2) + f(v_2)$ (mod m). For $v \in V$, the value $f(v)$ is the (vertex) "label" of v. For an edge $e = uv$, the value $f(u) + f(v)$ (mod m) is the (edge) label of e. Thus, in a harmonious labeling of G (if such a labeling exists), each vertex label occurs at most once and each edge label occurs

A. Abueida (✉)
University of Dayton, Dayton, OH 45469-2316, USA
e-mail: aabueida1@udayton.edu

K. Roblee
Troy University, Troy, AL 36082, USA
e-mail: kroblee@troy.edu

F. Hoffman (ed.), *Combinatorics, Graph Theory and Computing*,
Springer Proceedings in Mathematics & Statistics 388,
https://doi.org/10.1007/978-3-031-05375-7_10

exactly once. In the case when G is a tree, we are permitted to use exactly one vertex label two times. If a harmonious labeling of G exists, then G is said to be harmonious.

Harmonious labelings of graphs were first introduced by Graham and Sloan in [6]. There are many fundamental results from that foundational paper. For example, the complete graph K_n is harmonious if and only if $n \neq 4$; the complete bipartite graph $K_{m,n}$ is harmonious if and only if $m = 1$ or $n = 1$; the cycle graph C_n is harmonious if and only if $n \geq 3$ is odd. Other results in [6] include that wheels W_n are harmonious for $n \geq 3$, ladders $L_n = P_2 \times P_n$ are harmonious for $n \geq 3$ are harmonious, and the Petersen Graph is harmonious. For a comprehensive survey of results on harmonious and other types of graph labelings, please refer to [3].

It is not difficult to see that paths P_n and stars $S_n = K_{1,n}$ (referenced above) are harmonious. It has been shown that all trees with 31 or fewer vertices are harmonious in [2]. In [6], it was shown that all caterpillars are harmonious.

Relevant to the results here and regarding disjoint unions of graphs involving trees, it was shown in [7] that $C_s \cup P_3$ is harmonious for all odd $s \geq 3$. Additionally, one of the results by Gallian and Stewart implies that certain disjoint unions of odd cycles and paths are harmonious; see [5]. Abueida and Roblee re-proved this result (among other things) in [1]; namely, they showed that for odd $s \leq 3$ and even $t \geq 2$, the graph $C_s \cup P_{st+1}$ is harmonious. The labeling there involved labeling the cycle with the elements of the cyclic subgroup H generated by $1 + t$ of $\mathbf{Z}_{s(1+t)}$; different sections of the path were carefully labeled using the elements of the cosets of H. That labeling idea—which used the fact that $s \mid s(1 + t)$ and fundamental results about finite cyclic groups and their subgroups—was extended in [1] to show that for the same values of s and t, the graph $C_s \cup T_{st+1}$ is harmonious; here, T_{st+1} is a "starlike tree" consisting of a central vertex adjacent to one endpoint of each of t-many paths each on s vertices.

In these latter results, to accommodate the tree and to agree with how trees are permitted to be harmoniously labeled, exactly one vertex label was used twice. Moreover, by the method used in labeling these, one of the labels on the cycle was used again for a vertex label on the tree; hence, this was the one recycled label permitted.

Here, we consider how to approach harmonious labeling of and extension of the aforementioned problem of labeling a disjoint union of a tree and cycle. In particular, we propose a general definition of harmonious labeling of the disjoint union of a cycle with multiple trees and show how to extend labelings to new disjoint unions from the authors' previous work. Then we consider a more balanced version of this labeling, in the sense that the number of times each label is used on the vertices is as equitable as possible. Then we show how to extend labelings from previous work to these disjoint unions.

2 Main Results

The first definition we consider is one that allows disconnected graphs in which $|V| > |E|$.

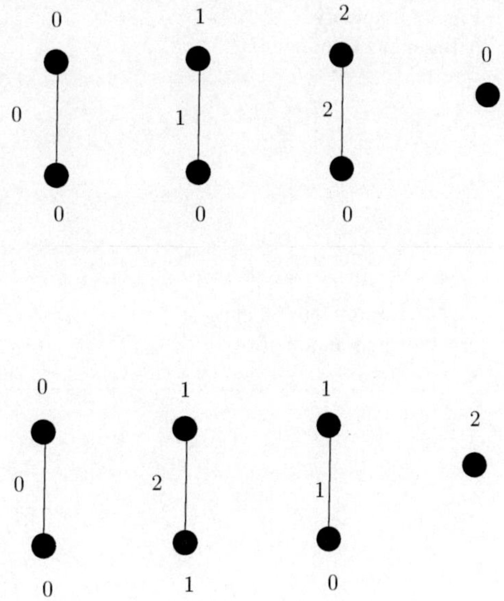

Fig. 1 Two harmonious labelings of a forest

Definition 1 Let $G = (V, E)$ be a non-edgeless simple graph with $|V| = n > m = |E|$. An onto function $f : V \to \mathbf{Z}_m$ is said to be a *harmonious labeling of* G provided that if $e_1 = u_1v_1$ and $e_2 = u_2v_2$ are distinct edges, then $f(u_1) + f(v_1) \not\equiv f(u_2) + f(v_2) \pmod{m}$. If such a function exists, then G is said to be harmonious.

Observe that this agrees with the common usage of the term "harmonious labeling" when a single tree is considered. Moreover, it agrees with other works (such as the authors' previous works) when considering the disjoint union of a cycle and a tree. In particular, precisely one vertex label was used twice, and so the function was onto. The onto condition ensures that each vertex label is used at least once. However, it is rather permissive beyond this: For example, let G be the disjoint union of three copies of P_2 and one isolated vertex; so $n = 6$ and $m = 3$, we could construct rather different harmonious labelings, such as using 0 to label both vertices of the first component; use 0 and 1 to label the second component; label 0 and 2 to label the third component; and label the isolated vertex 0. This is clearly an onto function and the respective edge labels would be 0, 1, and 2; thus it is a harmonious labeling of the graph. Moreover, once you have cleared the onto hurdle of the labeling, you are not restricted as to how many more times you could use a particular vertex label, as long as you keep the edge labels distinct. See Fig. 1 for this labeling as well as another harmonious labeling of the same graph.

For the same graph G, we could alternatively label the first component's vertices using 0 and 1; label the second component's vertices with 1 and 2; and label the third component's vertices using 1 and 2; then label the isolated vertex with 0. This is also an onto function and the respective edge labels would be 1, 2, and 3. This labeling is more balanced in the sense that each label is used the same number of times. Of course, this may not always be possible, but something that Libras and others who

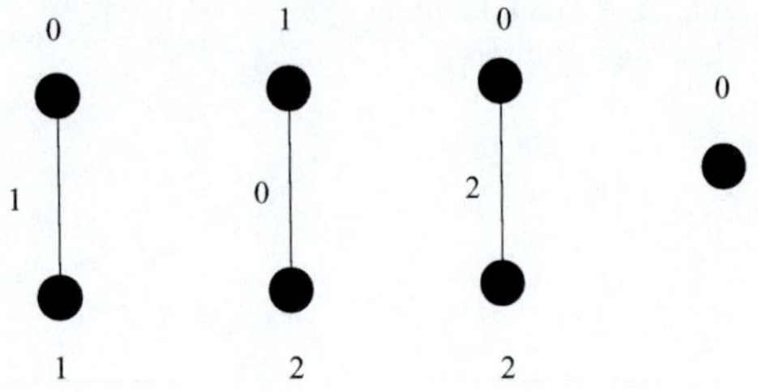

Fig. 2 Equitably harmonious labeling of a forest

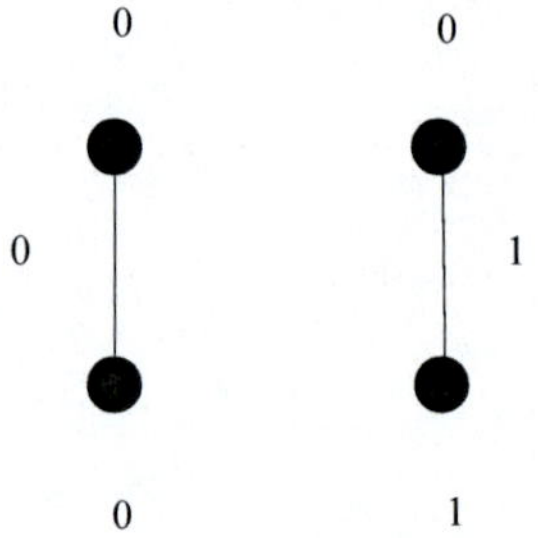

Fig. 3 Harmonious labeling of the disjoint union of two P_2's

like balance would hope for. See Fig. 2 for a picture of this labeling. Thus, we present a new definition.

Definition 2 Let G be as in the previous definition. An equitably harmonious labeling of G is a harmonious labeling of G such that for each $i, j \in \mathbf{Z}_m$, we have $||f^{-1}(i)| - |f^{-1}(j)|| \leq 1$.

With this definition, the vertex labels must be used as close to the same number of many times as possible. In Fig. 1, the labelings presented are not equitably harmonious, whereas in Fig. 2 we see an example of the same disconnected graph together with an equitably harmonious labeling of it. In Fig. 3, we have the disjoint union of two P_2's, which is interestingly harmonious but not equitably harmonious. One can refer to Fig. 4 to convince oneself that this is the case.

Now, we show how to extend some results in [1] to new graphs with harmonious and equitably harmonious labelings.

We let $s \geq 3$ be odd and $t \geq 2$ be even. A "starlike tree" $T = T_{st+1}$ on $st + 1$ vertices consists of a central vertex adjacent to one endpoint of each of t-many paths each on $s \geq 3$ vertices, where $1 \leq i \leq t$. Figure 5 shows an example to illustrate with $t = 4$ and $s = 3$; as in the definition, it is a central vertex adjacent to the ends of a given number of fixed-length paths.

The authors showed in [1] that for such values of s and t, then $C_s \cup T$ is harmonious, among other things. We re-state the result here formally as well as the method of labeling; we re-use some of this method to prove our first theorem a little later.

Theorem 1 (Abueida, Roblee, 2019) *Let $s \geq 3$ be odd, $t \geq 2$ be even, and $T = T_{st+1}$ be a starlike tree. Then $C_s \cup T$ is harmonious.*

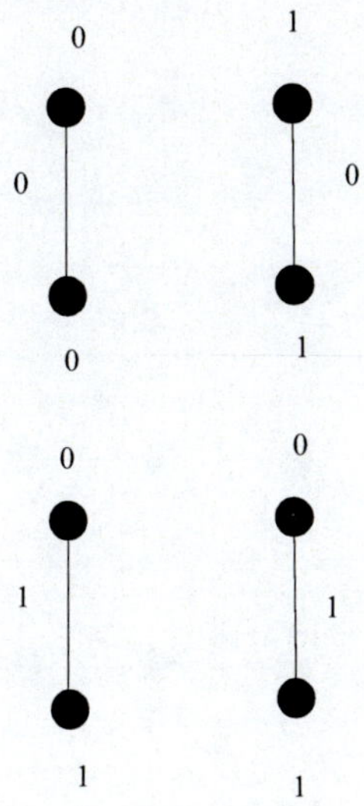

Fig. 4 Two failed attempts at equitably harmonious labelings of the disjoint union of two P_2's

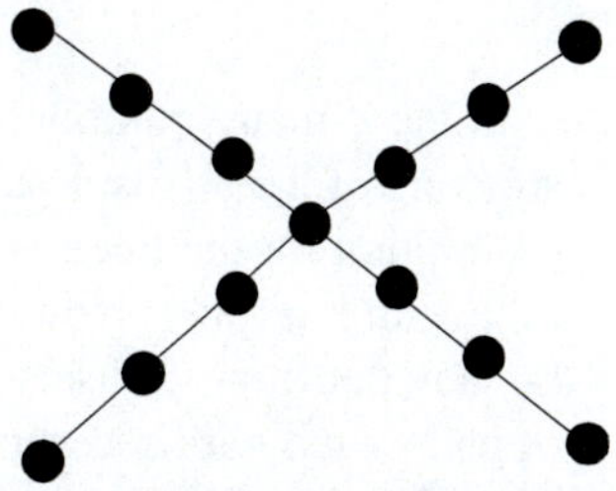

Fig. 5 Starlike tree consisting of a central vertex adjacent to $t = 4$ paths each on $s = 3$ vertices

For ease of notation, we denote the t-many paths each with s vertices in T by P_s^i, where $1 \leq i \leq t$. For convenience, we give the labeling of that graph again here.

For the cycle component, we label its vertices using the cyclic subgroup H of $\mathbf{Z}_{s+st}$ generated by the element $1 + t$; we denote this by writing $H = \langle 1 + t \rangle$. Thus, $H = \langle 1 + t \rangle = \{0, 1 + t, 2(1 + t), \ldots, (s - 1)(1 + t)\}$. In particular, choose a vertex and label it with an element (we will pick 0) of H; then proceed to consecutive vertices and label them with consecutive elements of H (so, $1 + t,\ 2(1 + t), \ldots (s - 1)(1 + t)$). See [4] for more on cyclic groups and other group-theoretic concepts.

Next, we denote the central vertex of the starlike tree T by v_0 and label it 0 (observe this vertex label was already used on the cycle—it is our only repeated vertex label). Now, for all i, where $1 \leq i \leq t$, we label the vertices of P_s^i starting from one end to another consecutively with the elements (in order) of the coset $i + H = \{i, i + (1 + t), i + 2(1 + t), \ldots, i + (s - 1)(1 + t)\}$. For odd i, we connect the endpoint of P_s^i with the smallest label to v_0; for even i, we connect the endpoint of P_s^i with the largest label to v_0. This was shown to be a harmonious labeling in [1]; Fig. 6 gives an example from that paper shown again here.

Moving to a different family of disconnected graphs, consider the graph $G = C_s \cup F_{r+st+1}$ with s, t as indicated above. Here, $F = F_{r+st+1}$ is a forest consisting

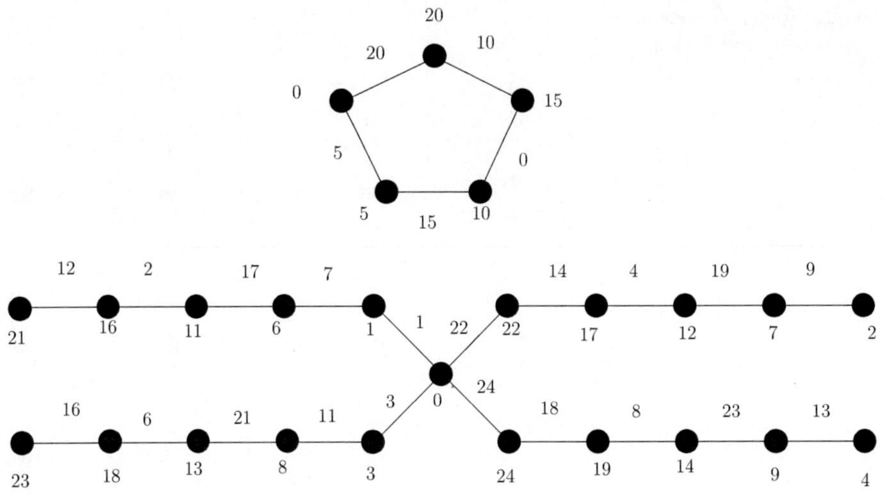

Fig. 6 Example of Odd Cycle Union Star-like tree

of some r-many components, namely, $T_1, \ T_2, \ldots, T_r$, where $1 \leq r \leq t$, and the structure of the T_i's is described in the next paragraph. To facilitate the notation and description for the trees, we denote by $P = \{P_s^1, P_s^2, \ldots, P_s^t\}$ a collection of paths each with s vertices. Now, let $\mathcal{P} = \{S_1, S_2, \ldots, S_r\}$ be a partition of P consisting of r-many nonempty subsets of P; thus, each $S_i \in \mathcal{P}$, where $1 \leq i \leq r$ is a collection of paths each with s vertices. For each such i we could more precisely write $S_i = \{P_s^{i_1}, P_s^{i_2}, \ldots, P_s^{k_i}\}$ for some $1 \leq k_i \leq t$ and such that $\sum_{i=1}^{r} k_i = t$.

With this notation, we specify the structure of the trees $T_1, T_2, \ldots, T_r$. Each of the tree components T_i is a starlike tree each with a central vertex v_i adjacent to an endpoint of each of the k_i-many paths taken from S_i. Now, we claim that G is harmonious.

To prove the claim, let us observe that the harmonious labeling of $C_s \cup T$ presented earlier can be extended to create a harmonious labeling of G. Let us note that $|V| = s + r + st > |E| = s + st$ and the vertex labels will be elements of $\mathbb{Z}_{s(1+t)}$.

To see how the previous labeling extends to this case, we again label the vertices of C_s using the elements of the subgroup $H = \langle 1 + t \rangle$ of $\mathbb{Z}_{s(1+t)}$. We label the central vertices $v_1, v_2, \ldots, v_r$ of $T_1, T_2, \ldots, T_r$ each with 0. Observe that the vertex label 0 will then be used a total of $r + 1$ times. As such, this will be the only repeated vertex label.

For each $1 \leq i \leq r$, we label the vertices of the paths $P_s^{i_1}, P_s^{i_2}, \ldots, P_s^{k_i}$ starting at one end of the path consecutively and in order with the elements of their respective cosets $i_1 + H, i_2 + H, \ldots, k_i + H$. Note these values $i_1, \imath_2, \ldots, k_i$ will be all distinct and will be distinct for different values of i. This is due to the partitioning we did in advance. For all $1 \leq i \leq r$ and all $i_1, i_2, \ldots, k_i$, we connect the central vertex v_i to the least label of the paths labeled by the odd coset-labeled paths (odd values of i_* or k_i for labeling from $i_* + H$ or $k_i + H$) and by the greatest labeled element from the

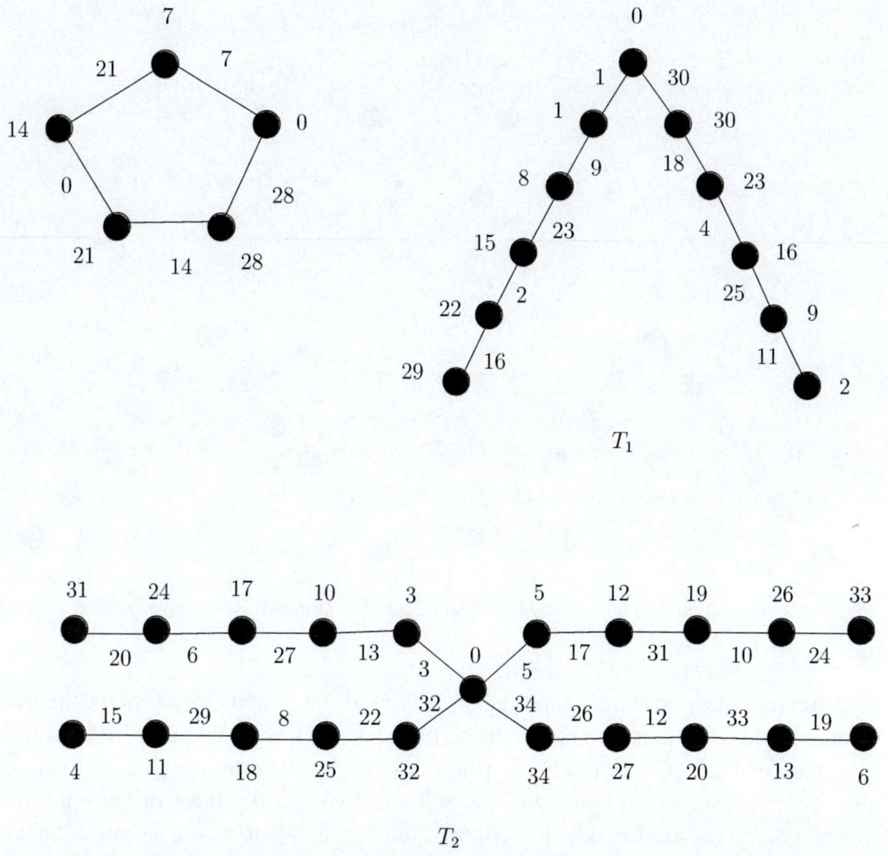

Fig. 7 Example of a harmonious labeling of an odd cycle union two nonisomorphic starlike trees

even coset-labeled paths (even i_* or k_i for labeling from $i_* + H$ or $k_i + H$). As this just "splits" the labeling from [1], still no repeated edge labels exist. In Fig. 7, we have $G = C_5 \cup F_{2+5\cdot 6}$. Note how the vertices in the left-side path in T_1 are labeled with elements (in order) of $1 + H$, where $H = \langle 7 \rangle \leq \mathbf{Z}_{35}$; the elements on the right-side path in T_1 are labeled with elements (in reverse order) of $2 + H$. The vertices in the 4 paths in T_2 are labeled with elements of $3 + H, 4 + H, 5 + H$, and $6 + H$, where the "even cosets" label in reverse order and the "odd cosets" label in the natural order.

This is essentially "splitting" the starlike tree in $C_s \cup T_{st+1}$ described earlier into some number of new starlike trees and keeping the same labeling, but just with the additional central vertices labels of 0. See Fig. 8 for another example. Thus, we have the following:

Theorem 2 *Let $s \geq 3$ be odd, $t \geq 2$ be even, and $r \geq 1$. Let $F = F_{r+st}$ be a forest of r-many starlike trees, with each central vertex v_i adjacent to an end of k_i-many paths each on s vertices. Then $C_s \cup F$ is harmonious.*

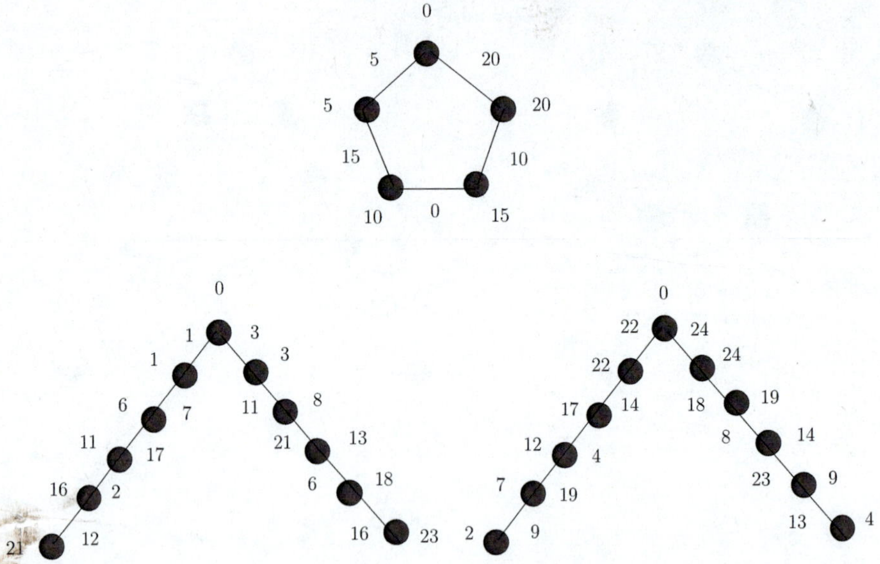

Fig. 8 Example of a harmonious labeling of an odd cycle union two isomorphic starlike trees

Although the harmonious labeling procedure for the graph $C_s \cup F$ in the theorem is not an equitably harmonious labeling (unless $r = 1$), we do ponder for the future whether for $C_s \cup F$ is equitably harmonious. In the meantime, we now focus on the specific case of this issue when $r = 2$ and two starlike trees in the union are isomorphic. That is, for odd $s \geq 3$ and even $t \geq 2$, what would be an equitably harmonious labeling of $C_s \cup T_1 \cup T_2$, where both T_1 and T_2 are both copies of the starlike tree $T_{s \cdot t/2+1}$ consisting of a central vertex adjacent to one end of $\frac{t}{2}$-many paths each on s vertices. We show that this graph is equitably harmonious. For example, see Fig. 9.

Thus, we have the following result.

Theorem 3 *Let $s \geq 3$ be odd, $t \geq 2$ be even. Then the graph $C_s \cup T_1 \cup T_2$, where $T_1 \cong T_2 \cong T_{s \cdot t/2+1}$ is equitably harmonious.*

Proof The vertex labels are elements of the group $\mathbf{Z}_{s(s+t)}$ and there are clearly $s(s+t)+2$ vertices. Thus, we would need the labeling function to be onto such that one pair of vertices use some label twice and exactly one more pair use a different label twice. As before, we label the vertices of the cycle (starting with any vertex, and consecutively around the cycle) with the elements of the subgroup $H = \langle 1+t \rangle$ of $\mathbf{Z}_{s(s+t)}$. As in the case of harmonious labelings of C_s a from [1], no edge labels on the cycle are repeated.

For T_1, we label its central vertex v_1 with 0. Let us denote the paths in T_1 by $P_s^1, P_s^2, \ldots, P_s^{t/2}$. For each $i = 1, 2, \ldots, \frac{t}{2}$, label the vertices of P_s^i from one end to another consecutively by the elements of $2i - 1 + H$ in increasing order; then make the central vertex adjacent to the least-labeled vertex in each path.

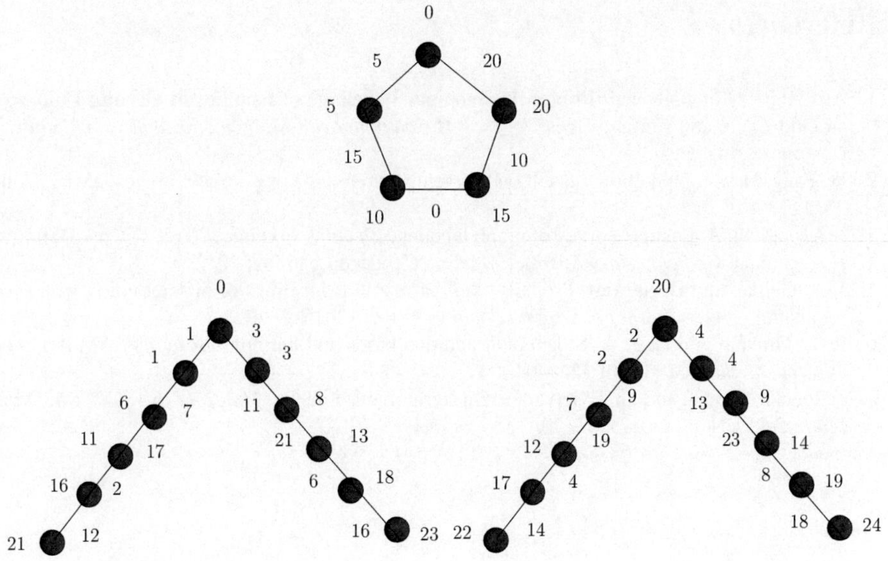

Fig. 9 Example of an equitably harmonious labeling of an odd cycle union two starlike trees

For T_2, label its central vertex v_2 with $(s-1)(1+t)$; denote the paths in T_2 by $Q_s^1, Q_s^2, \ldots, Q_s^{t/2}$. Then label the vertices of Q_s^i consecutively from one endpoint to another by the elements of $2i+H$ in increasing order for all $i = 1, 2, \ldots, \frac{t}{2}$. Finally, we make v_2 adjacent to the end with the *greatest* label in each path. Thus, we labeled the vertices of T_2 so to not use the vertex label of 0 three times as in the harmonious labeling of the forest part of our previous theorem. With T_2, we changed the label of the central vertex v_2 from the case of harmonious labelings to $(s-1)(1+t)$ and reversed the labeling of its paths from greatest to least labels in $2i+H$ starting from the v_2-connected end. This gives label of the edge with endpoints v_2 and its adjacent vertex in Q_s^i to be $(s-1)(1+t)+2i \pmod{s(1+t)}$ for $i = 1, 2, \ldots, \frac{t}{2}$.

The edge labels connecting v_1 to the paths in T_1 would clearly be $0+i=i$ for $i = 1, 2, \ldots, \frac{t}{2}$. Clearly, $2i-1 \not\equiv (s-1)(t+1)+2j \pmod{s(s+t)}$ for $1 \le i, j \le \frac{t}{2}$.

As the other edge labels coincide with the harmonious labeling in [1], then this is indeed a harmonious labeling. As only 0 and $(s-1)(1+t)$ are the only repeated vertex labels, then this is an equitably harmonious labeling. □

The authors wish to thank Professor Peter Johnson, Jr. of Auburn University for his valuable questions as to how harmonious labelings extend to more severely disconnected graphs such as these.

References

1. Atif Abueida and Kenneth Roblee, Harmonious Labelings of Families of Disjoint Unions of an Odd Cycle and Certain Trees, *Journal of Combinatorial Mathematics and Combinatorial Computing*, to appear.
2. W. Fang, New computational result on harmonious trees, arXiv:1106.3490v1 [cs.DM] 17 June 2011.
3. J. A. Gallian, A dynamic survey of graph labeling, *Electr. J. Combin.*, 20 (2017), p.# DS6.
4. J. A. Gallian, *Contemporary Abstract Algebra*, Cengage Learning, 2012.
5. J. A. Gallian and D. Stewart, Properly even harmonious labelings of disjoint union with even sequnetial graphs, *Journal of Graph Labeling*, 1 No.1 (2015), 1–10.
6. R. L. Graham and N. J. A. Sloane, On additive bases and harmonious graphs, *SIAM J. Alg. Discrete Methods*, **1** (1980) 382–404.
7. J. Renuka, P. Balaganesan, P. Selvaraju, On Harmonious Labeling, *Internat. J. Advance in Math. Math. Sci.*, **1** No. 2 (2012), 65–70.

On Rainbow Mean Colorings of Trees

James Hallas, Ebrahim Salehi, and Ping Zhang

Abstract A mean coloring of a connected graph G of order 3 or more is an edge coloring of G with positive integers such that the mean of the colors of the edges incident with every vertex is an integer. The associated color of a vertex is its chromatic mean. If distinct vertices have distinct chromatic means, then the edge coloring is a rainbow mean coloring of G. The maximum vertex color in a rainbow mean coloring c is the rainbow mean index of c, while the rainbow mean index of the graph G is the minimum rainbow mean index among all rainbow mean colorings of G. In this paper, rainbow mean colorings of trees are investigated.

Keywords Chromatic mean · Rainbow mean colorings · Rainbow chromatic mean index

AMS Subject Classification: 05C05, 05C07, 05C15, 05C78

1 Introduction

During the past several decades, there have been many studies of edge labelings or edge colorings of graphs that have given rise to vertex labelings or colorings where no two vertices have the same color (see [1, 3, 4, 6, 7], for example). One of the early examples of this occurred in 1986 when at the 250th Anniversary of Graph Theory Conference held at Indiana University-Purdue University Fort Wayne

J. Hallas
Department of Mathematics and Computer Science, Concord University, Athens, WV 24712, USA

E. Salehi (✉)
Department of Mathematical Sciences, University of Nevada Las Vegas, Las Vegas, NV 89154-4020, USA
e-mail: ebrahim.salehi@univ.edu

P. Zhang
Department of Mathematics, Western Michigan University, Kalamazoo, MI 49008-5248, USA

F. Hoffman (ed.), *Combinatorics, Graph Theory and Computing*,
Springer Proceedings in Mathematics & Statistics 388,
https://doi.org/10.1007/978-3-031-05375-7_11

(now called Purdue University Fort Wayne), Chartrand introduced a concept, often called the irregularity strength of graphs. The *irregularity strength* of a graph G is the smallest positive integer k for which there exists a coloring of the edges of G from the set $[k] = \{1, 2, \ldots, k\}$ resulting in a vertex coloring where the color of a vertex is the sum of the colors of its incident edges, such that no two vertices have the same color. The goal then was for the vertices to have distinct colors, regardless of how large these colors may be. For vertices of large degree, this requires some vertex colors to be large, possibly much larger than the order of the graph. However, in [2] an edge coloring of a graph was introduced in which each edge was colored with a positive integer in a manner so that each vertex is assigned a positive integer color that is the average of the colors of its incident edges and no two vertices have the same color. If the order of G is n, then the number of vertex colors must therefore be at least n. With all the conditions required for a graph to have such an edge coloring, one might anticipate that for some graphs at least, the largest vertex color may exceed the order of the graph, possibly by a large amount. The goal of this paper is to investigate this topic where the graphs in question are trees.

2 Rainbow Mean Colorings

A *mean coloring* of a connected graph G of order 3 or more is an edge coloring $c : E(G) \to \mathbb{N}$ of G such that for every vertex v of G, its vertex color

$$\text{cm}(v) = \frac{\sum_{e \in E_v} c(e)}{\deg v}, \text{ where } E_v \text{ is the set of edges incident with } v,$$

is an integer, called the *chromatic mean* of v. Clearly, every nontrivial connected graph G has mean colorings. For example, if every edge of G is assigned the same positive integer a, the resulting edge coloring is a mean coloring in which $\text{cm}(v) = a$ for every vertex v of G. If distinct vertices have distinct chromatic means, then the edge coloring c is called a *rainbow mean coloring* of G. The following result was obtained in [2].

Theorem 1 *Every connected graph of order* 3 *or more has a rainbow mean coloring.*

For a rainbow mean coloring c of a graph G, the maximum vertex color is the *rainbow chromatic mean index* (or simply, the *rainbow mean index*) $\text{rm}(c)$ of c. That is,

$$\text{rm}(c) = \max\{\text{cm}(v) : v \in V(G)\}.$$

The *rainbow chromatic mean index* (or the *rainbow mean index*) $\text{rm}(G)$ of the graph G itself is defined as

$$\text{rm}(G) = \min\{\text{rm}(c) : c \text{ is a rainbow mean coloring of } G\}.$$

Two immediate observations were also made in [2].

Observation 1 If G is a connected graph of order $n \geq 3$, then $\mathrm{rm}(G) \geq n$.

Observation 2 If c is a rainbow mean coloring of a connected graph G, then

$$\sum_{v \in V(G)} \deg v \cdot \mathrm{cm}(v) = 2 \sum_{e \in E(G)} c(e).$$

Furthermore, if the order of G is n and $\mathrm{rm}(c) = n$, then $\sum_{v \in V(G)} \mathrm{cm}(v) = \binom{n+1}{2}$.

The rainbow mean index was obtained in [2] for paths, cycles, and complete graphs.

Theorem 2 *For an integer $n \geq 3$,*

$$\mathrm{rm}(P_n) = \begin{cases} n & \text{if } n \neq 4 \\ 5 & \text{if } n = 4. \end{cases}$$

Theorem 3 *For an integer $n \geq 3$,*

$$\mathrm{rm}(C_n) = \begin{cases} n & \text{if } n \equiv 0, 1 \pmod 4 \\ n+1 & \text{if } n \equiv 2, 3 \pmod 4 \end{cases}$$

Theorem 4 *For an integer $n \geq 4$,*

$$\mathrm{rm}(K_n) = \begin{cases} n & \text{if } n \not\equiv 2 \pmod 4 \\ n+1 & \text{if } n \equiv 2 \pmod 4. \end{cases}$$

The rainbow mean index was determined for the complete bipartite graphs $K_{s,t}$, $1 \leq s \leq t$ and $s + t \geq 3$, in [5], with the case $s = 1$ observed in [2].

Theorem 5 *Let s and t be integers with $1 \leq s \leq t$ and $n = s + t \geq 3$. Then*

$$\mathrm{rm}(K_{s,t}) = \begin{cases} n & \text{if } st \text{ is even} \\ n+1 & \text{if } st \text{ is odd and } s \geq 3 \\ n+2 & \text{if } t \text{ is odd and } s = 1. \end{cases}$$

In a rainbow mean coloring of a connected graph G of order at least 3, each edge of G is assigned a positive integer color in such a way that every vertex color is an integer and all vertex colors are distinct. Hence, it may be anticipated that at least in some cases, vertex colors would be considerably larger than the order of the graph. However, no such graph has yet been found. Indeed, from the results obtained on the rainbow mean index of many connected graphs G of order $n \geq 3$, the value of $\mathrm{rm}(G)$ has always been either n or $n + 1$ with the one exception of stars of even order $n \geq 4$, which have rainbow mean index $n + 2$. In fact, the following conjecture was stated in [2].

Conjecture 1 For every connected graph G of order $n \geq 3$,

$$n \leq \mathrm{rm}(G) \leq n + 2.$$

Since only stars of even order $n \geq 4$ have been shown to have rainbow mean index different from n or $n+1$, this suggests studying the rainbow mean index of trees related to stars in some manner. In this paper, we determine the rainbow mean index of three classes of trees, namely cubic caterpillars, subdivided stars, and double stars.

3 The Rainbow Mean Index of Trees

Let c be a rainbow mean coloring of a connected graph G. For a vertex v of G, the *chromatic sum* $\mathrm{cs}(v)$ of v is defined as the sum of the colors of the edges incident with v. Hence, $\mathrm{cs}(v) = \sum_{e \in E_v} c(e) = \deg v \cdot \mathrm{cm}(v)$.

Observation 3 Let G be a connected bipartite graph with partite sets U and W. If c is an edge coloring of G, then $\sum_{u \in U} \mathrm{cs}(u) = \sum_{w \in W} \mathrm{cs}(w)$.

A connected graph of order 3 or more with a rainbow mean coloring is referred to as a *mean colored-graph*. A vertex v of a mean colored-graph G is called *chromatically odd* if $\mathrm{cs}(v) = \deg v \cdot \mathrm{cm}(v)$ is an odd integer; otherwise, v is *chromatically even*. The following are consequences of Observation 2.

Corollary 1 *Every mean colored-graph contains an even number of chromatically odd vertices.*

Corollary 2 *If G is a connected graph of order $n \geq 6$ with $n \equiv 2 \pmod 4$ all of whose vertices are odd, then $\mathrm{rm}(G) \geq n + 1$.*

By Theorem 5, for each integer $n \geq 3$,

$$\mathrm{rm}(K_{1,n-1}) = \begin{cases} n & \text{if } n \text{ is odd} \\ n+2 & \text{if } n \text{ is even.} \end{cases} \tag{1}$$

Consequently, if $n \not\equiv 0, 2 \pmod 4$, then $\mathrm{rm}(K_{1,n-1}) = n$. Of course, by Corollary 2, if $n \equiv 2 \pmod 4$, then $\mathrm{rm}(K_{1,n-1}) \neq n$. This brings up the question of determining $\mathrm{rm}(T)$ for those trees of order 5 or more which is neither a path nor a star. Figure 1 shows trees T (that are not paths or stars) of order n where $n \in \{5, 6\}$. With one exception, $\mathrm{rm}(T) = n$ for all these trees T. For this one exception, the tree T has order 6 and all vertices have odd degree. Of course, by Corollary 2, the rainbow mean index of this tree is at least 7. As shown in Fig. 1, $\mathrm{rm}(T) = 7$ for this tree T.

Figure 2 shows all trees T (that are not paths or stars) of order 7 together with a rainbow mean coloring for each of these trees. Thus, every tree of order 7 has rainbow mean index 7.

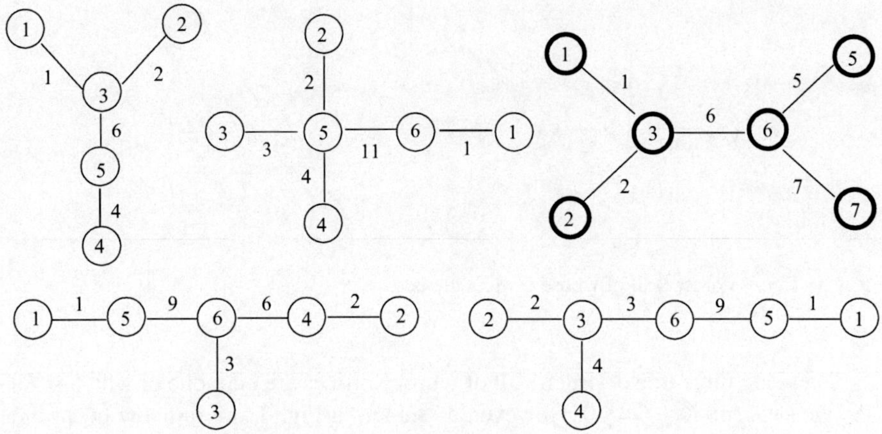

Fig. 1 Rainbow mean colorings of trees of order 5 and 6

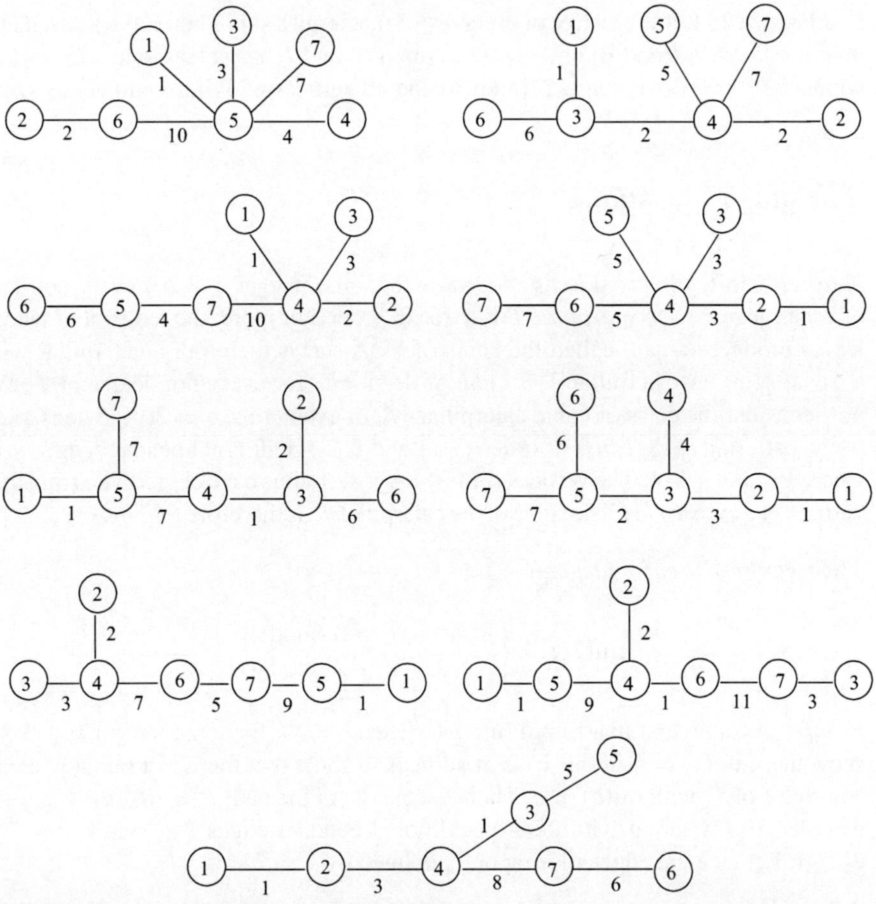

Fig. 2 Rainbow mean colorings of trees of order 7

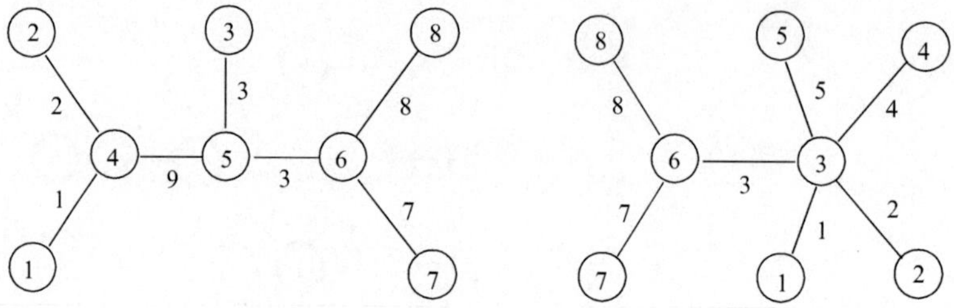

Fig. 3 Trees of order 8 all of whose vertices are odd

There are three tree of order 8 all of whose vertices are odd, one of which is $K_{1,7}$. As we saw, $\mathrm{rm}(K_{1,7}) = 10$. However, as shown in Fig. 3, the rainbow mean index of the two remaining such trees of order 8 is 8.

These observations lead us to the following conjecture.

Conjecture 2 Let T be a tree of order $n \geq 5$ that is not a star. Then $\mathrm{rm}(T) = n$ if and only if (i) $n \not\equiv 2 \pmod 4$ or (ii) $n \equiv 2 \pmod 4$ and T has at least one even vertex; while $\mathrm{rm}(T) = n + 1$ if $n \equiv 2 \pmod 4$ and all vertices of T have odd degrees.

4 Cubic Caterpillars

A tree T is often referred to as r *-regular* for some integer $r \geq 2$ if every non-leaf of T has degree r. A *caterpillar* T is a tree of order 3 or more, the removal of whose leaves produces a path called the *spine* of T. A star is therefore a caterpillar with a trivial spine. A caterpillar T is *cubic* if $\deg v = 3$ for every non-leaf v of T. We now consider the class of cubic caterpillars T_n of even order $n = 2\ell \geq 6$ consisting of the path $(u_0, u_1, \ldots, u_\ell)$ of order $\ell + 1$ and $\ell - 1$ additional pendant edges $u_i v_i$ where $1 \leq i \leq \ell - 1$. The vertices u_i, $1 \leq i \leq \ell - 1$, thus have degree 3 and all other vertices of T_n are leaves. The spine of the caterpillar T_n is therefore $(u_1, u_1, \ldots, u_{\ell-1})$.

Theorem 6 *For each integer $n \geq 6$,*

$$\mathrm{rm}(T_n) = \begin{cases} n & \text{if } n \equiv 0 \pmod 4 \\ n+1 & \text{if } n \equiv 2 \pmod 4. \end{cases}$$

Proof Assume first that $n \equiv 0 \pmod 4$. Then $n = 4k$ for some integer $k \geq 2$. To show that $\mathrm{rm}(T_n) = n$ in this case, it suffices to show that there is a rainbow mean coloring c of T_n with $\mathrm{rm}(c) = n$. Then T_n consists of the path $P = (u_0, u_1, \ldots, u_{2k})$ of order $2k + 1$ along with $2k - 1$ additional pendant edges $u_i v_i$ where $1 \leq i \leq 2k - 1$. Let c be the edge coloring of T_n defined by

$$c(e) = \begin{cases} 2i & \text{if } e = u_i v_i \text{ for } 1 \le i \le 2k-2 \\ 4k-3 & \text{if } e = u_{2k-1}v_{2k-1} \\ 1 & \text{if } e = u_0 u_1 \\ 2i+4 & \text{if } e = u_i u_{i+1} \text{ where} 1 \le i \le 2k-3 \text{ and } i \text{ is odd} \\ 2i+1 & \text{if } e = u_i u_{i+1} \text{ where} 2 \le i \le 2k-4 \text{ and } i \text{ is even} \\ 4k & \text{if } e = u_{2k-2}u_{2k-1}, u_{2k-1}u_{2k}. \end{cases}$$

Then the chromatic means of the vertices of T_n are given by

$$\text{cm}(u_i) = \begin{cases} 2i+1 & \text{if } 0 \le i \le 2k-3 \text{ or } i = 2k-1 \\ 2i+2 & \text{if } i = 2k-2 \\ 2i & \text{if } i = 2k \end{cases}$$

$$\text{cm}(v_i) = \begin{cases} 2i & \text{if } 1 \le i \le 2k-2 \\ 2i-1 & \text{if } i = 2k-1. \end{cases}$$

Hence, c is a rainbow mean coloring with $\text{rm}(c) = n$ and so $\text{rm}(T_n) = n$ if $n \equiv 0 \pmod 4$.

Next, suppose that $n \equiv 2 \pmod 4$. Then $n = 4k+2$ for a positive integer k. Then T_n consists of the path $P = (u_0, u_1, \ldots, u_{2k+1})$ of order $2k+2$ and $2k$ additional pendant edges $u_i v_i$ where $1 \le i \le 2k$. Since $n \equiv 2 \pmod 4$ and each vertex of T_n is odd, it follows by Corollary 2 that $\text{rm}(T_n) \ge n+1$. It therefore suffices to show that there is a rainbow mean coloring c of T_n with $\text{rm}(c) = n+1$. Let c be the edge coloring of T_n defined by

$$c(e) = \begin{cases} 2 & \text{if } e = u_1 v_1 \\ 2i+1 & \text{if } e = u_i v_i \text{ for } 2 \le i \le 2k \\ 1 & \text{if } e = u_0 u_1 \\ 2i+4 & \text{if } e = u_i u_{i+1} \text{ where } 1 \le i \le 2k-1 \text{ and } i \text{ is odd} \\ 2i+3 & \text{if } e = u_i u_{i+1} \text{ where } 2 \le i \le 2k \text{ and } i \text{ is even.} \end{cases}$$

Then the chromatic means of the vertices of T_n are given by

$$\text{cm}(u_i) = \begin{cases} 2i+1 & \text{if } i = 0, 1, 2k+1 \\ 2i+2 & \text{if } 2 \le i \le 2k \end{cases}$$

$$\text{cm}(v_i) = \begin{cases} 2 & \text{if } i = 1 \\ 2i+1 & \text{if } 2 \le i \le 2k. \end{cases}$$

Hence, c is a rainbow mean coloring with $\text{rm}(c) = n+1$ and so $\text{rm}(T_n) = n+1$ if $n \equiv 2 \pmod 4$. □

5 Subdivided Stars

The *subdivision graph* $S(G)$ of a graph G is that graph obtained from G by subdividing each edge of G exactly once (that is, by replacing each edge $e = uv$ of G by a new vertex w_e and the two new edges uw_e and vw_e, where w_e is called the *subdivision vertex* of e). If G is a graph of order n and size m, then the order of $S(G)$ is $n + m$ and its size is $2m$.

Theorem 7 *For each integer $t \geq 3$, $\text{rm}(S(K_{1,t})) = 2t + 1$.*

Proof Let $G = S(K_{1,t})$ be the subdivision graph of the star $K_{1,t}$, where $t \geq 3$. Then the order of G is $n = 2t + 1$. By Observation 1, it suffices to show that there is a rainbow mean coloring c of G with $\text{rm}(c) = n$. We consider two cases, according to whether t is even or t is odd.

Case 1. *$t \geq 4$ is even.* Then $t = 2k$ for some integer $k \geq 2$. Let

$$V(K_{1,2k}) = \{u_1, u_2, \ldots, u_k\} \cup \{x_1, x_2, \ldots, x_k\} \cup \{w\},$$

where w is the central vertex of $K_{1,2k}$. For each integer i with $1 \leq i \leq k$, let v_i be the subdivision vertex of $u_i w$ and let y_i be the subdivision vertex of $x_i w$. Define the edge coloring $c : E(G) \to [4k + 1]$ as follows: For $1 \leq i \leq k$,

$$c(u_i v_i) = 2i - 1, c(v_i w) = 2i + 1,$$
$$c(x_i y_i) = 2k + 2i + 1, \text{ and } c(y_i w) = 2k + 2i - 1.$$

Then the chromatic means of the vertices of G are given by

$$\text{cm}(u_i) = 2i - 1 \text{ and } \text{cm}(v_i) = 2i \text{ for } 1 \leq i \leq k,$$
$$\text{cm}(w) = 2k + 1,$$
$$\text{cm}(x_i) = 2k + 2i + 1 \text{ and } \text{cm}(y_i) = 2k + 2i \text{ for } 1 \leq i \leq k.$$

Thus, c is a rainbow mean coloring of G with $\text{rm}(c) = 4k + 1$.

Case 2. *$t \geq 3$ is odd.* Then $t = 2k + 1$ for some positive integer k. Let

$$V(K_{1,2k+1}) = \{u_1, u_2, \ldots, u_k\} \cup \{x_1, x_2, \ldots, x_{k-1}\} \cup \{w_1, z_1\} \cup \{w\},$$

where w is the central vertex of $K_{1,2k+1}$. For each integer i with $1 \leq i \leq k$, let v_i be the subdivision vertex of $u_i w$ for $1 \leq i \leq k$, let y_i be the subdivision vertex of $x_i w$ for $1 \leq i \leq k - 1$, let w_2 be the subdivision vertex of $w_1 w$, and let z_2 be the subdivision vertex of $z_1 w$. Define the edge coloring $c : E(G) \to [4k + 3]$ by

$$c(u_i v_i) = 2i - 1 \text{ and } c(v_i w) = 2i + 1 \text{ for } 1 \leq i \leq k$$
$$c(w_1 w_2) = 2k + 3, c(w_2 w) = 2k - 1,$$
$$c(z_1 z_2) = 2k + 4, c(z_2 w) = 2k + 6,$$
$$c(x_i y_i) = 2k + 2i + 5, \text{ and } c(y_i w) = 2k + 2i + 3 \text{ for } 1 \leq i \leq k - 1.$$

Then the chromatic means of the vertices of G are given by

$$\text{cm}(u_i) = 2i - 1 \text{ and } \text{cm}(v_i) = 2i \text{ for } 1 \leq i \leq k,$$
$$\text{cm}(w) = 2k + 2, \text{cm}(w_1) = 2k + 3, \text{cm}(w_2) = 2k + 1,$$
$$\text{cm}(z_1) = 2k + 4, \text{cm}(z_2) = 2k + 5$$
$$\text{cm}(x_i) = 2k + 2i + 5 \text{ and } \text{cm}(y_i) = 2k + 2i + 4 \text{ for } 1 \leq i \leq k - 1.$$

Thus, c is a rainbow mean coloring of G with $\text{rm}(c) = 4k + 3$. □

6 Double Stars

We saw in Theorem 5 that the rainbow mean index of the star $K_{1,t}$, $t \geq 2$, is $t+1$ if t even and is $t+3$ if t is odd. In fact, the stars of even order 4 or more are the only connected graphs whose rainbow mean index has been shown to be neither the order nor one plus the order of the graph. This suggests investigating the rainbow mean index of the related double stars class of graphs. For integers a and b with $2 \leq a \leq b$, the double star $S_{a,b}$ is that tree of order $a+b$ (and size $a+b-1$) and diameter 3 whose central vertices u and v have degrees a and b, respectively. The vertex u is thus adjacent to $a-1$ end-vertices, denoted by $u_1, u_2, \ldots, u_{a-1}$, while v is adjacent to $b-1$ end-vertices, denoted by $v_1, v_2, \ldots, v_{b-1}$. First, we determine $\text{rm}(S_{a,a})$ where $a \geq 2$. Since $\text{rm}(S_{2,2}) = \text{rm}(P_4) = 5$ by Theorem 2, we may assume that $a \geq 3$.

Theorem 8 *For each integer $a \geq 3$,*

$$\text{rm}(S_{a,a}) = \begin{cases} 2a & \text{if } a \text{ is even} \\ 2a+1 & \text{if } a \text{ is odd.} \end{cases}$$

Proof Suppose that u and v are the central vertices of $G = S_{a,a}$ where u is adjacent to the $a-1$ end-vertices $u_1, u_2, \ldots, u_{a-1}$ and v is adjacent to the $a-1$ end-vertices $v_1, v_2, \ldots, v_{a-1}$. We consider two cases, according to whether a is even or a is odd.

Case 1. *$a \geq 4$ is even*. Then $a = 2k$ for some integer $k \geq 2$. Since the order of G is $4k$, it suffices to show that there is a rainbow mean coloring c of G with $\text{rm}(c) = 4k$ by Observation 1. Define the edge coloring c such that

$$\begin{aligned} \{c(uu_i) : 1 \leq i \leq 2k\} &= [k] \cup [3k+1, 4k-1] \\ c(uv) &= k \\ \{c(vv_i) : 1 \leq i \leq 2k\} &= ([k+1, 3k] \cup \{4k\}) - \{2k-1, 2k+1\}. \end{aligned}$$

Then the chromatic means of the vertices of G are given by

$$\text{cm}(u_i) = c(uu_i) \text{ and } \text{cm}(v_i) = c(vv_i) \text{ for } 1 \leq i \leq 2k,$$
$$\text{cm}(u) = 2k-1 \text{ and } \text{cm}(v) = 2k+1.$$

Thus, c is a rainbow mean coloring of G with $\text{rm}(c) = 4k$.

Case 2. *$a \geq 3$ is odd*. Then $a = 2k+1$ for some positive integer k. Since the order of G is $4k+2$ and every vertex of G is odd, it follows by Corollary 2 that $\text{rm}(G) \geq 4k+3$. Thus, it remains to show that there is a rainbow mean coloring c of G with $\text{rm}(c) = 4k+3$. An edge coloring c is defined as follows:

$$c(u_i u) = 2i-1 \text{ for } 1 \leq i \leq k \text{ and } c(u_i u) = 2i+1 \text{ for } k+1 \leq i \leq 2k$$
$$c(v_i v) = 2i \text{ for } 1 \leq i \leq k \text{ and } c(v_i v) = 2i+2 \text{ for } k+1 \leq i \leq 2k-1,$$
$$c(uv) = 2k+1 \text{ and } c(v_{2k} v) = 4k+3.$$

Then the chromatic means of the vertices of G are given by

$$\text{cm}(u_i) = c(u_i u) \text{ for } 1 \le i \le 2k \text{ and } \text{cm}(v_i) = c(v_i v) \text{ for } 1 \le i \le k,$$
$$\text{cm}(u) = 2k + 1 \text{ and } \text{cm}(v) = 2k + 2.$$

Thus, c is a rainbow mean coloring of G with $\text{rm}(c) = 4k + 3$. □

If $a, b \ge 3$ are odd and $a \equiv b \pmod 4$, it then follows by Corollary 2 that $\text{rm}(S_{a,b}) \ge a + b + 1$. In fact, $\text{rm}(S_{a,b}) = a + b + 1$ as we show next.

Theorem 9 *If a and b are odd integers with $a, b \ge 3$ and $a \equiv b \pmod 4$, then*

$$\text{rm}(S_{a,b}) = a + b + 1.$$

Proof By Theorem 8, we may assume that $a < b$. Since a and b are odd integers and $a \equiv b \pmod 4$, it follows that either a and b are both congruent to 1 modulo 4 or a and b are both congruent to 3 modulo 4. In each case, $a + b \equiv 2 \pmod 4$ and every vertex of G is odd. Hence, $\text{rm}(G) \ge a + b + 1$ by Corollary 2. Thus, it remains to show that there is a rainbow mean coloring c of G with $\text{rm}(c) = a + b + 1$. We consider these two cases.

Case 1. $a \equiv 1 \pmod 4$ *and* $b \equiv 1 \pmod 4$. Then $a = 4j + 1$ and $b = 4k + 1$ for some integers j, k with $1 \le j < k$. Let u and v be the central vertices of $G = S_{4j+1,4j+1}$ where u is adjacent to the $a - 1 = 4j$ end-vertices $u_1, u_2, \ldots u_{4j}$ and v is adjacent to the $b - 1 = 4k$ end-vertices $v_1, v_2, \ldots, v_{4k}$. Define the edge coloring c by

$$\{c(uu_i) : 1 \le i \le 4j\} = [4j + 1] - \{2j + 1\},$$
$$c(uv) = 2j + 1$$
$$\{c(vv_i) : 1 \le i \le 4k\} = [4j + 2, 4j + 4k + 3] - \{2k + 2j + 2, 2k + 4j + 2\}.$$

Then the chromatic means of the vertices of G are given by

$$\text{cm}(u_i) = c(uu_i) \text{ for } 1 \le i \le 4j,$$
$$\text{cm}(u) = 2j + 1, \text{cm}(v) = 2k + 4j + 2.$$
$$\text{cm}(v_i) = c(vv_i) \text{ for } 1 \le i \le 4k.$$

Thus, c is a rainbow mean coloring of G with $\text{rm}(c) = 4j + 4k + 3$.

Case 2. $a \equiv 3 \pmod 4$ *and* $b \equiv 3 \pmod 4$. Then $a = 4j + 3$ and $b = 4k + 3$ for some integers j, k with $0 \le j < k$. Let u and v be the central vertices of $G = S_{4j+3,4j+3}$ where u is adjacent to the $a - 1 = 4j + 2$ end-vertices $u_1, u_2, \ldots u_{4j+2}$ and v is adjacent to the $b - 1 = 4k + 2$ end-vertices $v_1, v_2, \ldots, v_{4k+2}$. Define the edge coloring c by

$$\{c(uu_i) : 1 \le i \le 4j + 2\} = [4j + 3] - \{2j + 2\},$$
$$c(uv) = 2j + 2$$
$$\{c(vv_i) : 1 \le i \le 4k + 2\} =$$
$$[4j + 4, 4j + 4k + 7] - \{2k + 2j + 4, 2k + 4j + 5\}.$$

Then the chromatic means of the vertices of G are given by

$$\text{cm}(u_i) = c(uu_i) \text{ for } 1 \le i \le 4j+2,$$
$$\text{cm}(u) = 2j+2, \text{cm}(v) = 2k+4j+5.$$
$$\text{cm}(v_i) = c(vv_i) \text{ for } 1 \le i \le 4k+2.$$

Thus, c is a rainbow mean coloring of G with $\text{rm}(c) = 4j + 4k + 7$. □

We now turn our attention to the double stars $S_{a,b}$ where $2 \le a < b$ and at least one of a and b is even.

Theorem 10 *If a and b are integers with $2 \le a < b$ such that ab is even, then*

$$\text{rm}(S_{a,b}) = a + b.$$

Proof Let $G = S_{a,b}$ where $2 \le a < b$ and ab is even. By Observation 1, it suffices to show that there is a rainbow mean coloring c of G with $\text{rm}(c) = a + b$. We consider three cases, according to the parities of a and b.

Case 1. *a and b are both even.* Then $a = 2j$ and $b = 2k$ where j and k are integers and $1 \le j < k$. Let u and v be the central vertices of $G = S_{2j,2k}$ where u is adjacent to the $a - 1 = 2j - 1$ end-vertices $u_1, u_2, \ldots u_{2j-1}$ and v is adjacent to the $b - 1 = 2k - 1$ end-vertices $v_1, v_2, \ldots, v_{2k-1}$. It suffices to show that there exists a rainbow mean coloring c with $\text{rm}(c) = a + b$. Define the edge coloring c by

$$\{c(uu_i) : 1 \le i \le 2j-1\} = [j+1, 3j-1],$$
$$c(uv) = 2j(j+1)$$
$$\{c(vv_i) : 1 \le i \le 2k-1\} = [j] \cup [3j+1, 2j+2k] - \{2j+k\}.$$

Then the chromatic means of the vertices of G are given by

$$\text{cm}(u_i) = c(uu_i) \text{ for } 1 \le i \le 2j-1,$$
$$\text{cm}(u) = 3j, \text{cm}(v) = 2j+k.$$
$$\text{cm}(v_i) = c(vv_i) \text{ for } 1 \le i \le 2k-1.$$

Since $j + 1 \le k$, it follows that $\text{cm}(u) \ne \text{cm}(v)$. Thus, c is a rainbow mean coloring of G with $\text{rm}(c) = 2j + 2k$.

Case 2. *$a \ge 3$ is odd and $b \ge 4$ is even.* Then $a = 2j + 1$ and $b = 2k$ for some integers j, k with $1 \le j < k$. Let u and v be the central vertices of G where u is adjacent to the $a - 1 = 2j$ end-vertices $u_1, u_2, \ldots, u_{2j}$ and v is adjacent to the $b - 1 = 2k - 1$ end-vertices $v_1, v_2, \ldots, v_{2k-1}$. Define the edge coloring c by

$$c(uu_i) = i \text{ for } 1 \le i \le 2j, c(uv) = 2jk + 2j + k + 1$$
$$\{c(vv_i) : 1 \le i \le 2k-1\} = [2j+1, 2k+2j+1] - \{k+j+1, k+3j+1\}.$$

Then the chromatic means of the vertices of G are given by

$$\text{cm}(u_i) = c(uu_i) \text{ for } 1 \le i \le 2j, \text{cm}(u) = k+j+1, \text{cm}(v) = k+3j+1.$$
$$\text{cm}(v_i) = c(vv_i) \text{ for } 1 \le i \le 2k-1.$$

Thus, c is a rainbow mean coloring of G with $\text{rm}(c) = 2j + 2k + 1$.

Case 3. *$a \ge 2$ is even, and $b \ge 3$ is odd.* Then $a = 2j$ and $b = 2k + 1$ where $1 \le j \le k$. Let u and v be the central vertices of G where u is adjacent to the $a - 1 = 2j - 1$ end-vertices $u_1, u_2, \ldots u_{2j-1}$ and v is adjacent to the $b - 1 = 2k$ end-vertices $v_1, v_2, \ldots, v_{2k}$. Define the edge coloring c by

$$\{c(uu_i) : 1 \le i \le 2j-1\} = [j+k+2, 3j+k],$$
$$c(uv) = 2j(j+1)+k+1$$
$$\{c(vv_i) : 1 \le i \le 2k\} = [j+k] \cup [3j+k+2, 2j+2k+1].$$

Then the chromatic means of the vertices of G are given by

$$\text{cm}(u_i) = c(uu_i) \text{ for } 1 \le i \le 2j-1,$$
$$\text{cm}(u) = 3j+k+1, \text{cm}(v) = j+k+1.$$
$$\text{cm}(v_i) = c(vv_i) \text{ for } 1 \le i \le 2k.$$

Thus, c is a rainbow mean coloring of G with $\text{rm}(c) = 2j+2k+1$. □

The one remaining class of double stars $S_{a,b}$ for which the rainbow mean index has not yet been determined is that where a and b are both odd and $a \not\equiv b \pmod 4$. In order to present a result dealing with this class, it is convenient to establish the following two lemmas.

Lemma 1 *For positive integers a and b with $a \le b$ and the set*

$$X = [4a+4b+4] - \{2a+2b+1, 2a+2b+3\},$$

let $s_1 = \sum_{i=1}^{4a} i$ and $s_2 = \sum_{i=1}^{4a}(4b+4+i)$. For every integer s with $s_1 \le s \le s_2$, there exists a $(4a)$-element subset S of X such that $\sum_{x \in S} x = s$.

Proof First, we show that there exists a $(4a)$-element subset $S \subseteq [4a+4b+4]$ such that $\sum_{x\in S} x = s$. If $s = s_1$ or $s = s_2$, then the result holds. Thus, we may assume that $s_1 < s < s_2$. Let m be the minimum integer in $[4b+4]$ such that

$$[m+(m+1)+\cdots+(m+4a-1)] < s <$$
$$[(m+1)+(m+2)+\cdots+(m+4a)].$$

Let $t = (m+1)+(m+2)+\cdots+(m+4a-1)$. Therefore, $m+t < s < t+(m+4a)$. Thus, $s = m+t+r$ for some integer r with $1 \le r \le 4a-1$. Consequently, by adding 1 to the last r terms in the sum $m+(m+1)+\cdots+(m+4a-1)$, we obtain the $(4a)$-element set

$$T = \{m, m+1, \ldots, m+4a-r-1\} \cup \{m+4a-r+1, m+4a-r+2, \ldots, m+4a\}$$

such that $\sum_{x\in T} x = s$.

It remains to show that there are $4a$ distinct integers in X whose sum is s. Of course, if neither $2a+2b+1$ nor $2a+2b+3$ belongs to T, then T has the desired property. Thus, we may assume that at least one of $2a+2b+1$ and $2a+2b+3$ belongs to T, say $2a+2b+1 \in T$.

- ⋆ If $2a+2b+3 \in T$ as well, then we remove $2a+2b+1$ and $2a+2b+3$ from T and replace them by 1 and $4a+4b+3$, obtaining the set $T_1 \subseteq X$ such that the sum of elements in T_1 is s.

- ⋆ If $2a + 2b + 3 \notin T$, then either $2a + 2b \in T$ or $2a + 2b + 2 \in T$, say the former. Hence, we remove $2a + 2b$ and $2a + 2b + 1$ from T and replace them by 1 and $4a + 4b$, obtaining the set $T_2 \subseteq X$ such that the sum of elements in T_2 is s. □

Lemma 2 *For positive integers a and b with $a \le b$ and the set*

$$X = [4a + 4b + 4] - \{2a + 2b + 1, 2a + 2b + 3\},$$

let $s_1 = \sum_{i=1}^{4a+2} i$ *and* $s_2 = \sum_{i=1}^{4a+2} (4b + 2 + i)$. *For every integer s with $s_1 \le s \le s_2$, there exists a $(4a + 2)$-element subset S of X such that* $\sum_{x \in S} x = s$.

Proof First, we show that there exists a $(4a + 2)$-element subset $S \subseteq [4a + 4b + 4]$ such that $\sum_{x \in S} x = s$. If $s = s_1$ or $s = s_2$, then the result holds. Thus, we may assume that $s_1 < s < s_2$. Let m be the minimum integer in $[4b + 2]$ such that

$$[m + (m + 1) + \cdots + (m + 4a + 1)] < s <$$
$$[(m + 1) + (m + 2) + \cdots + (m + 4a + 2)].$$

Let $t = (m + 1) + (m + 2) + \cdots + (m + 4a + 1)$. Therefore, $m + t < s < t + (m + 4a + 2)$. Thus, $s = m + t + r$ for some integer r with $1 \le r \le 4a + 1$. Consequently, by adding 1 to the last r terms in the sum $m + (m + 1) + \cdots + (m + 4a + 1)$, we obtain the $(4a + 2)$-element set

$$T = \{m, m + 1, \ldots, m + 4a - r + 1\} \cup \{m + 4a - r + 3, m + 4a - r + 4, \ldots, m + 4a + 2\}$$

such that $\sum_{x \in T} x = s$.

It remains to show that there are $4a + 2$ distinct integers in X whose sum is s. Of course, if neither $2a + 2b + 1$ nor $2a + 2b + 3$ belongs to T, then T has the desired property. Thus, we may assume that at least one of $2a + 2b + 1$ and $2a + 2b + 3$ belongs to T, say $2a + 2b + 1 \in T$.

- ⋆ If $2a + 2b + 3 \in T$ as well, then we remove $2a + 2b + 1$ and $2a + 2b + 3$ from T and replace them by 1 and $4a + 4b + 3$, obtaining the set $T_1 \subseteq X$ such that the sum of elements in T_1 is s.
- ⋆ If $2a + 2b + 3 \notin T$, then either $2a + 2b \in T$ or $2a + 2b + 2 \in T$, say the former. Hence, we remove $2a + 2b$ and $2a + 2b + 1$ from T and replace them by 1 and $4a + 4b$, obtaining the set $T_2 \subseteq X$ such that the sum of elements in T_2 is s. □

We are now prepared to present the following result.

Theorem 11 *If a and b are odd integers with $3 \le a < b$ such that $a \not\equiv b \pmod 4$, then $\mathrm{rm}(S_{a,b}) = a + b$.*

Proof Let $G = S_{a,b}$. We show that there is a rainbow mean coloring $c : E(G) \to [a + b]$ of G with $\mathrm{rm}(c) = a + b$ such that $\mathrm{cm}(u)$ and $\mathrm{cm}(v)$ have certain prescribed values. We consider two cases. In each case, we let

$$A = \sum_{i=1}^{a-1} c(uu_i) = \sum_{i=1}^{a-1} \text{cm}(u_i)$$
$$B = \sum_{i=1}^{b-1} c(vv_i) = \sum_{i=1}^{b-1} \text{cm}(v_i)$$
$$x = c(uv).$$

Observe that $A + x = \text{cm}(u) \cdot a$ and $B + x = \text{cm}(v) \cdot b$. Furthermore,

$$A + B + \text{cm}(u) + \text{cm}(v) = 1 + 2 + \cdots + (a + b) = \binom{a+b+1}{2}.$$

Case 1. $a \equiv 3 \pmod 4$ *and* $b \equiv 1 \pmod 4$. Then $a = 4j + 3$ and $b = 4k + 1$ where $0 \le j < k$. We show that there is a rainbow mean coloring $c : E(G) \to [4k + 4j + 4]$ of G with $\text{rm}(c) = 4j + 4k + 4$ such that $\text{cm}(u) = 2k + 2j + 1$ and $\text{cm}(v) = 2k + 2j + 3$. For such an edge coloring c of G, we have

$$\begin{aligned} A + x &= (2k + 2j + 1)(4j + 3) = 8kj + 8j^2 + 6k + 10j + 3 \\ B + x &= (2k + 2j + 3)(4k + 1) = 8kj + 8j^2 + 14k + 2j + 3 \\ A + B &= 1 + 2 + \cdots + (4k + 4j + 5) - (\text{cm}(u) + \text{cm}(v)) \\ &= (16kj + 8k^2 + 8j^2 + 18k + 18j + 10) - (4k + 4j + 4) \\ &= 16kj + 8k^2 + 8j^2 + 14k + 14j + 6. \end{aligned}$$

Hence,

$$\begin{aligned} A &= 8kj + 8j^2 + 3k + 9j + 3 \\ B &= 8kj + 8k^2 + 11k + 3j + 3 \\ x &= 3k - j. \end{aligned}$$

Therefore, such an edge coloring c of G exists if there are $4a + 2$ distinct elements in the set $X = [4k + 4j + 4] - \{2k + 2j + 1, 2k + 2j + 3\}$ whose sum is $A = 8kj + 8j^2 + 3k + 9j + 3$. The sum of the $4j + 2$ smallest integers in the set $[4k + 4j + 4]$ is

$$\binom{4j+3}{2} = (2j + 1)(4j + 3) = 8j^2 + 10j + 3;$$

while the sum of the $4j + 2$ largest integers in the set $[4k + 4j + 4]$ is

$$(2j + 1)(8k + 4j + 7) = 16kj + 8j^2 + 8k + 18j + 7.$$

Since

$$8j^2 + 10j + 3 \le A \le 16kj + 8j^2 + 8k + 18j + 7,$$

it follows by Lemma 2 that there is a $(4a + 2)$-element subset S of X such that $\sum_{x \in S} x = S$. Observe that the sum of integers in $X - S$ is therefore B.

Case 2. $a \equiv 1 \pmod 4$ *and* $b \equiv 3 \pmod 4$. Then $a = 4j + 1$ and $b = 4k + 3$ where $1 \le j \le k$. We show that there is a rainbow mean coloring $c : E(G) \to$

$[4k+4j+4]$ of G with $\mathrm{rm}(c) = 4j+4k+4$ such that $\mathrm{cm}(u) = 2k+2j+1$ and $\mathrm{cm}(v) = 2k+2j+3$. For such an edge coloring c of G, we have

$$\begin{aligned} A+x &= (2k+2j+1)(4j+1) = 8kj+8j^2+2k+6j+1 \\ B+x &= (2k+2j+3)(4k+3) = 8kj+8j^2+18k+6j+9 \\ A+B &= 16kj+8k^2+8j^2+14k+14j+6. \end{aligned}$$

Hence,

$$\begin{aligned} A &= 8kj+8j^2-k+7j-1 \\ B &= 8kj+8k^2+15k+7j+7 \\ x &= 3k-j+2. \end{aligned}$$

Therefore, such an edge coloring c of G exists if there are $4a$ distinct elements in the set $X = [4k+4j+4] - \{2k+2j+1, 2k+2j+3\}$ whose sum is $A = 8kj + 8j^2 - k + 7j - 1$. The sum of the $4j$ smallest integers in the set $[4k+4j+4]$ is

$$\binom{4j+1}{2} = 2j(4j+1) = 8j^2+2j;$$

while the sum of the $4j$ largest integers in the set $[4k+4j+4]$ is

$$2j(8k+4j+9) = 16kj+8j^2+18j.$$

Since $8j^2+2j \le A \le 16kj+8j^2+18j$, it follows by Lemma 1 that there is a $4a$-element subset S of X such that $\sum_{x\in S} x = A$. Again, the sum of integers in $X - S$ is therefore B. □

In summary, we have the following result.

Theorem 12 *For integers a and b where $a, b \ge 2$,*

$$\mathrm{rm}(S_{a,b}) = \begin{cases} a+b & \text{if } ab \text{ is even or } ab \text{ is odd and } a+b \not\equiv 2 \pmod 4 \\ a+b+1 & \text{if } ab \text{ is odd and } a+b \equiv 2 \pmod 4. \end{cases}$$

References

1. G. Chartrand, C. Egan, and P. Zhang, *How to Label a Graph.* Springer, New York (2019).
2. G. Chartrand, J. Hallas, E. Salehi, and P. Zhang, Rainbow mean colorings of graphs. *Discrete Math. Lett.* **3** (2020), 1–8.
3. G. Chartrand and P. Zhang, *Chromatic Graph Theory.* Second Edition. Chapman & Hall/CRC Press, Boca Raton (2020).
4. J. Hallas, *Extremal Theory on Induced Graph Colorings*. Doctoral Dissertation, Western Michigan University (2020).
5. J. Hallas, E. Salehi, and P. Zhang, Rainbow mean colorings of bipartite graphs. *Bull. Insti. Combin. Appl.*. **88** (2020), 78–97.
6. P. Zhang, *Color-Induced Graph Colorings*. Springer, New York (2015).
7. P. Zhang, *A Kaleidoscopic View of Graph Colorings*. Springer, New York (2016).

Examples of Edge Critical Graphs in Peg Solitaire

Robert A. Beeler and Aaron D. Gray

Abstract Peg solitaire is a game in which pegs are placed in every hole but one and the player jumps over pegs along rows or columns to remove them. Usually, the goal is to remove all but one peg. In a 2011 paper, this game is generalized to graphs. In this paper, we examine graphs in which any single edge addition changes solvability. In order to do this, we introduce a family of graphs and provide necessary and sufficient conditions for the solvability for this family. We show that infinite subsets of this family are edge critical. We also determine the maximum number of pegs that can be left on this family with the condition that a jump is made whenever possible. Finally, we give a list of graphs on eight vertices that are edge critical.

Keywords Games on graphs · Peg solitaire · Critical graphs

AMS Subject Classification 05C57 (91A43, 05C35)

1 Introduction

Peg solitaire is a table game which traditionally begins with "pegs" in every space except for one which is left empty (i.e., a "hole"). If in some row or column two adjacent pegs are next to a hole (as in Fig. 1), then the peg in x can jump over the peg in y into the hole in z. The peg in y is then removed. The goal is to remove every peg but one. If this is achieved, then the board is considered solved [1, 12].

R. A. Beeler
Department of Mathematics and Statistics, East Tennessee State University, Johnson City, TN 37614-1700, USA
e-mail: beelerr@etsu.edu

A. D. Gray (✉)
Mathematics Division, Northeast State Community College, Blountville, TN 37617-0246, USA
e-mail: adgray@northeaststate.edu

F. Hoffman (ed.), *Combinatorics, Graph Theory and Computing*,
Springer Proceedings in Mathematics & Statistics 388,
https://doi.org/10.1007/978-3-031-05375-7_12

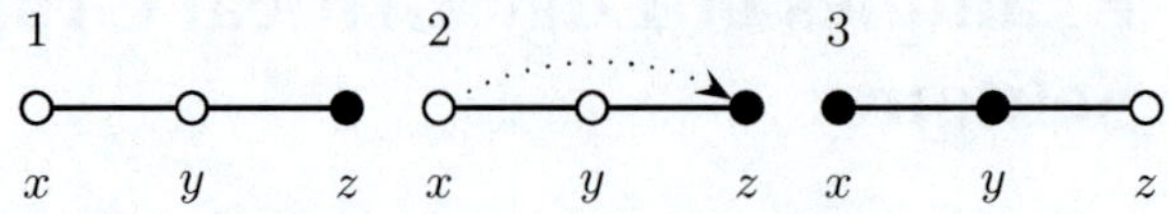

Fig. 1 A Typical Jump in Peg Solitaire, $x \cdot \overrightarrow{y} \cdot z$

In [6], this notion is generalized to graphs. A graph, $G = (V, E)$, is a set of vertices, V, and a set of edges, E. Because of the restrictions of peg solitaire, we will assume that all graphs are finite, undirected, connected graphs with no loops or multiple edges. For all undefined graph theory terminology, refer to West [18]. If there are pegs in vertices x and y and a hole in z, then we allow the peg in x to jump over the peg in y into the hole in z provided that $xy, yz \in E$. The peg in y is then removed. This jump is denoted $x \cdot \overrightarrow{y} \cdot z$.

In general, the game begins with a *starting state* $S \subset V$ which is a set of vertices with holes. A *terminal state* $T \subset V$ is a set of vertices that have pegs at the end of the game. A terminal state T is *associated* with starting state S if T can be obtained from S by a series of jumps. Unless otherwise noted, we will assume that S consists of a single vertex. A graph G is *solvable* if there exists some vertex s so that, starting with $S = \{s\}$, there exists an associated terminal state consisting of a single peg. A graph G is *freely solvable* if for all vertices s so that, starting with $S = \{s\}$, there exists an associated terminal state consisting of a single peg. A graph G is *k-solvable* if there exists some vertex s so that, starting with $S = \{s\}$, there exists an associated terminal state consisting of k nonadjacent pegs. In particular, a graph is *distance 2-solvable* if there exists some vertex s so that, starting with $S = \{s\}$, there exists an associated terminal state consisting of two pegs that are distance 2 apart.

We now include several results from previous studies of peg solitaire on graphs that will aid us in our results.

Theorem 1 ([6, 7])

(i) *The cycle C_n is freely solvable if and only if n is even or $n = 3$; C_n is distance 2-solvable in all other cases.*
(ii) *For $n \geq 2$, the complete graph K_n is freely solvable.*
(iii) *For $n \geq 2$ and $m \geq 2$, the complete bipartite graph $K_{n,m}$ is freely solvable.*
(iv) *The double star $K_{1,1}(a_1; b_1)$ is freely solvable if and only if $a_1 = b_1$ and $b_1 \neq 1$. It is solvable if and only if $a_1 \leq b_1 + 1$. It is distance 2-solvable if and only if $a_1 = b_1 + 2$. It is $(a_1 - b_1)$-solvable if $a_1 \geq b_1 + 3$.*

The following proposition from [6] is also useful.

Proposition 1 ([6])

(i) *If a graph G is k-solvable with the initial hole in s and a jump is possible, then there is a first jump; say $s'' \cdot \overrightarrow{s'} \cdot s$. Hence, if there are holes in s'' and s' and pegs elsewhere, then G can be k-solved from this configuration.*
(ii) (Inheritance Principle) *Suppose that H is a k-solvable graph and G is a spanning subgraph of H, then G is (at best) k-solvable.*

The following theorem allows the completion of the game in reverse by exchanging the roles of pegs and holes. Beeler and Rodriguez [9] define the *dual* of a configuration of pegs on a graph as the arrangement of pegs obtained by reversing the roles of pegs and holes. The dual of a configuration is particularly useful in determining which initial holes can be used to solve the graph.

Theorem 2 (Duality Principle [6, 9]) *Suppose that S is a starting state of G with associated terminal state T. Let S' and T' be the duals of S and T, respectively. It follows that T' is a starting state of G with associated terminal state S'.*

2 The Hairy Complete Bipartite Graph

In this section, we consider a family of graphs that generalize both the complete bipartite graph and the double star. The *hairy complete bipartite graph* is the graph on $n + m + a_1 + \cdots + a_n + b_1 + \cdots + b_m$ vertices obtained from the complete bipartite graph $K_{n,m}$ by appending a_i pendant vertices to x_i for $i = 1, \ldots, n$ and appending b_j pendant vertices to y_j for $j = 1, \ldots, m$. We denote this graph $K_{n,m}(a_1, \ldots, a_n; b_1, \ldots, b_m)$. Note that if $n = m = 1$, then the graph is the double star. If $n = 1$ and $m \geq 2$, then the graph is a tree of diameter four. As the solvability of double stars and trees of diameter four has already been determined [7, 11], we assume that $n \geq 2$ and $m \geq 2$. Further, we will assume that $a_1 \geq \cdots \geq a_n$, $b_1 \geq \cdots \geq b_m$, $a_1 \geq 1$, and $\sum_{i=1}^{n} a_i \geq \sum_{j=1}^{m} b_j$. We denote the a_i pendants adjacent to x_i by $x_{i,1}, \ldots, x_{i,a_i}$. We denote the b_j pendants adjacent to y_j by $y_{j,1}, \ldots, y_{j,b_j}$. Let $X_i = \{x_{i,1}, \ldots, x_{i,a_i}\}$, let $Y_j = \{y_{j,1}, \ldots, y_{j,b_j}\}$, let $X = \{x_1, \ldots, x_n\}$, and let $Y = \{y_1, \ldots, y_m\}$. For $S \subset V(G)$, the function $\rho(S)$ gives the current number of pegs in the set S. Figure 2 shows an example of the hairy complete bipartite graph.

Berlekamp, Conway, and Guy [12] explore a helpful device for the elimination of pegs. They define a *package* as a known configuration of pegs that may be eliminated with a predetermined series of jumps. The elimination of these pegs is called a *purge*. A purge acts as a type of "shortcut" that can be used to efficiently solve the game. While not explicitly stated in [7], a proof in that paper extends the notion of a purge to peg solitaire on graphs. In addition, [10] also discusses packages and purges in peg solitaire on graphs.

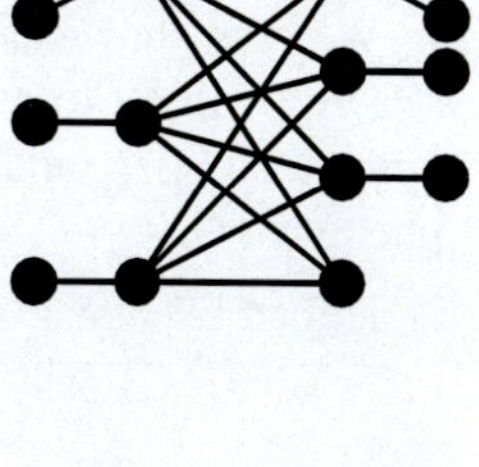

Fig. 2 The hairy complete bipartite graph $K_{3,4}(3, 1, 1; 2, 1, 1, 0)$

We will use packages and purges to aid in our results. Suppose that the graph G has a double star subgraph, $K_{1,1}(a_1; b_1)$, with a peg in x_1 and a hole in y_1. Assume that $\rho(X_1) \geq d$ and $\rho(Y_1) \geq d$. We can remove d pegs from both X_1 and Y_1 by performing the jumps $x_{1,i} \cdot \overrightarrow{x_1} \cdot y_1$ and $y_{1,i} \cdot \overrightarrow{y_1} \cdot x_1$ for $i = 1, \ldots, d$. Note that both before and after this sequence there is a peg in x_1 and a hole in y_1. This sequence is called a *double star purge* and is denoted $\mathcal{DS}(X_1, Y_1, d)$.

With this purge in mind, we now give necessary and sufficient conditions for the solvability of the hairy complete bipartite graph. These conditions will be dependent on a property $\mathcal{P}$ defined as: (i) $n = 2$, m is even, $a_1 \geq 2$, and $a_2 \leq \sum_{j=1}^{m} b_j$ *or* (ii) $n = 2$, m is odd, $a_1 = a_2 = 1$, and $\sum_{j=1}^{m} b_j = 0$. We define $(\sim\mathcal{P})$ as the negation of property $\mathcal{P}$.

Theorem 3 *For the hairy complete bipartite graph $G = K_{n,m}(a_1, \ldots, a_n;\ b_1, \ldots, b_m)$:*

(i) If $\mathcal{P}$, then the graph G is solvable if and only if $\sum_{i=1}^{n} a_i \leq \sum_{j=1}^{m} b_j + n - 1$. If $(\sim\mathcal{P})$, then the graph G is solvable if and only if $\sum_{i=1}^{n} a_i \leq \sum_{j=1}^{m} b_j + n$.
(ii) If $\mathcal{P}$, then the graph G is freely solvable if and only if $\sum_{i=1}^{n} a_i \leq \sum_{j=1}^{m} b_j + n - 2$. If $(\sim\mathcal{P})$, then the graph G is freely solvable if and only if $\sum_{i=1}^{n} a_i \leq \sum_{j=1}^{m} b_j + n - 1$.
(iii) If $\mathcal{P}$, then the graph G is distance 2-solvable if and only if $\sum_{i=1}^{n} a_i = \sum_{j=1}^{m} b_j + n$. If $(\sim\mathcal{P})$, then the graph G is distance 2-solvable if and only if $\sum_{i=1}^{n} a_i = \sum_{j=1}^{m} b_j + n + 1$.
(iv) If $\mathcal{P}$, then graph G is $(\sum_{i=1}^{n} a_i - \sum_{j=1}^{m} b_j - n + 2)$-solvable if $\sum_{i=1}^{n} a_i \geq \sum_{j=1}^{m} b_j + n$. If $(\sim\mathcal{P})$, then the graph G is $(\sum_{i=1}^{n} a_i - \sum_{j=1}^{m} b_j - n + 1)$-solvable if $\sum_{i=1}^{n} a_i \geq \sum_{j=1}^{m} b_j + n + 1$.

Proof We begin by establishing the necessary conditions for (i), (iii), and (iv). We first examine the optimal method for solving the graph. The pegs in each cluster must be eliminated. Hence all pegs in each X_i must be removed. To do so, a peg must first be in x_i. For this to occur, one of two jumps must be made, namely, $y_{j,k} \cdot \overrightarrow{y_j} \cdot x_i$ or $x_\ell \cdot \overrightarrow{y_j} \cdot x_i$, where $\ell \neq i$. Therefore, one of two double star purges is necessary, namely $\mathcal{DS}(X_i, Y_j, d)$or $\mathcal{DS}(X_i, X - \{x_i\}, d)$. In the first, each Y_j can "exchange" b_j pegs with X_i. In the second, each x_ℓ, where $\ell \in \{1, \ldots, i-1, i+1, \ldots, n\}$, can "exchange" one peg with X_i. Hence $\sum_{i=1}^{n} a_i \leq \sum_{j=1}^{m} b_j + n$ is necessary for the graph to be solvable (and also freely solvable). Moreover, if $\sum_{i=1}^{n} a_i \geq \sum_{j=1}^{m} b_j + n + 1$, then, at best, $\sum_{i=1}^{n} a_i - \sum_{j=1}^{m} b_j - n$ pegs remain in the graph. Note that this bound is only achievable in the $(\sim \mathcal{P})$ case, as we will show in the coming paragraphs. Adjustments for the case of $\mathcal{P}$ will be discussed at the end of the proof.

We now show that the conditions given in (i), (iii), and (iv) are sufficient. Our strategy will be to reduce the number of pegs in $X_1, \ldots, X_n$ by exchanging pegs with $Y_1, \ldots, Y_m$ and, if necessary, X. To this end, we define a graph homomorphism $\phi : G \to G'$, where $G' = K_{n,1}(a_1, \ldots, a_n; \sum_{j=1}^{m} b_j)$. The homomorphism ϕ is defined by $\phi(y_j) = y'$, $\phi(y_{j,\ell}) = y'_{s_j+\ell}$, with $s_j = \sum_{k=1}^{j-1} b_k$, and $\phi(v) = v$ for all other vertices. Let Y' denote the set of all $y'_{s_j+\ell}$.

This homomorphism has the effect of collapsing the support vertices of $Y_1,\ldots,$ Y_m. In addition, it allows the movement of a hole along each of the y_j. This occurs because as each Y_j empties, the jumps $x_{i,1}\cdot\overrightarrow{x_i}\cdot y_j$ and $y_{j-1,1}\cdot\overrightarrow{y_j}\cdot x_i$, for $k \neq j$, results in the hole in y_j being moved to y_{j-1}.

Begin with the initial hole in y'. This corresponds to beginning with the initial hole in y_j for some j. Perform the double star purge $\mathcal{DS}(X_{n-i+1}, Y', \min\{\rho(Y'), a_{n-i+1}\})$, for $i = 1, \ldots, n$. Now $\rho(Y_j) = 0$ for $j = 1, \ldots, m$, $\rho(X) = n$, $\rho(Y) = m - 1$, and $\rho(X_1 \cup \cdots \cup X_n) = \sum_{i=1}^{n} a_i - \sum_{j=1}^{m} b_j$. Without loss of generality, assume that the hole in Y is in y_m.

If $\sum_{i=1}^{n} a_i = \sum_{j=1}^{m} b_j$, then this reduces G to the complete bipartite graph with a hole in a single vertex. This is solvable with the final two pegs in x_i and y_j, for any i and j. Thus, the graph may be solved with the final peg in x_i, y_j, $x_{i,1}$, or $y_{j,1}$ for any i and j. Hence, G is freely solvable by the Duality Principle. This provides part of the sufficient conditions in (ii).

If $\sum_{i=1}^{n} a_i \geq \sum_{j=1}^{m} b_j + 1$, then let ℓ be the greatest integer such that $\rho(X_\ell) \geq 1$. If $\ell = 1$, then perform the double star purge $\mathcal{DS}(X_1, X - \{x_1, x_2\}, \min\{\rho(X_1), \rho(X - \{x_1, x_2\})\})$. If $\ell \geq 2$, then for $i = 1, \ldots, \ell$, perform the double star purge $\mathcal{DS}(X_{\ell-i+1}, X - \{x_{\ell-i+1}, x_1\}, \min\{\rho(X_{\ell-i+1}), \rho(X - \{x_{\ell-i+1,x_1}\})\})$ until two pegs remain in X. We note that if $n = 2$, then we omit these purges. In any case, we then jump $x_{1,1}\cdot\overrightarrow{x_1}\cdot y_m$. If $\sum_{i=1}^{n} a_i \geq \sum_{j=1}^{m} b_j + n$, then $\rho(X_1 \cup \cdots \cup X_n) = \sum_{i=1}^{n} a_i - \sum_{j=1}^{m} b_j - n + 1$, $\rho(Y_j) = 0$ for $j = 1, \ldots, m$, $\rho(X) = 1$, and $\rho(Y) = m$.

If $\sum_{j=1}^{m} b_j + 1 \leq \sum_{i=1}^{n} a_i \leq \sum_{j=1}^{m} b_j + n - 1$, then this reduces the graph to the complete bipartite graph with a hole in x_1. This is solvable with the final two pegs in x_i and y_j, for any i and j. Thus, the graph may be solved with the final peg in x_i, y_j, $x_{i,1}$, or $y_{j,1}$ for any i and j. Hence, G is freely solvable by the Duality Principle. This provides part of the sufficient conditions in (ii).

If $\sum_{i=1}^{n} a_i \geq \sum_{j=1}^{m} b_j + n$, then let x_q be the vertex in X with a peg and let q' be an integer in $\{1, \ldots, n\}$ such that $q' \neq q$. For $k = 1, \ldots, \lfloor \frac{m-2}{2} \rfloor$, jump $x_q\cdot\overrightarrow{y_{2k-1}}\cdot x_{q'}$ and $x_{q'}\cdot\overrightarrow{y_{2k}}\cdot x_q$. We note that if $m = 2$ or $m = 3$, then we omit these jumps. Let ℓ be the greatest integer such that $\rho(X_\ell) \geq 1$

Assume that m is even. Then $\rho(X_1 \cup \cdots \cup X_n) = \sum_{i=1}^{n} a_i - \sum_{j=1}^{m} b_j - n + 1$, $\rho(Y_j) = 0$ for $j = 1, \ldots, m$, $\rho(X) = 1$, and $\rho(Y) = 2$. If $\ell \neq q$ and $n = 2$ (we note that this occurs if $a_1 \geq 2$ and $a_2 \leq \sum_{j=1}^{m} b_j$), then jump $x_q\cdot\overrightarrow{y_m}\cdot x_\ell$ and $y_{m-1}\cdot\overrightarrow{x_\ell}\cdot y_m$ to end the game with $\rho(X_1 \cup \cdots \cup X_n) = \sum_{i=1}^{n} a_i - \sum_{j=1}^{m} b_j - n + 1$, $\rho(X) = 0$, and $\rho(Y) = 1$. In particular if $\sum_{i=1}^{n} a_i = \sum_{j=1}^{m} b_j + n$, then the graph is distance 2-solvable since one peg remains in $X_1 \cup \cdots \cup X_n$ and one peg remains in Y. If $\ell \neq q$ and $n \geq 3$, then let q'' be an integer in $\{1, \ldots, n\}$ such that $q'' \neq q$ and $q'' \neq \ell$. Jump $x_q\cdot\overrightarrow{y_{m-1}}\cdot x_{q''}$, $x_{q''}\cdot\overrightarrow{y_m}\cdot x_\ell$, and $x_{\ell,1}\cdot\overrightarrow{x_\ell}\cdot y_m$ to end the game with $\rho(X_1 \cup \cdots \cup X_n) = \sum_{i=1}^{n} a_i - \sum_{j=1}^{m} b_j - n$, $\rho(X) = 0$, and $\rho(Y) = 1$. If $\ell = q$, then jump $x_\ell\cdot\overrightarrow{y_m}\cdot x_{q'}$, $x_{q'}\cdot\overrightarrow{y_{m-1}}\cdot x_\ell$, and $x_{\ell,1}\cdot\overrightarrow{x_\ell}\cdot y_m$ to end the game with $\rho(X_1 \cup \cdots \cup X_n) = \sum_{i=1}^{n} a_i - \sum_{j=1}^{m} b_j - n$, $\rho(X) = 0$, and $\rho(Y) = 1$. We note that in the two previous cases if $\sum_{i=1}^{n} a_i = \sum_{j=1}^{m} b_j + n + 1$, then the graph is distance 2-solvable since one peg remains in $X_1 \cup \cdots \cup X_n$ and one peg remains in Y.

Assume that m is odd. Then $\rho(X_1 \cup \cdots \cup X_n) = \sum_{i=1}^{n} a_i - \sum_{j=1}^{m} b_j - n + 1$, $\rho(Y_j) = 0$ for $j = 1, \ldots, m$, $\rho(X) = 1$, and $\rho(Y) = 3$. If $\ell = q, n = 2$, and $\rho(X_{q'}) = 0$ (we note that this occurs if $a_1 = a_2 = 1$, and $\sum_{j=1}^{m} b_j = 0$), then jump $x_\ell \cdot \overrightarrow{y_{m-2}} \cdot x_{q'}$, $x_{q'} \cdot \overrightarrow{y_m} \cdot x_\ell$, and $y_{m-1} \cdot \overrightarrow{x_\ell} \cdot y_m$ to end the game with $\rho(X_1 \cup \cdots \cup X_n) = \sum_{i=1}^{n} a_i - \sum_{j=1}^{m} b_j - n + 1$, $\rho(X) = 0$, and $\rho(Y) = 1$. In particular if $\sum_{i=1}^{n} a_i = \sum_{j=1}^{m} b_j + n$, then the graph is distance 2-solvable since one peg remains in $X_1 \cup \cdots \cup X_n$ and one peg remains in Y. If $\ell = q$, $n = 2$, and $\rho(X_{q'}) \geq 1$, then jump $x_\ell \cdot \overrightarrow{y_{m-2}} \cdot x_{q'}$, $x_{q'} \cdot \overrightarrow{y_{m-1}} \cdot x_\ell$, $x_\ell \cdot \overrightarrow{y_m} \cdot x_{q'}$, and $x_{q',1} \cdot \overrightarrow{x_{q'}} \cdot y_m$ to end the game with $\rho(X_1 \cup \cdots \cup X_n) = \sum_{i=1}^{n} a_i - \sum_{j=1}^{m} b_j - n$, $\rho(X) = 0$, and $\rho(Y) = 1$. If $\ell = q$ and $n \geq 3$, then let ℓ' and ℓ'' be integers in $\{1, \ldots, n\}$ such that ℓ, ℓ', and ℓ'' are distinct. Jump $x_\ell \cdot \overrightarrow{y_{m-2}} \cdot x_{\ell'}$, $x_{\ell'} \cdot \overrightarrow{y_{m-1}} \cdot x_{\ell''}$, $x_{\ell''} \cdot \overrightarrow{y_m} \cdot x_\ell$, and $x_{\ell,1} \cdot \overrightarrow{x_\ell} \cdot y_m$ to end the game with $\rho(X_1 \cup \cdots \cup X_n) = \sum_{i=1}^{n} a_i - \sum_{j=1}^{m} b_j - n$, $\rho(X) = 0$, and $\rho(Y) = 1$. If $\ell \neq q$, then jump $x_q \cdot \overrightarrow{y_{m-2}} \cdot x_\ell$, $x_\ell \cdot \overrightarrow{y_{m-1}} \cdot x_q$, $x_q \cdot \overrightarrow{y_m} \cdot x_\ell$, and $x_{\ell,1} \cdot \overrightarrow{x_\ell} \cdot y_m$ to end the game with $\rho(X_1 \cup \cdots \cup X_n) = \sum_{i=1}^{n} a_i - \sum_{j=1}^{m} b_j - n$, $\rho(X) = 0$, and $\rho(Y) = 1$. We note that in the three previous cases if $\sum_{i=1}^{n} a_i = \sum_{j=1}^{m} b_j + n + 1$, then the graph is distance 2-solvable since one peg remains in $X_1 \cup \cdots \cup X_n$ and one peg remains in Y.

In the above arguments we have established the necessary conditions for (i), (iii), and (iv), as well as the sufficient conditions for (i), (ii), (iii), and (iv). We now establish the necessary conditions for (ii). Note that any graph that is not solvable is also not freely solvable. For this reason, and because if $\mathcal{P}$ and $\sum_{i=1}^{n} a_i \leq \sum_{j=1}^{m} b_j + n - 2$, then $G = K_{n,m}(a_1, \ldots, a_n; b_1, \ldots, b_m)$ is solvable by (i), we need only show that if $\mathcal{P}$ and $\sum_{i=1}^{n} a_i \leq \sum_{j=1}^{m} b_j + n - 1$, then G is not freely solvable. To do so we need only show that such a graph cannot be solved for a particular choice of the initial hole.

Assume that the initial hole is in x_i for some $i \in \{1, \ldots, n\}$, and let $j \in \{1, \ldots, m\}$ and $k \in \{1, \ldots, i-1, i+1, \ldots, n\}$. Without loss of generality, one of two first jumps may occur. If we jump $y_{j,1} \cdot \overrightarrow{y_j} \cdot x_i$ (which is only possible if $\sum_{j=1}^{m} b_j \geq 1$), then $\rho(X_1, \ldots, X_n) = \sum_{i=1}^{n} a_i$ but $\rho(Y_1 \cup \cdots \cup Y_m) = \sum_{j=1}^{m} b_j - 1$. Since one fewer peg in $Y_1, \ldots, Y_m$ can be used to purge the pegs in $X_1, \ldots, X_n$, the graph is not solvable by (i). If we jump $x_k \cdot \overrightarrow{y_j} \cdot x_i$, then $\rho(X_1, \ldots, X_n) = \sum_{i=1}^{n} a_i$ but $\rho(X) = n - 1$. Since one fewer peg in X can be used to purge the pegs in $X_1, \ldots, X_n$, the graph is not solvable by (i). Similar arguments are used to show that if $(\sim\mathcal{P})$ and $\sum_{i=1}^{n} a_i \leq \sum_{j=1}^{m} b_j + n$, then G is not freely solvable. Note the effect of property $\mathcal{P}$ is that it allows the removal of one fewer peg from $X_1 \cup \cdots \cup X_n$ than in the $(\sim\mathcal{P})$ case, as described in the previous paragraph. For this reason, the remaining cases follow analogously. ■

There are several variants of peg solitaire on graphs [4, 5, 13–15]. One notable variant is fool's solitaire. In *fool's solitaire*, the objective of the game is to leave the maximum number of pegs on the graph G under the caveat that a jump must be made whenever possible. This maximum number of pegs is the *fool's solitaire number* of G and is denoted $Fs(G)$. For completeness, we include the fool's solitaire number for the hairy complete bipartite graph. A sharp upper bound on $Fs(G)$ is $\alpha(G)$, the independence number of G. For this reason, our strategy will be to attempt to solve the dual of the maximum independent set. If this is possible, then $Fs(G) = \alpha(G)$ by Theorem 2. More information on the fool's solitaire problem can be found in [9, 17]. Note that for this theorem, we ignore the assumption that $\sum_{i=1}^{n} a_i \geq \sum_{j=1}^{m} b_j$.

Theorem 4 *For $G = K_{n,m}(a_1, \ldots, a_n; b_1, \ldots, b_m)$ with $a_i = 0$ for $i \geq n - \ell + 1$ and $b_j = 0$, where $j \geq m - \lambda + 1$ and $\ell \geq \lambda$:*

(i) If $\ell = 0$, then $Fs(G) = \sum_{i=1}^{n} a_i + \sum_{j=1}^{m} b_j = \alpha(G)$;
(ii) If $1 \leq \ell \leq n - 1$, then $Fs(G) = \sum_{i=1}^{n} a_i + \sum_{j=1}^{m} b_j + \ell = \alpha(G)$;
(iii) If $\ell = n$, then $Fs(G) = \sum_{i=1}^{n} a_i + \sum_{j=1}^{m} b_j + \ell - 1 = \alpha(G) - 1$.

Proof *(i)* Suppose that $\ell = 0$. This implies that $\lambda = 0$. Therefore $a_i \geq 1$ for all i and $b_j \geq 1$ for all j. A maximum independent set is $A = X_1 \cup \cdots \cup X_n \cup Y_1 \cup \cdots \cup Y_m$. The dual of A is $X \cup Y$. Jump $y_1 \cdot \overrightarrow{x_1} \cdot x_{1,1}$, $x_2 \cdot \overrightarrow{y_2} \cdot x_1$, and $x_{1,1} \cdot \overrightarrow{x_1} \cdot y_2$. If $n = 2$ and $m = 2$, then the dual is solved with the final peg in y_2. If $n = 3$ and $m = 2$, then jump $x_3 \cdot \overrightarrow{y_2} \cdot x_1$ to solve the dual. If $n \geq 4$ and $m \geq 2$, then the subgraph induced by $(X - \{x_1, x_2\}) \cup Y$ is isomorphic to $K_{n-2,m}$ with a hole in y_1. Hence it is solvable.

(ii) Suppose that $1 \leq \ell \leq n - 1$. A maximum independent set is $A = X_1 \cup \cdots \cup X_n \cup Y_1 \cup \cdots \cup Y_m \cup \{x_{n-\ell+1}, \ldots, x_n\}$. The dual of A is $\{x_1, \ldots, x_{n-\ell}\} \cup Y$. The subgraph induced by $\{x_1, \ldots, x_{n-\ell+1}\} \cup Y \cup \{x_{n-\ell+1}\}$ is isomorphic to $K_{n-\ell+1,m}$ with a hole in $x_{n-\ell+1}$. Hence it is solvable.

(iii) Suppose that $\ell = n$. A maximum independent set is $A = X \cup Y_1 \cup \cdots \cup Y_m$. The dual of A is Y. Since no pegs are adjacent in the dual, it is not solvable. Thus at least one peg must be added to the dual. We add x_1 to the dual to obtain $Y \cup \{x_1\}$. The subgraph induced by $Y \cup \{x_1, x_2\}$ is isomorphic to $K_{2,m}$ with a hole in x_2. Hence it is solvable. ∎

3 Edge Critical Results

In [6], Beeler and Hoilman present an open problem considering the set of connected graphs on n vertices and k edges, which they denote $G_{n,k}$. The problem is, given a fixed n, determine the minimum k such that all graphs in $G_{n,k}$ are solvable. In [3], Beeler and Gray explore this problem by considering edge critical graphs. A graph G is *edge critical* if the addition of *any* single edge to G changes the solvability of G. We are particularly interested in the case when the addition of any single edge to an unsolvable (solvable but not freely solvable) graph results in a solvable (freely solvable) graph. An example of an edge critical graph is the cycle on an odd number of vertices [2]. The odd cycle C_{2k+1} is distance 2-solvable by Theorem 1. However, the addition of a single edge results in a solvable graph as shown in [2]. An additional family of edge critical graphs is explored in [3]. We now present families of edge critical hairy complete bipartite graphs. We denote the addition of edge uv to graph G by $G + uv$. As a reminder, property $\mathcal{P}$ is when either: (i) $n = 2$, m is even, $a_1 \geq 2$, and $a_2 \leq \sum_{j=1}^{m} b_j$ *or* (ii) $n = 2$, m is odd, $a_1 = a_2 = 1$, and $\sum_{j=1}^{m} b_j = 0$. When $n \geq 3$, the analogous property $\mathcal{Q}$ is: (i) $n \geq 3$, $a_1 \geq n$ and $\sum_{i=2}^{n} a_i \leq \sum_{j=1}^{m} b_j$ *or* (ii) $n \geq 3$, $a_1 = \cdots = a_n = 1$, and $\sum_{j=1}^{m} b_j = 0$.

Theorem 5 *The hairy complete bipartite graph $G = K_{2,m}(a_1, a_2; 0, \ldots, 0)$ is an edge critical graph if $\mathcal{P}$. For $n \geq 3$, the hairy complete bipartite graph $H = K_{n,m}(a_1, \ldots, a_n; 0, \ldots, 0)$ is edge critical if $\mathcal{Q}$(ii).*

Proof By Theorem 3 if $\mathcal{P}$, then G is $(a_1 + a_2)$-solvable with $a_1 + a_2 - 1$ pegs remaining in $X_1 \cup X_2$ and one peg remaining in Y. We now show that if $\mathcal{P}$, then any single edge addition to G allows the removal of at least one additional peg. Likewise, if $\mathcal{Q}$(ii), then H is solvable, but not freely solvable. Among these cases, an additional edge can be inserted in one of six places (up to automorphism on the vertices):

(1) An edge is inserted between $x_{1,1}$ and $x_{1,2}$. We note that this is only possible if $\mathcal{P}$(i). With the initial hole in x_1, jump $x_{1,1}\cdot\overrightarrow{x_{1,2}}\cdot x_1$, $y_m\cdot\overrightarrow{x_1}\cdot x_{1,2}$, and $x_2\cdot\overrightarrow{y_{m-1}}\cdot x_1$. For $k = 1, \ldots, \frac{m-2}{2}$, jump $x_1\cdot\overrightarrow{y_{2k-1}}\cdot x_2$ and $x_2\cdot\overrightarrow{y_{2k}}\cdot x_1$. Then jump $x_{1,2}\cdot\overrightarrow{x_1}\cdot y_m$ to end the game with $\rho(X_1) = a_1 - 2$ and $\rho(Y) = 1$.
(2) An edge is inserted between $x_{1,1}$ and $x_{2,1}$. We note that this is only possible if $\mathcal{P}$(ii) or $\mathcal{Q}$(ii). With the initial hole in $x_{2,1}$, jump $y_m\cdot\overrightarrow{x_2}\cdot x_{2,1}$ and $x_{1,1}\cdot\overrightarrow{x_{2,1}}\cdot x_2$. If $\mathcal{P}$(ii), then we end the game by solving the $K_{2,m}$ subgraph with a hole in y_m. If $\mathcal{Q}$(ii), then we solve the $K_{n,m}(1, \ldots, 1, 0, 0; 0, \ldots, 0)$ subgraph with a hole in y_m, which is solvable with the final two pegs in x_1 and y_1 by Theorem 3. Hence, it is freely solvable.
(3) An edge is inserted between $x_{1,1}$ and x_2. We relabel $x_{1,1}$ as y_{m+1}. If $\mathcal{P}$, then the graph is isomorphic to $K_{2,m+1}(a_1 + a_2 - 1, 0; 0, \ldots, 0)$, which is $(a_1 + a_2 - 2)$-solvable by Theorem 3. If $\mathcal{Q}$(ii), then the graph is isomorphic to $K_{n,m+1}(1, \ldots, 1, 0; 0, \ldots, 0)$ which is freely solvable.
(4) An edge is inserted between x_1 and x_2. Place the initial hole in x_1. Assume $\mathcal{P}$(i). Jump $y_m\cdot\overrightarrow{x_2}\cdot x_1$, $x_{1,1}\cdot\overrightarrow{x_1}\cdot x_2$, $y_{m-1}\cdot\overrightarrow{x_2}\cdot x_1$, and $x_{1,2}\cdot\overrightarrow{x_1}\cdot x_2$. Then for $k = 1, \ldots, \frac{m-2}{2}$, jump $x_2\cdot\overrightarrow{y_{2k-1}}\cdot x_1$ and $x_1\cdot\overrightarrow{y_{2k}}\cdot x_2$ to end the game with $\rho(X_1) = a_1 - 2$ and a peg in x_2.
Assume $\mathcal{P}$(ii) or $\mathcal{Q}$(ii). Jump $x_{2,1}\cdot\overrightarrow{x_2}\cdot x_1$ and $x_{1,1}\cdot\overrightarrow{x_1}\cdot x_2$. If $\mathcal{P}$(ii), then we finish the game by solving the $K_{2,m}$ subgraph with a hole in x_1. If $\mathcal{Q}$(ii), then we finish the game by solving the $K_{n,m}(1, \ldots, 1, 0, 0; 0, \ldots, 0)$ subgraph with a hole in x_1, which is freely solvable.
(5) An edge is inserted between y_1 and y_2. Place the initial hole in y_m. Assume $\mathcal{P}$(i). Jump $x_{1,1}\cdot\overrightarrow{x_1}\cdot y_m$, $y_1\cdot\overrightarrow{y_2}\cdot x_1$, and $x_{1,2}\cdot\overrightarrow{x_1}\cdot y_2$. If $n = m = 2$ and $a_1 = 2$, then jump $x_2\cdot\overrightarrow{y_2}\cdot x_1$ to end the game with a single peg in x_1. If $n = m = 2$ and $a_1 \geq 3$, then jump $x_2\cdot\overrightarrow{y_2}\cdot x_1$ and $x_{1,3}\cdot\overrightarrow{x_1}\cdot y_2$ to end the game with $\rho(X_1) = a_1 - 3$ and $\rho(Y) = 1$. If $n = 2$ and $m \geq 3$, then finish the game by solving the remaining $K_{2,m-1}(a_1 - 2, 0; 0, \ldots, 0)$ subgraph after the first jump, which results in at most $a_1 + a_2 - 2$ pegs remaining in the graph at the end of the game.

Assume $\mathcal{P}$(ii) or $\mathcal{Q}$(ii). Jump $x_{2,1}\cdot\overrightarrow{x_2}\cdot y_m$, $y_1\cdot\overrightarrow{y_2}\cdot x_2$, and $x_{1,1}\cdot\overrightarrow{x_1}\cdot y_2$. If $n = 2$, then we finish the game by solving the $K_{2,m-1}$ subgraph with a hole in x_1. If $n \geq 3$, then we finish the game by solving the $K_{n-1,m}(1, \ldots, 1, 0, 0; 0, \ldots, 0)$ subgraph with a hole in y_1, which is solvable by Theorem 3.

(6) An edge is inserted between $x_{1,1}$ and y_1. Assume $\mathcal{P}$(i). With the initial hole in y_1, jump $x_{1,2}\cdot\overrightarrow{x_1}\cdot y_1$ and $y_1\cdot\overrightarrow{x_{1,1}}\cdot x_1$. We finish the game by solving the $K_{2,m}(a_1 - 2, 0; 0, \ldots, 0)$ subgraph with a hole in y_1, which results in at most $a_1 + a_2 - 2$ pegs remaining in the graph at the end of the game.
Assume $\mathcal{P}$(ii) or $\mathcal{Q}$(ii). With the initial hole in x_1, jump $x_{1,1}\cdot\overrightarrow{y_1}\cdot x_1$ and $x_{2,1}\cdot\overrightarrow{x_2}\cdot y_1$. If $n = 2$, then we finish the game by solving the $K_{2,m}$ subgraph with a hole in x_2. If $n \geq 3$, then we finish the game by solving the $K_{n,m}(1, \ldots, 1, 0, 0; 0, \ldots, 0)$ subgraph with a hole in x_2, which is freely solvable.

The arguments are similar for when G is solvable, but not freely solvable, and the addition of a single edge results in a freely solvable graph. ∎

Note that graphs in the family $\mathcal{Q}$(i) are *not* edge critical. To see this, consider $G = K_{n,m}(a_1, 0, \ldots, 0; 0, \ldots, 0) + x_2x_3$, where $n \geq 3$ and $a_1 \geq n$. This graph is $(a_1 - n + 1)$-solvable by Theorem 3. Essentially, the only jumps that utilize the "new" edge x_2x_3 are $x_2\cdot\overrightarrow{x_3}\cdot y_j$ and $y_j\cdot\overrightarrow{x_2}\cdot x_3$. Because neither of these jumps result in an additional peg in X, neither jump will allow us to remove an additional peg from X_1. By adapting the $\mathcal{P}$(i) jumps, it is easy to see that adding any other edge to $K_{n,m}(a_1, 0, \ldots, 0; 0, \ldots, 0)$ for $n \geq 3$ and $a_1 \geq n$ will improve its solvability. The existence of additional edge critical hairy complete bipartite graph families is left as an open problem.

4 Graphs on Eight Vertices

The solvability of all 996 non-isomorphic connected graphs with at most seven vertices is given in [2]. We now extend that catalog by giving the solvability of all 11,117 non-isomorphic connected graphs with eight vertices. The graphs are from [16] and an exhaustive computer search [8] is used to determine solvability. Note that all connected graphs on eight vertices and at least 12 edges are freely solvable. Of the remaining 11,117 non-isomorphic graphs on eight vertices and at most eleven edges, 94 are not freely solvable. In the interest of space, Figs. 3, 4, and 5 list only those graphs that are *not* freely solvable.

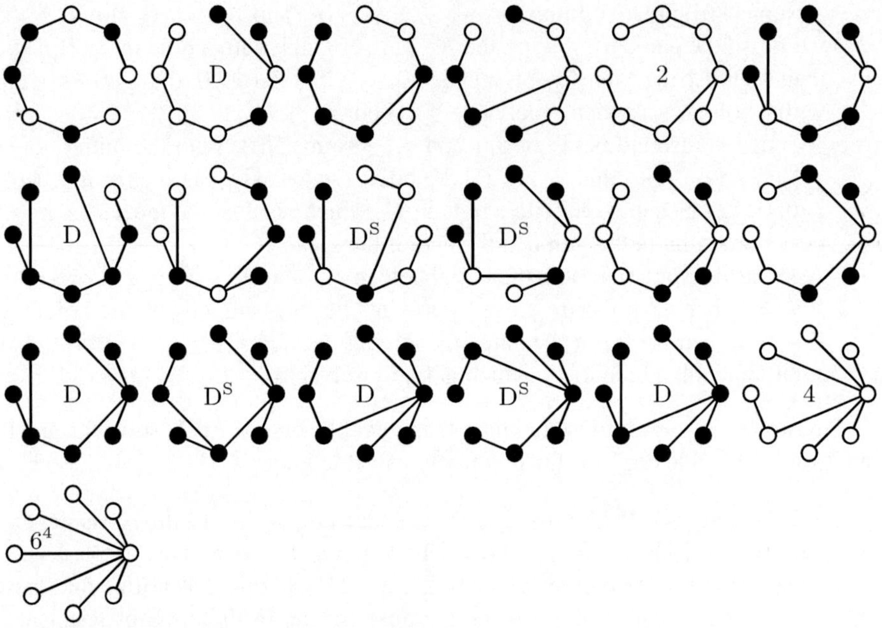

Fig. 3 Graphs with eight vertices and seven edges that are not freely solvable

In each figure, a black vertex indicates that the graph can be solved with the initial hole in that vertex. If the graph is not solvable, then we list the minimum number of pegs that can be obtained in a terminal state associated with a single vertex starting state. If the graph is distance 2-solvable, then this is indicated with a 'D', and a black vertex indicates that the graph can be distance 2-solved with the initial hole in that vertex.

If the graph is edge critical, then we denote the maximum solvability after any single edge addition with a superscript, where 'D' indicates distance 2-solvable, 'S' indicates solvable, but not freely solvable, and 'F' indicates freely solvable. If none of these three apply, then we use the minimum number of pegs that can be obtained in a terminal state associated with a single vertex starting state instead.

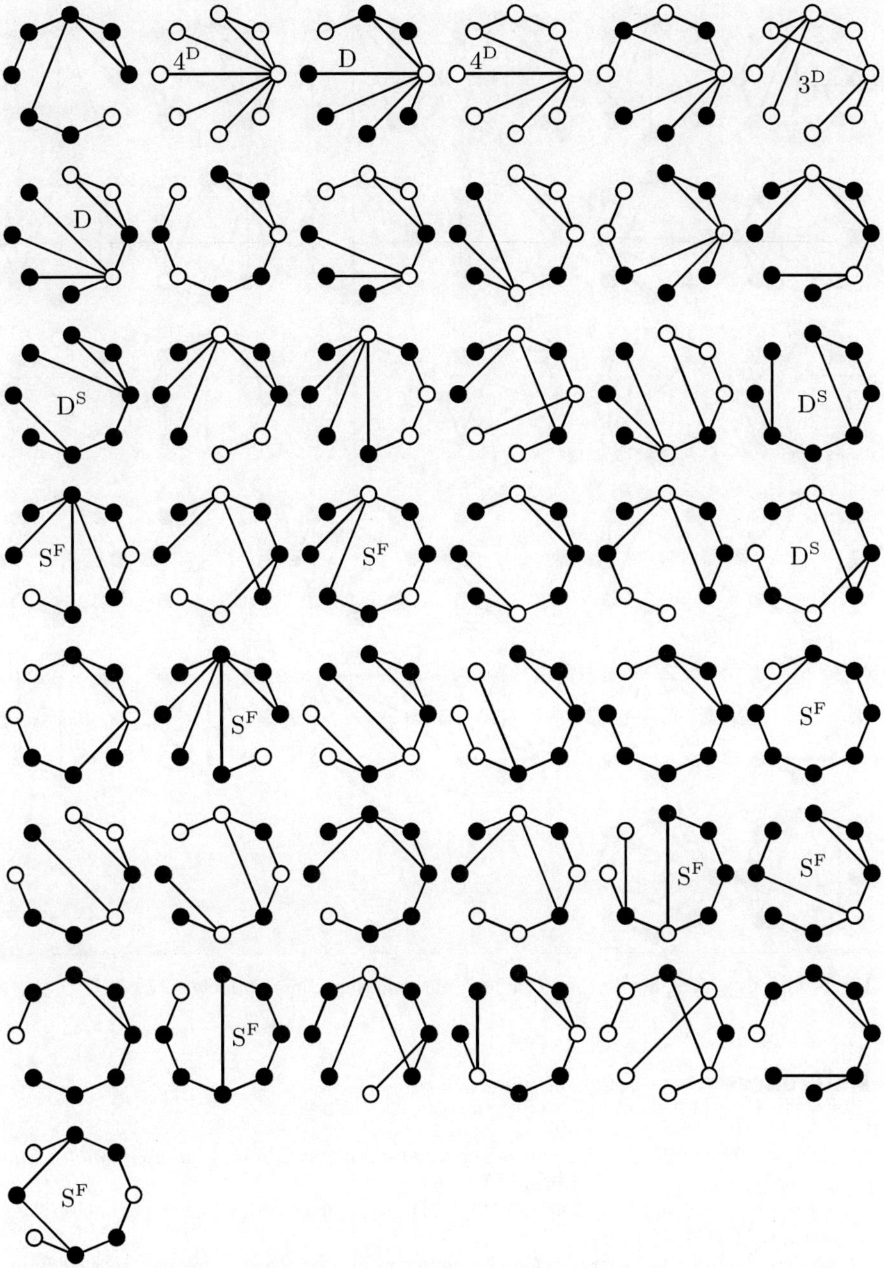

Fig. 4 Graphs with eight vertices and eight edges that are not freely solvable

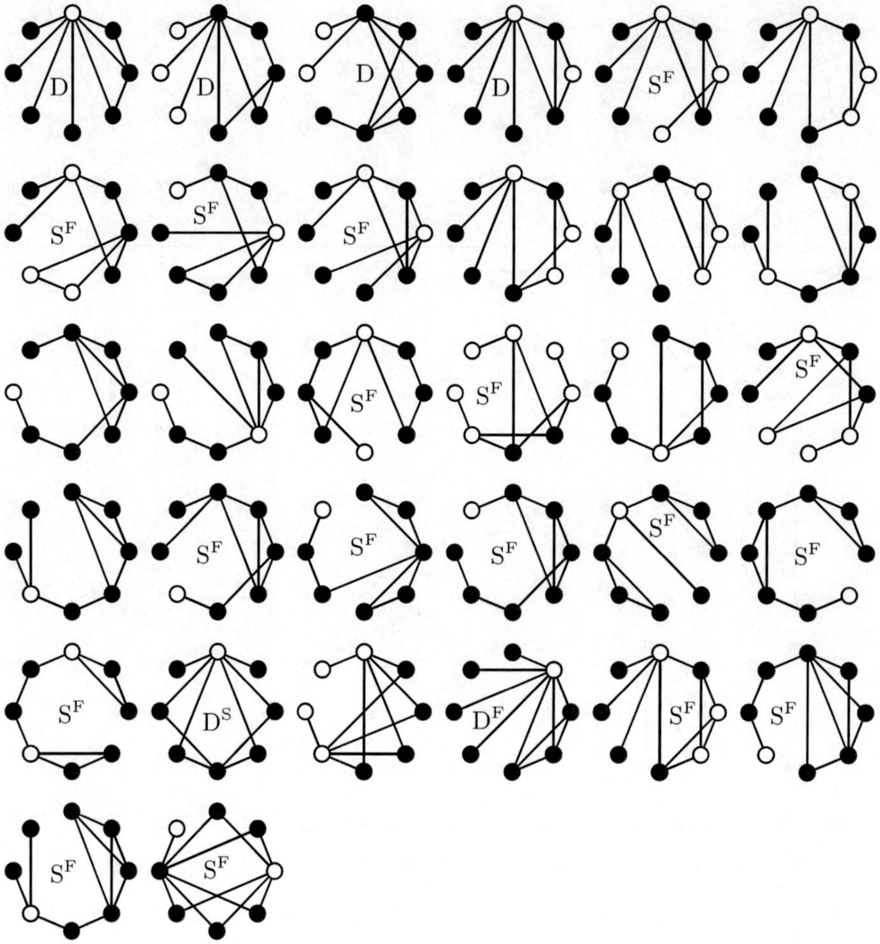

Fig. 5 Graphs with eight vertices and at least nine edges that are not freely solvable

References

1. John D. Beasley. *The ins and outs of peg solitaire*, volume 2 of *Recreations in Mathematics*. Oxford University Press, Eynsham, 1985.
2. Robert A. Beeler and Aaron D. Gray. Peg solitaire on graphs with seven vertices or less. *Congr. Numer.*, 211:151–159, 2012.
3. Robert A. Beeler and Aaron D. Gray. Extremal results for peg solitaire on graphs. *Bull. Inst. Combin. Appl.*, 77:30–42, 2016.
4. Robert A. Beeler and Aaron D. Gray. An introduction to peg duotaire on graphs. *J. Combin. Math. Combin. Comput.*, 104:171–186, 2018.
5. Robert A. Beeler and Aaron D. Gray. Double jump peg solitaire on graphs. *The Bulletin of the ICA*, 91:80–93, 2021.
6. Robert A. Beeler and D. Paul Hoilman. Peg solitaire on graphs. *Discrete Math.*, 311(20):2198–2202, 2011.

7. Robert A. Beeler and D. Paul Hoilman. Peg solitaire on the windmill and the double star. *Australas. J. Combin.*, 52:127–134, 2012.
8. Robert A. Beeler and Hal Norwood. Peg solitaire on graphs solver applet. http://faculty.etsu.edu/BEELERR/solitaire/.
9. Robert A. Beeler and Tony K. Rodriguez. Fool's solitaire on graphs. *Involve*, 5(4):473–480, 2012.
10. Robert A. Beeler and Clayton A. Walvoort. Packages and purges for peg solitaire on graphs. *Congr. Numer.*, 218:33–42, 2013.
11. Robert A. Beeler and Clayton A. Walvoort. Peg solitaire on trees with diameter four. *Australas. J. Combin.*, 63:321–332, 2015.
12. Elwyn R. Berlekamp, John H. Conway, and Richard K. Guy. *Winning ways for your mathematical plays. Vol. 2*. A K Peters Ltd., Natick, MA, second edition, 2003.
13. Grady D. Bullington. Peg solitaire: "burn two bridges, build one". *Congr. Numer.*, 223:187–191, 2015.
14. John Engbers and Christopher Stocker. Reversible peg solitaire on graphs. *Discrete Math.*, 338(11):2014–2019, 2015.
15. John Engbers and Ryan Weber. Merging peg solitaire on graphs. *Involve*, 11(1):53–66, 2018.
16. Jason Grout. Small graph database. http://artsci.drake.edu/grout/graphs/.
17. Sarah Loeb and Jennifer Wise. Fool's solitaire on joins and Cartesian products of graphs. *Discrete Math.*, 338(3):66–71, 2015.
18. Douglas B. West. *Introduction to graph theory*. Prentice Hall Inc., Upper Saddle River, NJ, 1996.

Regular Tournaments with Minimum Split Domination Number and Cycle Extendability

Kim A. S. Factor, Larry Langley, and Sarah Merz

Abstract A set of vertices, S, in a strongly connected digraph D, is split dominating provided it is: (1) dominating and (2) $D - S$ is trivial or not strongly connected. The split domination number of a strongly connected digraph is the minimum cardinality of a split dominating set for that digraph. We show that for any k-regular tournament, the split domination number is at least $\lceil \frac{2k+3}{3} \rceil$ and this bound is tight. We explore properties of regular tournaments with split domination number equal to the lower bound, including sufficient conditions for $\{1\}$-extendability.

Keywords Domination · Separating set · Tournament · Split domination · Cycle extendability

A set of vertices, S, in a graph is considered *dominating* when, for each vertex v in the graph, either $v \in S$ or there is an edge $\{s, v\}$ in the graph for some $s \in S$. In a digraph, a set of vertices, S, is considered *dominating* provided every vertex v in the digraph is either an element of S, or there is an arc (s, v) in the digraph for some $s \in S$. For a thorough introduction to graph theoretic domination, see [7]. For more advanced topics, including an overview of domination in digraphs, see [6].

Variations on domination in both graphs and digraphs are well-studied. In this paper, the variation considered is split domination. In a digraph D, a set of vertices S is *split dominating* provided the following two conditions hold. First, S is a dominating set. Second, removal of S results in a digraph, denoted by $D - S$, that is either trivial or has a reduced level of connectedness. In this paper, we focus on strongly connected digraphs. So S will be split dominating provided S is dominating and $D - S$ is either trivial or not strongly connected. Split domination was introduced in a (non-directed) graph context in 1979 by Kulli and Janakiram [10]. More recently, this problem was

K. A. S. Factor
Department of Mathematical and Statistical Sciences, Marquette University, Milwaukee WI 53233, USA

L. Langley · S. Merz (✉)
Department of Mathematics, University of the Pacific, Stockton CA, 95211, USA
e-mail: smerz@pacific.edu

F. Hoffman (ed.), *Combinatorics, Graph Theory and Computing*,
Springer Proceedings in Mathematics & Statistics 388,
https://doi.org/10.1007/978-3-031-05375-7_13

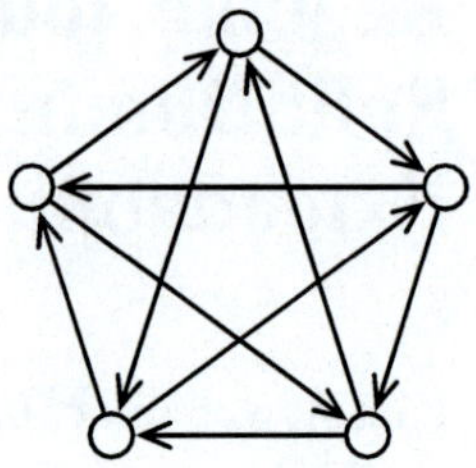

Fig. 1 A 2-regular tournament with split domination number equal to three

explored in graphs by Hedetniemi, Knoll, and Laskar [8] and in digraphs by Factor and Merz [5] and Factor, Langley, and Merz [4].

A tournament is a directed graph with the property that for each pair of vertices u and v, either (u, v) or (v, u) is an arc, but not both. Domination in tournaments has been considered in [11–13]. The *split domination number* of a tournament T is the minimum integer, denoted $\gamma_s(T)$, such that there is a split dominating set of size $\gamma_s(T)$. For example, the tournament shown in Fig. 1 has split domination number three. Dominating pairs occur in the tournament (any consecutive pair around the outer cycle is dominating), but no such pair is separating. Add any third vertex to a dominating pair and we have a split dominating set of size three. For more on this particular example and its generalization to a tournament with any odd number of vertices, see [5].

In a directed graph D, $V(D)$ and $A(D)$ denote the vertex and arc sets respectively. If $v \in V(D)$, the *out-set* of v and *in-set* of v are respectively

$$N^+(v) = \{u : (v, u) \in A(D)\} \text{ and } N^-(v) = \{u : (u, v) \in A(D)\}.$$

If $S \subseteq V(D)$, then $N_S^+(v) = N^+(v) \cap S$ and $N_S^-(v) = N^-(v) \cap S$.

A *regular tournament* is a tournament such that $|N^+(v)| = |N^+(u)|$ for all vertices v and u. A k-regular tournament is one in which $|N^+(v)| = k$ for each vertex v. It is well-known that k-regular tournaments are strongly connected. The consideration of the split domination number of a k-regular tournament is a natural consequence. In [5], Factor and Merz prove the following.

Proposition 1 *If T is a k-regular tournament, then $\gamma_s(T) \leq k + 1$.*

Furthermore, for all $k \geq 1$, they provide a k-regular tournament T with $\gamma_s(T) = k + 1$. Their example for $k = 2$ is shown in Fig. 1. In Sect. 1, we provide a tight lower bound for the split domination number of a regular tournament. In Sect. 2, we discuss properties of k-regular tournaments with split domination number equal to the lower bound.

1 The Lower Bound

Since a tournament on n vertices has $n(n-1)/2$ arcs, the average out-degree (and average in-degree) of the vertices is $(n-1)/2$. This means there is at least one vertex of out-degree greater than or equal to $(n-1)/2$ in any tournament, and that relationship is strict if n is even. Likewise, in any tournament with n vertices, there is a vertex with in-degree greater than or equal to $(n-1)/2$ and if n is even we know that there is a vertex with in-degree at least $n/2$.

Lemma 1 *If T is a k-regular tournament, then $\gamma_s(T) \geq \frac{2k+1}{3}$.*

Proof Let V be the vertex set of T. Assume S is a minimum size split dominating set of T. Since the induced tournament on $V - S$ is not strong, we may partition the vertices of $V - S$ into X and Y such that (x, y) is an arc for all $x \in X$ and all $y \in Y$. Observe that $|S| + |X| + |Y| = 2k + 1$.

Consider the subtournament induced by X. There must be at least one vertex x' in X with $|N_X^+(x')| \geq (|X| - 1)/2$. Since x' is directed toward every vertex in Y, $k = |N^+(x')| \geq \frac{|X|-1}{2} + |Y|$. Therefore, $(|X| - 1)/2 + |Y| \leq k$. By a similar argument, there is some vertex y' in Y such that $k = |N^-(y')| \geq \frac{|Y|-1}{2} + |X|$ so $(|Y| - 1)/2 + |X| \leq k$. Adding the two inequalities yields $|X| + |Y| \leq 2(2k + 1)/3$. Since $|S| + |X| + |Y| = 2k + 1$, we conclude that $|S| \geq (2k + 1)/3$. □

Theorem 1 *If T is a k-regular tournament, then $\gamma_s(T) \geq \lceil \frac{2k+3}{3} \rceil$.*

Proof By Lemma 1, we know that $\gamma_s(T) \geq (2k + 1)/3$. We will show that $\gamma_s(T) = (2k + 1)/3$ and $\gamma_s(T) = (2k + 2)/3$ are impossible. Let S be a split dominating set of minimum size. As in Lemma 1, partition the vertices of $T - S$ into X and Y so every vertex in X has an arc to every vertex in Y.

Case one: suppose $\gamma_s(T) = |S| = (2k + 1)/3$. Since $|S| + |X| + |Y| = 2k + 1$, $|X| + |Y| = 2(2k + 1)/3$. Suppose $|X| < |Y|$. That is,

$$|X| = \frac{2k+1}{3} - w \text{ and } |Y| = \frac{2k+1}{3} + w, \text{ for some integer } w > 0.$$

Since T is a tournament, there exists $x' \in X$ such that

$$|N^+(x')| \geq \frac{|X|-1}{2} + |Y| = \frac{(2k+1)/3 - w - 1}{2} + \frac{2k+1}{3} + w = k + \frac{w}{2}.$$

This is a contradiction since T is k-regular. An analogous argument with insets rules out the possibility that $|Y| < |X|$. Therefore, we know $|X| = |Y| = (2k + 1)/3$.

Since S is dominating, for all $y \in Y$, $N_S^-(y) \neq \emptyset$. Furthermore, there must be a vertex y' in Y such that $|N_Y^-(y')| \geq (|Y| - 1)/2$. Thus,

$$|N^-(y')| \geq |N_S^-(y')| + \frac{|Y|-1}{2} + |X| \geq 1 + \frac{(2k+1)/3 - 1}{2} + \frac{2k+1}{3} = k + 1.$$

This is a contradiction since T is k-regular. Thus, $\gamma_s(T) > (2k+1)/3$.

Case two: suppose $\gamma_s(T) = (2k+2)/3$. Observe that $|X| + |Y| = (4k+1)/3$ and this is odd. Consequently, either $|X|$ or $|Y|$ is odd and the other is even. Thus $|X|$ and $|Y|$ differ by at least one. Since regular tournaments have an odd number of vertices, it follows that $|S|$ is even. Suppose $|X| < |Y| - 1$. Then $|X| = \frac{2k-1}{3} - w$ and $|Y| = \frac{2k+2}{3} + w$. Again, there must be some vertex x' in X such that

$$|N^+(x')| \geq \frac{|X|-1}{2} + |Y| = \frac{(2k-1)/3 - w - 1}{2} + \frac{2k+2}{3} + w = \frac{2k}{2} + \frac{w}{2} > k,$$

a contradiction since T is k regular. Similar arguments result in a contradiction when $|Y| < |X| - 1$. Thus $|X|$ and $|Y|$ differ by exactly 1.

Suppose $|X| = |Y| - 1$. Then

$$\frac{4k+1}{3} = |X| + |Y| = 2|Y| - 1 \Rightarrow |Y| = \frac{2k+2}{3} = |S| \Rightarrow |X| = \frac{2k-1}{3}.$$

On the other hand, suppose $|Y| = |X| - 1$. Then

$$|X| = \frac{2k+2}{3} = |S| \text{ and } |Y| = \frac{2k-1}{3}.$$

In either case, observe that the larger of the two sets is even. Since S is dominating, for each vertex $v \notin S$, $N^-(v) \cap S \neq \emptyset$. If $|Y|$ is odd, then there is some vertex y' in Y such that

$$|N^-(y')| \geq |N_S^-(y')| + \frac{|Y|-1}{2} + |X| \geq 1 + \frac{(2k-1)/3 - 1}{2} + \frac{2k+2}{3} = 1 + k,$$

a contradiction since T is k-regular. On the other hand, if $|Y|$ is even, then there is some vertex y' in Y such that

$$|N^-(y')| \geq |N_S^-(y')| + \frac{|Y|}{2} + |X| = 1 + \frac{2k+2}{6} + \frac{2k-1}{3} = 1 + k,$$

a contradiction making case two impossible. So $\gamma_s(T) \geq \lceil (2k+3)/3 \rceil$. □

Observe that the tournament shown in Fig. 1 is k-regular with split domination number equal to $\lceil (2k+3)/3 \rceil$. Indeed, the bound in Theorem 1 is tight. When every vertex in set X has an arc to every vertex in set Y, we write $X \to Y$. If S is a set of vertices in digraph D, $D[S]$ is the subdigraph of D induced by S.

Theorem 2 *For all natural numbers $k \geq 1$, there is a k-regular tournament such that $\gamma_s(T) = \lceil \frac{2k+3}{3} \rceil$.*

Proof A 3-cycle is an example for $k = 1$. The tournament in Fig. 1 is an example for $k = 2$. So assume $k \geq 3$. Observe that either $2k+3$, $2k+4$ or $2k+5$ must be a multiple of 3. We consider each of these three cases.

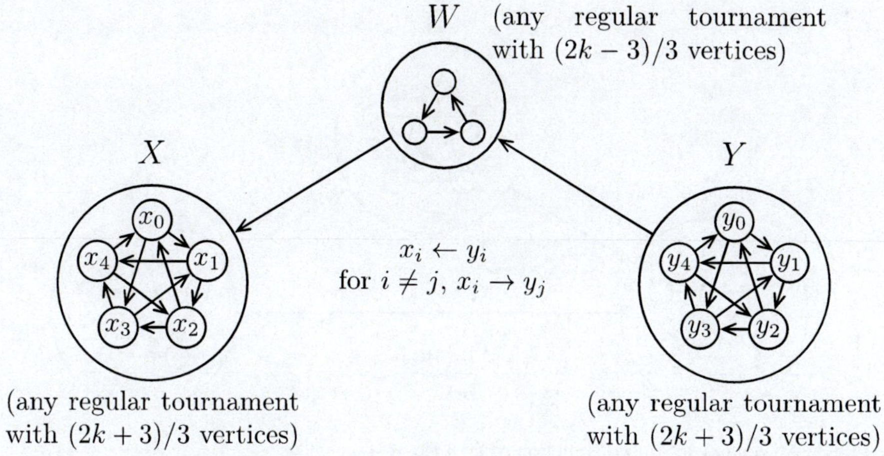

Fig. 2 An example of the construction when $3|(2k+3)$

Suppose $\lceil \frac{2k+3}{3} \rceil = \frac{2k+3}{3}$. Then k is a multiple of 3. Construct T as follows. Let X and Y be sets of vertices of size $(2k+3)/3$. Let W be a set of

$$(2k+1) - \frac{2(2k+3)}{3} = \frac{2k-3}{3}$$

vertices. Note that $2k+3$ and $2k-3$ are both odd, so $(2k+3)/3$ and $(2k-3)/3$ are also odd.

Create tournament T so that $T[W]$, $T[X]$ and $T[Y]$ are regular, $W \to X$, and $Y \to W$. It may be that W is a single vertex. Arcs between X and Y are oriented as follows. Label the vertices of X and Y by $x_0, x_1, \ldots, x_{2r}$ and $y_0, y_1, \ldots, y_{2r}$ respectively. Orient an arc from y_i toward x_i and, if $i \neq j$ orient an arc from x_j toward y_i. Figure 2 shows an example with $k = 6$.

Since T must be a regular tournament, every vertex must have in-degree of size k and out-degree of size k. By our construction, for all w in W,

$$|N^+(w)| = |N^-(w)| = |Y| + \frac{|W-1|}{2} = \frac{2k+3}{3} + \frac{(2k-3)/3-1}{2} = k.$$

Then the out-degree of each vertex of Y is

$$\frac{|Y-1|}{2} + |W| + 1 = \frac{(2k+3)/3-1}{2} + \frac{2k-3}{3} + 1 = k.$$

Because T is a tournament with $2k+1$ vertices in total, the in-degree of each vertex in Y must also equal k, so the tournament is k-regular. Analogous reasoning shows $|N^+(x)| = |N^-(x)| = k$ for all $x \in X$. Observe that Y is dominating and that $T - Y$ is not strong, since all vertices of W are directed toward all vertices of X. Since $|Y| = (2k+3)/3$, by Theorem 1, $\gamma_s(T) = (2k+3)/3$.

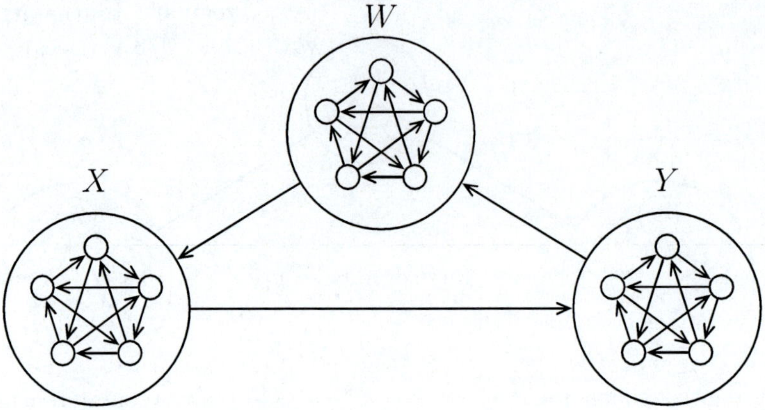

Fig. 3 An example of the construction when $3|(2k+4)$

Next, suppose $2k+4$ is a multiple of 3 with $k \geq 4$. Let $r = (2k+1)/3$. Let T consist of three regular r-tournaments W, X, and Y with $W \to X$, $X \to Y$, and $Y \to W$. For every vertex v, $|N^+(v)| = |N^-(v)| = (r-1)/2 + r = k$, T is regular. Figure 3 shows an example with $k = 7$.

Consider $S = W \cup \{x\}$ for some $x \in X$. Since S is a split dominating set size $r + 1 = (2k+4)/3$, by Theorem 1, $\gamma_s(T) = (2k+4)/3 = \lceil \frac{2k+3}{3} \rceil$.

The final case to consider is when $2k+5$ is a multiple of 3. So $k \geq 5$. Partition the $2k+1$ vertices of T into sets W, X and Y where $|X| = |Y| = (2k+5)/3$ and

$$|W| = 2k + 1 - \frac{2(2k+5)}{3} = \frac{2k-7}{3}.$$

Let each of the three sets induce a regular tournament, $W \to X$ and $Y \to W$. Let $r = (2k+5)/3$. Let $x_0, \ldots, x_r$ and $y_0, \ldots, y_r$ denote the vertices of X and Y respectively. For $i \in \{0, \ldots, r\}$, orient an arc from y_i to x_i and from y_i to x_{i+1}. All other arcs between X and Y are oriented from the vertex in X toward the vertex in Y. See Fig. 4.

Observe that for each w in W,

$$|N^+(w)| = |N^-(w)| = |Y| + \frac{|W-1|}{2} = \frac{2k+5}{3} + \frac{\frac{2k-7}{3} - 1}{2} = k.$$

Label the vertices of X and Y by $x_0, x_1, \ldots, x_{2r}$ and $y_0, y_1, \ldots, y_{2r}$ respectively. Orient an arc from y_i toward x_i and x_{i+1}. If $i \neq j$ and $i \neq j-1$, orient an arc from x_j toward y_i. Then the out-degree of each vertex of Y is

$$\frac{|Y-1|}{2} + |W| + 2 = \frac{(2k+5)/3 - 1}{2} + \frac{2k-7}{3} + 2 = k.$$

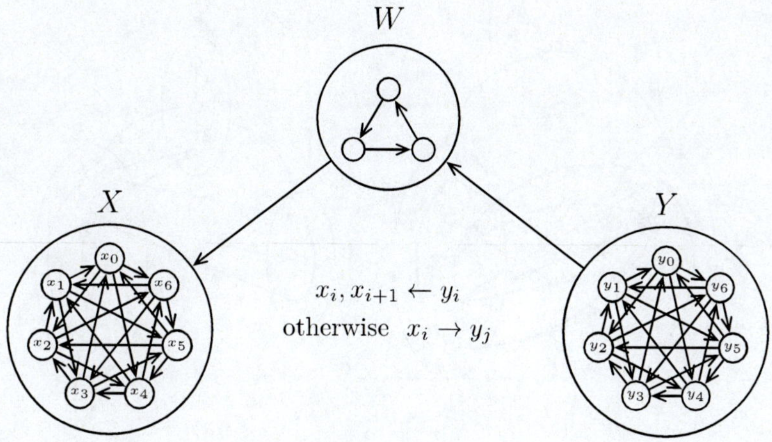

Fig. 4 An example of the construction when $3|(2k+5)$

An analogous calculation shows that $|N^-(x)| = k$ for all $x \in X$. Thus, T is regular. Observe that Y is a split dominating set of size $(2k+5)/3$. So by Theorem 1, $\gamma_s(T) = (2k+5)/3 = \lceil \frac{2k+3}{3} \rceil$. □

2 Properties of Tournaments That Meet the Bound

The examples given in Theorem 2 share several common properties, yielding the result that each of these tournaments is not $\{1\}$-extendable. In this paper, a cycle in a digraph is assumed to be directed. If D is a digraph with n vertices, we say a cycle C of length $m < n$ is *$\{1\}$-extendable* if there is a single vertex x such that the m vertices of C, together with x, induce a digraph that contains a cycle of length $m+1$. Cycle extendability has been considered in digraphs [1] and tournaments [2]. The connection between split domination and cycle extendability was introduced in [3]. A digraph is *$\{1\}$-extendable* if every cycle in D is $\{1\}$-extendable. The following result, due to Hendry, is relevant.

Proposition 2 *[9] If T is a regular tournament, then T is not $\{1\}$-extendable if and only if its vertex set can be partitioned into three non-empty sets W, X, and Y such that $T[W]$ is a nontrivial regular tournament, $W \rightarrow X$, $Y \rightarrow W$, and $|X| = |Y|$.*

Observe that each of the three constructions in Theorem 2 meets the conditions of Proposition 2, proving that the tournaments constructed are not $\{1\}$-extendable. This prompts us to wonder if there is any k-regular tournament satisfying the lower bound of Theorem 1 that is $\{1\}$-extendable. The tournament in Fig. 5a is such an example, where S is a split dominating set. You can see this tournament is $\{1\}$-extendable by inspection using the fact that any strongly connected subtournament will have a cycle containing all its vertices.

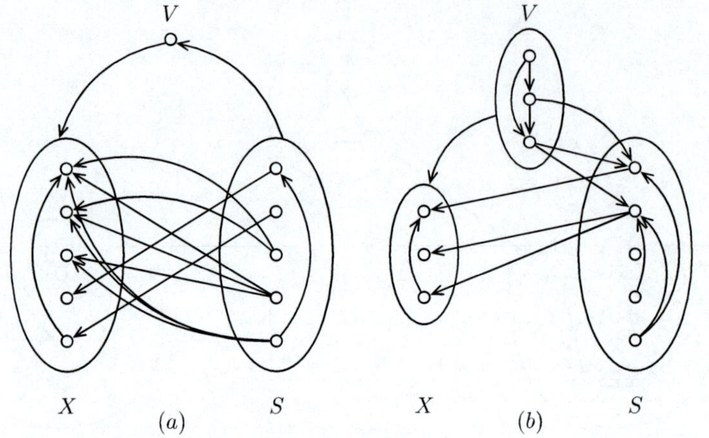

Fig. 5 Tournament (**a**) and **b** are 5-regular. Not all arcs are shown. Any arc not shown within sets X and S is directed from higher vertex to lower vertex. Any arc not shown between sets X and S is directed from X to S. In tournament **b**, arcs not shown between V and S are directed from S to V. $T[X]$ in tournament **a** and $T[V]$ in tournament **b** are not regular even though $|X|$ and $|V|$ are odd

A split dominating set in a tournament suggests a partition of the vertices, but what are the properties of this partition and what are the similarities to the partition of Proposition 2? Let S denote a split dominating set of size $\gamma_s(T)$. Let (V, X) denote the partition of $T - S$ into sets so that $V \to X$ and $T[X]$ is strong. Furthermore, assume that $\gamma_s(T)$ satisfies the lower bound of Theorem 1. For the examples in Figs. 2 and 4, $S = Y$, $V = W$, and X is as labeled. For the example in Fig. 3, let $w \in W$. Then $S = Y \cup \{w\}$, $V = W - \{w\}$, and X is as labeled. In each of these examples, if $|X|$ is odd, then $T[X]$ is regular. The same can be said for V. This is not always the case, as illustrated by the examples in Fig. 5. In a way, the examples of Fig. 5 are the only exceptions. In order to prove this, we use the following lemma.

Lemma 2 *Let T be a k-regular tournament with $\gamma_s(T) = \lceil \frac{2k+3}{3} \rceil$, $k \geq 3$. Let S be a minimum split dominating set and (V, X) the partition of $T - S$ so that $V \to X$ and X is strong. Then*

1. *if $\gamma_s(T) = \frac{2k+3}{3}$, then $|V| = \frac{2k}{3} - 1$,*
2. *if $\gamma_s(T) = \frac{2k+4}{3}$, then $|V| = \frac{2k-5}{3}$ or $\frac{2k-5}{3} + 1$, and*
3. *if $\gamma_s(T) = \frac{2k+5}{3}$, then $|V| = \frac{2k-7}{3}$, $\frac{2k-7}{3} + 1$, or $\frac{2k-7}{3} + 2$.*

Proof Let S be a split dominating set of size $\gamma_s(T)$. Then $|V| + |X| = 2k + 1 - \gamma_s(T)$. Since for all $x \in X$, $|N^-(x)| = k$ and $V \subseteq N^-(x)$, the fact that S dominates T, means that $|V| \leq k - 1$. Thus

$$|X| \geq 2k + 1 - \gamma_s(T) - (k - 1) = k + 2 - \gamma_s(T).$$

Furthermore, for all $v \in V$, $X \subseteq N^+(v)$ so $|X| \leq k$. Therefore

$$|V| \geq 2k + 1 - \gamma_s(T) - k = k + 1 - \gamma_s(T).$$

This gives us the following bounds:

$$k + 1 - \gamma_s(T) \leq |V| \leq k - 1 \text{ and } k + 2 - \gamma_s(T) \leq |X| \leq k. \tag{1}$$

For the smallest cases, $k = 3, 4$, and 5, $\gamma_s(T) = 3, 4$, and 5 respectively. When we observe that $T[X]$ strong implies that $|X| \neq 2$, the bounds of (1) prove the lemma for these values of k. Thus, we can assume that $k \geq 6$. There are three cases to consider: $\lceil \frac{2k+3}{3} \rceil = \frac{2k+3}{3}, \frac{2k+4}{3}$, or $\frac{2k+5}{3}$.

In each case, since $k \geq 6, 7$, and 8, respectively, $|V| \geq 2$ and $|X| \geq 3$. Thus, there exists $v \in V$ with $N_V^+(v) \geq 1$ and since $X \subseteq N^+(v)$, we conclude that $|X| \leq k - 1$. Thus,

$$|V| = 2k + 1 - \gamma_s(T) - |X| \Rightarrow |V| \geq k + 2 - \gamma_s(T).$$

Since $|X| \geq 3$, for all $x \in X$, $|N_X^-(x)| \geq 1$, making $|V| \leq k - 2$. Thus,

$$|X| = 2k + 1 - \gamma_s(T) - |V| \Rightarrow |X| \geq k + 3 - \gamma_s(T).$$

Thus the bounds in (1) are tightened to

$$k + 2 - \gamma_s(T) \leq |V| \leq k - 2 \text{ and } k + 3 - \gamma_s(T) \leq |X| \leq k - 1. \tag{2}$$

Assume $\gamma_s(T) = (2k + 3)/3$. Then (2), along with the fact that $|V| + |X| = 2k + 1 - \gamma_s(T)$, means that

$$|V| = \frac{k}{3} + w \text{ and } |X| = k - w \text{ where } 1 \leq w \leq \gamma_s(T) - 3. \tag{3}$$

There is a vertex $x \in X$ with arcs from at least half the other vertices in X, all of V, and at least one vertex in S. So

$$k = |N^-(x)| \geq |V| + \frac{|X| - 1}{2} + 1 = \frac{5k}{6} + \frac{w}{2} + \frac{1}{2}. \tag{4}$$

Then

$$k \geq \frac{5k}{6} + \frac{w}{2} + \frac{1}{2} \Rightarrow w \leq \frac{k-3}{3}.$$

So $|V| = k/3 + w \leq (2k/3) - 1$. Similarly, there exists $v \in V$ such that

$$k = |N^+(v)| \geq \frac{|V| - 1}{2} + |X| = \frac{7k}{6} - \frac{w}{2} - \frac{1}{2} \Rightarrow w \geq \frac{k-3}{3}. \tag{5}$$

Thus, $|V| = \frac{k}{3} + w \geq \frac{2k}{3} - 1$. Therefore, $|V| = \frac{2k}{3} - 1$ if $\gamma_s(T) = \frac{2k+3}{3}$.

Next, assume $\gamma_s(T) = (2k + 4)/3$. In this case, instead of (3), we have

$$|V| = \frac{k-1}{3} + w \text{ and } |X| = k - w \text{ where } 1 \leq w \leq \gamma_s(T) - 3.$$

Repeating the calculations in (4) and (5) yields $\frac{2k-5}{3} \leq |V| \leq \frac{2k-2}{3}$, thereby finishing case two.

Finally, assume $\gamma_s(T) = (2k+5)/5$. Then (3) becomes

$$|V| = \frac{k-2}{3} + w \text{ and } |X| = k - w \text{ where } 1 \leq w \leq \gamma_s(T) - 3.$$

Repeating the calculations in (4) and (5) yields $\frac{2k-7}{3} \leq |V| \leq \frac{2k-1}{3}$. □

Finally, we show that the instances shown in Fig. 5 are the only cases where either $|V|$ is odd and $T[V]$ is not regular or $|X|$ is odd and $T[X]$ is not regular.

Theorem 3 *Let T be a k-regular tournament with $\gamma_s(T) = \lceil \frac{2k+3}{3} \rceil$, $k \geq 3$. Let S be a minimum split dominating set and (V, X) the partition of $T - S$ so that $V \to X$ and X is strong. Then*

1. *if $|V|$ is odd and $|V| \neq \frac{2k-7}{3} + 2$ then $T[V]$ is regular, and*
2. *if $|X|$ is odd and $|X| \neq \frac{2k+5}{3}$, then $T[X]$ is regular.*

Proof We consider the three cases, $\lceil \frac{2k+3}{3} \rceil = \frac{2k+3}{3}, \frac{2k+4}{3}$, or $\frac{2k+5}{3}$. First assume $\gamma_s(T) = (2k+3)/3$. By Lemma 2, $|V| = (2k-3)/3$. Thus,

$$|X| = 2k + 1 - \gamma_s(T) - |V| = (2k+3)/3.$$

Since $3|(2k+3)$, $(2k-3)/3$ is odd so we must show that $T[V]$ is regular. Suppose not. Then there must be a vertex $v \in V$ with arcs to more than half the other vertices in V. Since $X \subseteq N^+(v)$, we find that

$$k = |N^+(v)| \geq \frac{|V|-1}{2} + 1 + |X| = \frac{(2k-3)/3 - 1}{2} + 1 + \frac{2k+3}{3} = k + 1, \tag{6}$$

a contradiction. Therefore, $T[V]$ is regular.

Note $|X|$ is also odd. Suppose $T[X]$ is not regular. Then some vertex $x \in X$ has arcs from more than half the other vertices in X, in addition to arcs from every vertex in V and at least one vertex in S. That is,

$$k = |N^+(x)| \geq |V| + \frac{|X|-1}{2} + 1 + 1 \geq = k + 1, \tag{7}$$

a contradiction. Thus $T[X]$ is regular.

Next, assume that $\gamma_s(T) = (2k+4)/3$. By Lemma 2, $|V| = (2k-8)/3 + w$ where $w \in \{1, 2\}$. Since $|X| = 2k + 1 - \gamma_s(T) - |V|$, if $|V| = (2k-8)/3 + w$, then $|X| = (2k+7)/3 - w$ for $w \in \{1, 2\}$.

Suppose, analogous to (6), that $|V|$ is odd and there is a vertex $v \in V$ with arcs to more than half the other vertices in V. Then

$$k = |N^+(v)| > \frac{|V|-1}{2} + 1 + |X| = k + \frac{3}{2} - \frac{w}{2} \text{ for } w \in \{1, 2\},$$

a contradiction. Thus, if $|V|$ is odd, then $T[V]$ is regular.

Suppose, analogous to (7), $|X|$ is odd and there is a vertex $x \in X$ such that

$$|N^-(x)| \geq |V| + \frac{|X|-1}{2} + 1 + 1 \geq k + \frac{w}{2} \text{ for } w \in \{1, 2\},$$

a contradiction. Thus, if $|X|$ is odd, then $T[X]$ is regular.

Lastly, suppose that $\gamma_s(T) = \frac{2k+5}{3}$. By Lemma 2, $|V| = (2k-10)/3 + w$ for $w \in \{1, 2, 3\}$. Since $|X| = 2k + 1 - \gamma_s(T) - |V|$, if $|V| = (2k-10)/3 + w$ where $w \in \{1, 2, 3\}$, then $|X| = (2k+8)/3 - w$. Suppose, analogous to (6), $|V|$ is odd and there is a vertex $v \in V$ with arcs to more than half the other vertices in V. Then

$$k = |N^+(v)| \geq \frac{|V|-1}{2} + 1 + |X| = k + \frac{3}{2} - \frac{w}{2} \text{ for } w \in \{1, 2, 3\}.$$

This is a contradiction so long as $w \neq 3$ ($|V| \neq (2k-7)/3 + 2$). Thus, if $|V|$ is odd and $|V| \neq (2k-7)/3 + 2$, then $T[V]$ is regular.

Suppose, analogous to (7),$|X|$ is odd and there is a vertex $x \in X$ with arcs from more than half the other vertices in X. Then

$$k = |N^-(x)| \geq |V| + \frac{|X|-1}{2} + 1 + 1 \geq k - \frac{1}{2} + \frac{w}{2} \text{ for } w \in \{1, 2, 3\}.$$

If $w \neq 1$ then we have a contradiction. So if $|X|$ is odd and $|X| \neq (2k+5)/3$, then $T[X]$ is regular. □

Corollary 1 *Let $k \geq 6$ and T be a k-regular tournament with $\gamma_s(T) = \lceil \frac{2k+3}{3} \rceil$. Let S be a minimum split dominating set and (V, X) the partition of $T - S$ so that $V \to X$ and X is strong. If*

1. $3|(2k+3)$, *or*
2. $3|(2k+4)$ *and* $|V|$ *is odd, or*
3. $3|(2k+5)$ *and* $|V| = \frac{2k-7}{3}$,

then T is not $\{1\}$-extendable.

Proof Assume $3|(2k+3)$. Then $|W| = (2k-3)/3$ and $|S| = |X|$. Since $|V| > 1$, by Theorem 3 and Proposition 2, T is not $\{1\}$-extendable. Assume $3|(2k+4)$ and $|V|$ is odd. Then by Lemma 2, $|V| = (2k-5)/3$. So $|S| = |X|$. Since $|V| > 1$, the result follows from Theorem 3 and Proposition 2. Finally, assume $3|(2k+5)$ and $|V| = (2k-7)/3$. Then $|V|$ is odd and $|S| = |X|$. Again, $|V| > 1$ so the result follows from Theorem 3 and Proposition 2. □

Independence and Domination of Chess Pieces on Triangular Boards and on the Surface of a Tetrahedron

Krista Munger, Joseph Petrone, and David R. Prier

Abstract Independence and domination results are given for six chess-like pieces on triangular boards with triangular spaces and triangular boards with hexagonal spaces. The question of independence and domination on for these same boards on the surface of a tetrahedron is introduced, and some initial results are given. …

Keywords Independence · Domination · Graph theory · Triangular chess boards

MSC Code 05C69

1 Introduction

Independence and domination on chessboards is a well-known topic of study. A graph BP can be made from a board B and a piece P where each space of the board is a vertex and there is an edge from vertex u to vertex v if and only if piece P can move from u to v. An independent set of vertices on BP is a set such that no two vertices share an edge. The independence number, denoted $\beta(BP)$, is the maximum size of an independent set of vertices. A dominating set of vertices on BP is one in which every vertex of the graph is either in the set or adjacent to some member of the set. The domination number, denoted $\gamma(BP)$, is the minimum size of a dominating set of vertices. For an undirected simple graph G, it is a well known fact that $\gamma(G) \leq \beta(G)$.

In this paper, two types of triangular chess boards are considered. Section 2 will examine T_n, a triangular board with n^2 triangular spaces as shown in Fig. 1. Section 3 will examine H_n, a triangular board with $\binom{n+1}{2}$ hexagonal spaces as shown in Fig. 2. Chess pieces that move analogously to those in standard chess have been defined in

K. Munger · J. Petrone · D. R. Prier (✉)
Gannon University, Erie PA, USA
e-mail: prier001@gannon.edu

F. Hoffman (ed.), *Combinatorics, Graph Theory and Computing*,
Springer Proceedings in Mathematics & Statistics 388,
https://doi.org/10.1007/978-3-031-05375-7_14

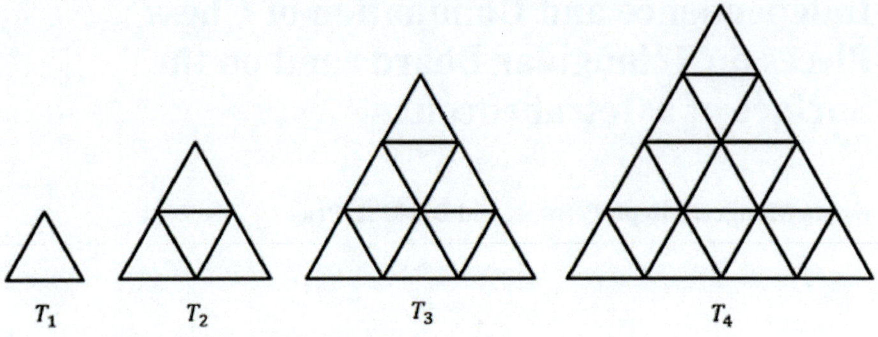

Fig. 1 T_n, triangular boards with triangular spaces

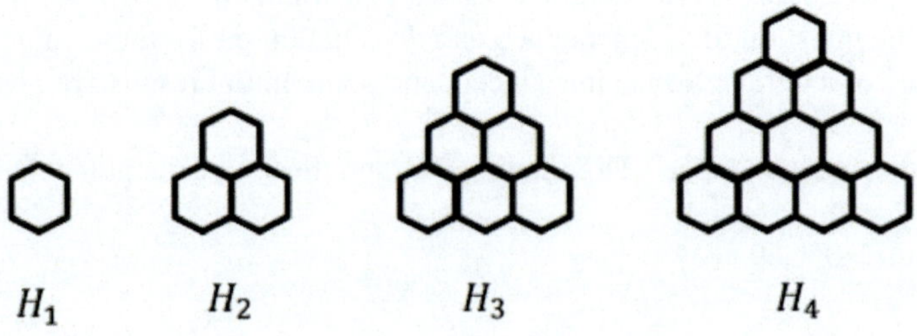

Fig. 2 H_n, triangular boards with hexagonal spaces

many papers, and six will be studied here. See [1]; and [2]; for examples. Section 4 will utilize the results from Sects. 2 and 3 and apply them to boards constructed by placing T_n or H_n on the four surfaces of a tetrahedron.

2 Triangular Boards with Triangular Spaces

In Fig. 3 we define the moves of six pieces on the triangular board with triangular spaces. The board with hexagonal spaces will be discussed in Sect. 3. The six pieces are the King denoted K, the Queen denoted Q, the Rook denoted R, the Bishop denoted B, the Knight denoted N, and the Grid denoted G. The Grid is similar to a pawn in that it can move only one space in some direction, however is is not limited in the direction that it can move.

2.1 *Independence on T_n*

Results were found concerning the independence numbers associated with these six pieces on T_n. After presenting some of this work at a conference, it was found

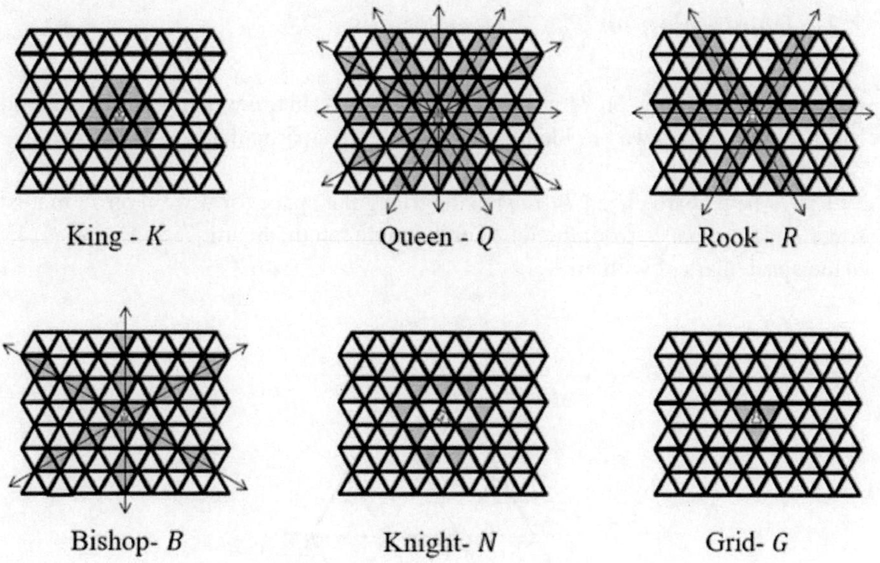

Fig. 3 The six pieces and their moves defined on a board with triangular spaces

that many of these results had already appeared in a German Mathematics Journal titled *Abhandlungen der Braunschweigischen Wissenschaftlichen Gesellschaft*. The pertinent results will be summarized here, and the full journal article can be found at [1].

Piece P	$\beta(T_n P)$	Conditions
K	$\frac{1}{3}\binom{n+2}{2} - 1$	$n \equiv 2, 4, 5, 7 (mod\ 12)$
K	$\lfloor \frac{1}{3}\binom{n+2}{2} \rfloor$	otherwise
Q	$\beta(T_n Q) \leq \beta(T_n R) \leq \lfloor \frac{2n+1}{3} \rfloor$	
R	$\lfloor \frac{2n+1}{3} \rfloor$	
B	$2n - 3$	$n \equiv 0 (mod 3)$
B	$2n - 1$	$n = \frac{3^k+1}{2}$ for k a nonnegative integer
B	$2n - 3 \leq \beta(T_n B) \leq 2n - 1$	otherwise
N	$\binom{n+1}{2}$	$n \neq 2$
N	4	$n = 2$
G	$\binom{n+1}{2}$	

Note that though many of these independence questions have been solved, there is still room for further work. The Bishop, B, is still not completely settled, and the Queen, Q, has only the trivial inequality that its independence number is less than the Rook's. This must be true since the Rook's moves are a subset of the Queen's.

2.2 Domination on T_n

Domination questions on T_n will be addressed next. In order to aid this explanation, it is useful to have a way to identify each space on a particular board. The following definition will do this.

Let T_n be a board. Then $T_n(i, j)$ is the triangular space (or vertex) on T_n that is in row i and is j spaces from the left. For example, on the board, T_4 in Fig. 4, $T_4(4, 3)$ is the space marked with an X.

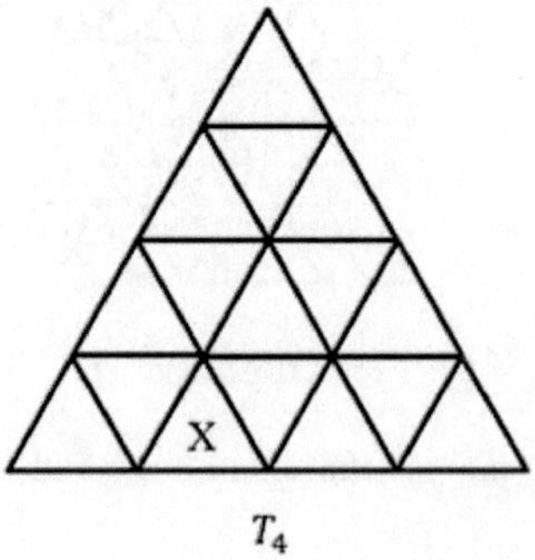

Fig. 4 $T_4(4, 3)$ is marked with an X

Grid

Theorem 1 $\gamma(T_nG) = \frac{n^2}{4}$ *if n is even.* $\lceil \frac{n^2}{4} \rceil \leq \gamma(T_nG) \leq \frac{(n-1)^2}{4} + \lfloor \frac{2n}{3} \rfloor$ *if n is odd.*

Proof The Grid, G, dominates at most 4 spaces. Since T_n has n^2 vertices $\gamma(T_nG) \geq \lceil \frac{n^2}{4} \rceil$. If n is even then the set of vertices $\{T_n(i, j) : i \text{ is even and } j \equiv 2 (\text{mod } 4)\} \cup \{T_n(i, j) : i \text{ is odd and } j \equiv 3 (\text{mod } 4)\}$ is a minimum dominating set of size $\frac{n^2}{4}$. If n is odd, then the minimum dominating set for T_{n-1} of size $\frac{(n-1)^2}{4}$ along with the set $[T_n(n, j) : j \equiv 2 (\text{mod } 3)]$ of size $\lfloor \frac{2n}{3} \rfloor$ is dominating. □

Rook

Theorem 2 $\lceil \frac{n^2}{4(n-1)} \rceil \leq \gamma(T_nR) \leq \lceil \frac{n}{2} \rceil$.

Proof Any one Rook, R, dominates at most $4(n-1)$ spaces, so $\lceil \frac{n^2}{4(n-1)} \rceil \leq \gamma(T_nR)$. A dominating set for T_n of size $\lceil \frac{n}{2} \rceil$ is $\{T_n(i, j) : i \equiv 0 (\text{mod } 2) \text{ and } j = \frac{i}{2}\}$. The above inequality seems to have a lot of room for improvement. The authors of this paper, however, have found no values of n where $\gamma(T_nR) < \frac{n}{2}$. It is conjectured that $\gamma(T_nR) = \frac{n}{2}$. □

Queen

Theorem 3 $\lceil \frac{n^2}{7(n-2)} \rceil \leq \gamma(T_n Q) \leq \gamma(T_n R) \leq \lceil \frac{n}{2} \rceil$.

Proof Any one Queen, Q, dominates at most $7(n-2)$ spaces, so $\lceil \frac{n^2}{7(n-2)} \rceil \leq \gamma(T_n Q)$. Because the Rook's moves are a subset of the Queen's, $\gamma(T_n Q) \leq \gamma(T_n R)$. However, as in the Rook's case, the authors of this paper have found no values of n where $\gamma(T_n Q) < \frac{n}{2}$. □

Bishop

Theorem 4 $\lceil \frac{n^2}{3n-2} \rceil \leq \gamma(T_n B) \leq 2n-1$ *if n is even, and* $\lceil \frac{n^2}{3n-4} \rceil \leq \gamma(T_n B) \leq 2n-1$ *if n is odd.*

Proof The maximum number of spaces of T_n dominated by one Bishop, B, is $3n-2$ if n is even and $3n-4$ if n is odd. The upper bound of $2n-1$ comes from the fact that the set $\{T_n(i, j) : i = n\}$ is a rather trivial dominating set of size $2n-1$. To see that this upper bound is far from sharp, one can notice that $\gamma(T_6 B) = 3$ with minimum dominating set $\{T_6(4, 4), T_6(5, 4), T_6(5, 6)\}$. □

King

Theorem 5 $\lceil \frac{n^2}{13} \rceil \leq \gamma(T_n K) \leq \lceil \frac{n^2+3n}{10} \rceil$.

Proof The maximum number of spaces one King, K, can dominate is 13, so this fact accounts for the lower bound. Now define the bottom five rows of T_n to be a new graph called D_n for all $n \geq 6$. It will be shown that $\gamma(D_n K) \leq n-1$ for all $n \geq 6$. Consider the set of vertices, S, defined as the following:
$S = \{T_n(n-1, j) : j \equiv 1 (\text{mod } 4)\} \cup \{T_n(n-4, j) : j \equiv 3 (\text{mod } 8)\} \cup \{T_n(n-5, j) : j \equiv 6 (\text{mod } 8)\}$ If $n \equiv 2 (\text{mod } 4)$, then S has $n-2$ vertices and dominates D_n. If $n \equiv 3 (\text{mod } 4)$, then S has $n-3$ vertices. S along with the vertices $T_n(n, 2n-1)$ and $T_n(n-5, 2n-9)$ dominates D_n with $n-1$ total vertices. If $n \equiv 0 (\text{mod } 4)$, then S has $n-2$ vertices. S along with the vertex $T_n(n-4, 2n-7)$ dominates D_n with $n-1$ total vertices. If $n \equiv 1 (\text{mod } 4)$, then S has $n-2$ vertices. S along with the vertex $T_n(n, 2n-1)$ dominates D_n with $n-1$ total vertices. In all four cases, $\gamma(D_n K) \leq n-1$.

Using the fact that $\gamma(D_n K) \leq n-1$, T_n can be dominated by dominating its subgraphs $D_n K, D_{n-5} K, D_{n-2\cdot 5} K$, and so on.

If $n \equiv 0 (\text{mod } 5)$, then $\gamma(T_n K) \leq \gamma(D_n K) + \gamma(D_{n-5} K) + \gamma(D_{n-2\cdot 5} K) + \cdots + \gamma(D_5 K) \leq (n-1) + (n-5-1) + (n-2\cdot 5-1) + \cdots + (5-1) = \frac{n^2+3n}{10}$. And thus $\gamma(T_n K) \leq \frac{n^2+3}{10} = \lceil \frac{n^2+3n}{10} \rceil$.

If $n \equiv 1 (\text{mod } 5)$, then $\gamma(T_n K) \leq \gamma(D_n K) + \gamma(D_{n-5} K) + \gamma(D_{n-2\cdot 5} K) + \cdots + \gamma(D_6 K) + \gamma(T_1 K) \leq (n-1) + (n-5-1) + (n-2\cdot 5-1) + \cdots + (6-1) + 1$

$= \frac{n^2+3n+6}{10}$. Note here that the fact $\gamma(T_1K) = 1$ is used, and thus $\gamma(T_nK) \leq \frac{n^2+3+6}{10} = \lceil\frac{n^2+3n}{10}\rceil$.

If $n \equiv 2(\text{mod } 5)$, then $\gamma(T_nK) \leq \gamma(D_nK) + \gamma(D_{n-5}K) + \gamma(D_{n-2\cdot5}K) + \cdots + \gamma(D_7K) + \gamma(T_2K) \leq (n-1) + (n-5-1) + (n-2\cdot5-1) + \cdots + (7-1) + 1 = \frac{n^2+3n}{10}$. Here the fact $\gamma(T_2K) = 1$ is used, and therefore $\gamma(T_nK) \leq \frac{n^2+3}{10} = \lceil\frac{n^2+3n}{10}\rceil$.

If $n \equiv 3(\text{mod } 5)$, then $\gamma(T_nK) \leq \gamma(D_nK) + \gamma(D_{n-5}K) + \gamma(D_{n-2\cdot5}K) + \cdots + \gamma(D_8K) + \gamma(T_3K) \leq (n-1) + (n-5-1) + (n-2\cdot5-1) + \cdots + (8-1) + 2 = \frac{n^2+3n+2}{10}$. Here the fact $\gamma(T_3K) = 2$ is used, and therefore $\gamma(T_nK) \leq \frac{n^2+3+2}{10} = \lceil\frac{n^2+3n}{10}\rceil$.

If $n \equiv 4(\text{mod } 5)$, then $\gamma(T_nK) \leq \gamma(D_nK) + \gamma(D_{n-5}K) + \gamma(D_{n-2\cdot5}K) + \cdots + \gamma(D_9K) + \gamma(T_4K) \leq (n-1) + (n-5-1) + (n-2\cdot5-1) + \cdots + (9-1) + 3 = \frac{n^2+3n+2}{10}$. Here the fact $\gamma(T_4K) = 3$ is used, and therefore $\gamma(T_nK) \leq \frac{n^2+3+2}{10} = \lceil\frac{n^2+3n}{10}\rceil$. □

Knight

Theorem 6 $\lceil\frac{n^2}{10}\rceil \leq \gamma(T_nN) \leq \frac{n^2}{4}$ *if* $n \equiv 0(mod\ 4)$.
$\lceil\frac{n^2}{10}\rceil \leq \gamma(T_nN) \leq \frac{n^2+7}{4}$ *if* $n \equiv 1(mod\ 4)$.
$\lceil\frac{n^2}{10}\rceil \leq \gamma(T_nN) \leq \frac{n^2+n+2}{4}$ *if* $n \equiv 2(mod\ 4)$.
$\lceil\frac{n^2}{10}\rceil \leq \gamma(T_nN) \leq \frac{n^2+2n+1}{4}$ *if* $n \equiv 3(mod\ 4)$.

Proof The maximum number of spaces on Knight, N, can dominate is 10, so this fact accounts for the lower bound of $\lceil\frac{n^2}{10}\rceil$. The upper bound for $n \equiv 0(\text{mod } 4)$ relies on the fact that $\gamma(T_4N) = 4$ with minimum dominating $\{T_4(3, 2), T_4(3, 3), T_4(3, 4), T_4(4, 4)\}$ and that a board of size $n \equiv 0(\text{mod } 4)$ can be tiled with $(\frac{n}{4})^2$ copies of T_4N. For the other three cases, the upper bounds are achieved by using this same tiling and then dominating the last one, two, or three rows.

If $n \equiv 1(\text{mod } 4)$, then after a tiling of the first $n-1$ rows of T_nN with $(\frac{n-1}{4})^2$ copies of T_4N, the set $\{T_n(n, j) : j \equiv 1(\text{mod } 8)\} \cup \{T_n(n-2, j) : j \equiv 7(\text{mod } 8)\} \cup \{T_n(n, 2), T_n(n, 2n-2)\}$ of size $\frac{n+3}{2}$ finishes dominating the nth row. The total size of this dominating set is then $\frac{(n-1)^2}{4} + \frac{n+3}{2} = \frac{n^2+7}{4}$.

If $n \equiv 2(\text{mod } 4)$, then after a tiling of the first $n-2$ rows of T_nN with $(\frac{n-2}{4})^2$ copies of T_4N, the set $\{T_n(n, j) : j \equiv 2, 6(\text{mod } 8)\} \cup \{T_n(n-1, j) : j \equiv 0, 1, 2 (\text{mod } 8)\}$ of size $\frac{5n-2}{4}$ finishes dominating the last two rows. The total size of this dominating set is then $\frac{(n-2)^2}{4} + \frac{5n-2}{4} = \frac{n^2+n+2}{4}$.

If $n \equiv 3(\text{mod } 4)$, then after a tiling of the first $n-3$ rows of T_nN with $(\frac{n-3}{4})^2$ copies of T_4N, the set $S = \{T_n(n, j) : j \equiv 4(\text{mod } 8)\} \cup \{T_n(n-1, j) : j \equiv 2, 3, 4 (\text{mod } 8)\} \cup \{T_n(n-2, j) : j \equiv 5, 6, 7(\text{mod } 8)\} \cup \{T_n(n-1, 2n-5), T_n(n-1, 2n-6)\} - \{T_n(n, 2n-2), T_n(n-1, 2n-3)\}$ of size $2n-2$ finishes dominating the last three rows. The total size of this dominating set is then $\frac{(n-3)^2}{4} + (2n-2) = \frac{n^2+2n+1}{4}$. It is worth mentioning that the set S, while always dominating, is not always the dominating set of smallest size. One instance to note is when

$n \equiv 7, 11 (\text{mod } 28)$ the set $S_0 = \{T_n(n, j) : j \equiv 4 (\text{mod } 14)\} \cup \{T_n(n-1, j) : j \equiv 2, 3, 4, 9, 10, 11 (\text{mod } 14)\} \cup \{T_n(n-2, j) : j \equiv 9 (\text{mod } 14)\}$ dominates the last three rows of T_n. And, for example, in the case where $n \equiv 11(\text{mod } 28)$, $|S_0| = \frac{8n-4}{7}$ implying that $\gamma(T_n N) \leq \frac{(n-3)^2}{4} + \frac{8n-4}{7} = \frac{7n^2-10n+47}{28}$. This upper bound is much less than $\frac{n^2+2n+1}{4}$ achieved by using S to dominate the last three rows. □

3 Triangular Boards with Hexagonal Spaces

In Fig. 5 we define the moves of six pieces on the triangular board with hexagonal spaces, H_n. As in Sect. 2, there are six pieces defined; the King denoted K, the Queen denoted Q, the Rook denoted R, the Bishop denoted B, the Knight denoted N, and the Grid denoted G. As with T_n, many results were found concerning the independence numbers associated with these six pieces on H_n. But, as before, a more complete list of independence results was found in [2]. These results are summarized here.

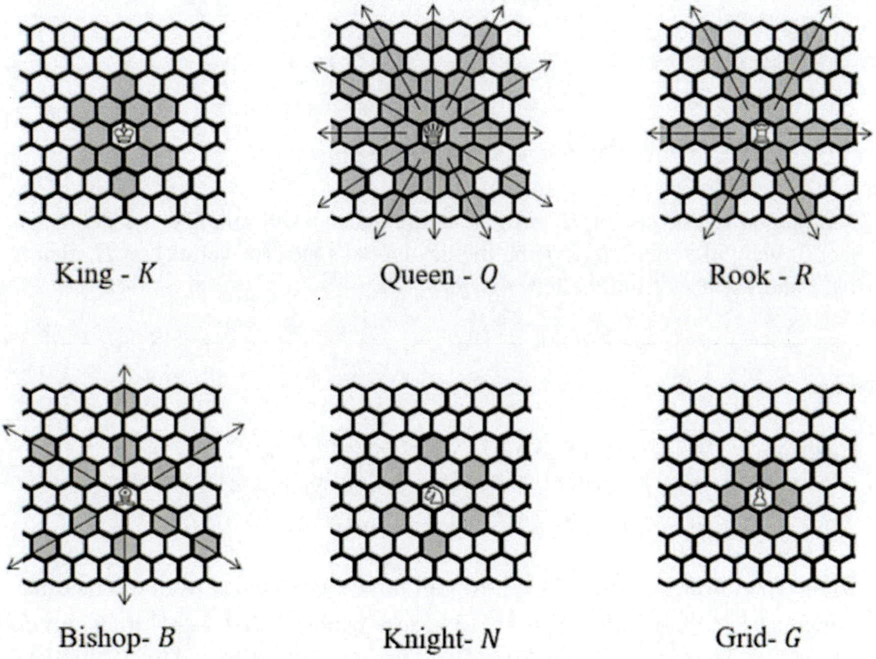

Fig. 5 The six pieces and their moves defined on a board with hexagonal spaces

3.1 Independence on H_n

Piece P	$\beta(H_n P)$	Conditions
K	$\lfloor \frac{n+2}{2} \rfloor \lfloor \frac{n+3}{2} \rfloor / 2$	
Q	$\beta(H_n Q) \leq \beta(H_n R) = \lfloor \frac{2n+1}{3} \rfloor$	
R	$\lfloor \frac{2n+1}{3} \rfloor$	
B	$2n-3$, $2n-6$, or $2n-9$	$n \equiv 0(mod 3)$
B	$2n-i$ for $3 \leq i \leq 9$	$n \equiv 1(mod 3)$
B	$2n-1$, $2n-4$, or $2n-7$	$n \equiv 2(mod 3)$
N	$\lfloor \frac{n^2+3n+2}{6} \rfloor$	$n \equiv 0, 1(mod 3)$
N	$\frac{n^2+5n+4}{6}$	$n \equiv 2(mod 3)$
G	$\lfloor \frac{n(n+1)+4}{6} \rfloor$	$n \neq 3, 5$
G	3 and 6	$n = 3$ and 5 respectively

Note that there is still work that could be done to complete this list. The Queen's independence number, for example, only has the rather trivial restriction that it is less than the Rook's. Harborth et al. in [2] noted that for $1 \leq n \leq 31$, $\beta(H_n Q) = \beta(H_n R)$ or $\beta(H_n R) - 1$.

3.2 Domination on H_n

Domination questions on H_n will be addressed next. Similar to the definition in Sect. 3, we will define $H_n(i, j)$ be the hexagonal space (or vertex) on H_n that is in row i and j spaces from the left.

King

Theorem 7 $\lceil \frac{n(n+1)}{26} \rceil \leq \gamma(H_n K) \leq \frac{n(n+3)}{18}$ *if* $n \equiv 0(mod\ 3)$,
$\lceil \frac{n(n+1)}{26} \rceil \leq \gamma(H_n K) \leq \frac{(n-1)(n+2)}{18}$ *if* $n \equiv 1(mod\ 3)$, $n \geq 4$,
$\lceil \frac{n(n+1)}{26} \rceil \leq \gamma(H_n K) \leq \frac{(n+1)(n+4)}{18}$ *if* $n \equiv 2(mod\ 3)$.

Proof The lower bound of $\lceil \frac{n(n+1)}{26} \rceil$ in all three cases comes from the fact that an individual King dominates at most 13 spaces while H_n has $\frac{n(n+1)}{2}$ total spaces. If $n \equiv 1(\text{mod } 3)$ and $n \geq 4$ then $S = \{H_n(i, j) : i \equiv 0(\text{mod } 3)$ and $j \equiv 2(\text{mod } 3)\}$ is a dominating set of size $\frac{(n-1)(n+2)}{18}$. If $n \equiv 0(\text{mod } 3)$, then S is a dominating set of size $\frac{(n)(n+3)}{18}$. And Finally, if $n \equiv 2(\text{mod } 3)$, then $S \cup \{H_n(n, j) : j \equiv 2(\text{mod } 3)\}$ is a dominating set of size $\frac{(n+1)(n+4)}{18}$. □

Rook

Theorem 8 $\lceil \frac{n(n+1)}{4n-2} \rceil \leq \gamma(H_n R) \leq \lceil \frac{n}{2} \rceil$.

Proof Any one Rook dominates exactly $2n - 1$ of the $\frac{n(n+1)}{2}$ spaces of H_n, thus giving the lower bound. $S = \{H_n(i, j) : i \equiv 1 (\mathrm{mod}\ 2)$ and $j = \frac{i+1}{2}\}$ is a dominating set of size $\lceil \frac{n}{2} \rceil$ giving the upper bound. □

Queen

Theorem 9 $\lceil \frac{n(n+1)}{7n-8} \rceil \leq \gamma(H_n Q) \leq \lceil \frac{n}{2} \rceil$.

Proof If n is odd, then a Queen dominates at most $\frac{n(n+1)}{7n-9}$ spaces of H_n. If n is even, then the maximum number is $\frac{n(n+1)}{7n-8}$. Since $\frac{n(n+1)}{7n-8}$ is the lesser number, this gives the lower bound. Since the Rook's moves are a subset of the Queens, trivially $\gamma(H_n Q) \leq \gamma(H_n R)$ giving the upper bound. □

Bishop

Theorem 10 $\lceil \frac{n(n+1)}{3n-4} \rceil \leq \gamma(H_n B) \leq 2n - 9$ *for* $n \geq 6$, $\gamma(H_1 B) = 1$, *and* $\gamma(H_k (B)) = 3$ *for* $2 \leq k < 6$.

Proof If $n > 1$ is odd, then a Bishop dominates at most $\frac{3n-5}{2}$ spaces of H_n. If n is even, then the maximum number is $\frac{3n-4}{2}$. These facts give the lower bound. If $n \geq 6$ then $S = \{H_n(n-1, j) : 3 \leq j \leq n-3\} \cup \{H_n(n-2, j) : 2 \leq j \leq n-3\}$ is a dominating set of size $2n - 9$ giving the upper bound. If $n = 6$, then $|S| = 3$. Sets of size 3 similar to this are minimum dominating sets for $H_k(B)$ for $2 \leq k < 6$. □

Knight

Theorem 11 $\lceil \frac{n(n+1)}{14} \rceil \leq \gamma(H_n N) \leq \frac{2n^2+5n}{25}$ *if* $n \equiv 0 (mod\ 5)$,
$\lceil \frac{n(n+1)}{14} \rceil \leq \gamma(H_n N) \leq \frac{2n^2+11n+12}{25}$ *if* $n \equiv 1 (mod\ 5), n > 5$,
$\lceil \frac{n(n+1)}{14} \rceil \leq \gamma(H_n N) \leq \frac{2n^2+17n+8}{25}$ *if* $n \equiv 2 (mod\ 5), n > 5$,
$\lceil \frac{n(n+1)}{14} \rceil \leq \gamma(H_n N) \leq \frac{2n^2+13n+18}{25}$ *if* $n \equiv 3 (mod\ 5), n > 5$,
$\lceil \frac{n(n+1)}{14} \rceil \leq \gamma(H_n N) \leq \frac{2n^2+9n-93}{25}$ *if* $n \equiv 4 (mod\ 5), n > 5$,
$\gamma(H_1 N) = 1$, *and* $\gamma(H_2 N) = \gamma(H_3 N) = \gamma(H_4 N) = 3$.

Proof A Knight dominates at most seven spaces of H_n giving the lower bounds in all five cases.

If $n \equiv 0 (\mathrm{mod}\ 5)$, then $S = \{H_n(i, j) : i \equiv 2 (\mathrm{mod}\ 5), j \equiv 4 (\mathrm{mod}\ 5)\} \cup \{H_n(i, j) : i \equiv 3 (\mathrm{mod}\ 5), j \equiv 2 (\mathrm{mod}\ 5)\} \cup \{H_n(i, j) : i \equiv 4 (\mathrm{mod}\ 5), j \equiv 2, 3 (\mathrm{mod}\ 5)\}$ is a dominating set of size $\frac{2n^2+5n}{25}$. If $n = 5$, this set of size 3 is minimum dominating, and similar sets of size three dominate H_2N, H_3N, and H_4N.

If $n \equiv 1 (\text{mod } 5)$, then $S \cup \{H_n(n-1, j) : j \equiv 3(\text{mod } 5)\} \cup \{H_n(n, j) : j \equiv 1 (\text{mod } 5)\}$ is a dominating set of size $\frac{2n^2+11n+12}{25}$.

If $n \equiv 2(\text{mod } 5)$, then $S \cup \{H_n(n-1, j) : j \equiv 2(\text{mod } 5)\} \cup \{H_n(n, j) : j \equiv 0, 2, 3(\text{mod } 5)\} \cup \{H_n(n, n-1)\}$ is a dominating set of size $\frac{2n^2+17n+8}{25}$.

If $n \equiv 3(\text{mod } 5)$, then $S \cup \{H_n(n-1, j) : j \equiv 1, 2, 4(\text{mod } 5)\} \cup \{H_n(n, j) : j \equiv 2(\text{mod } 5)\}$ is a dominating set of size $\frac{2n^2+13n+18}{25}$.

If $n \equiv 4(\text{mod } 5)$, then S is a dominating set of size $\frac{2n^2+9n-93}{25}$. □

Grid

Theorem 12 $\lceil \frac{n(n+1)}{14} \rceil \leq \gamma(H_nG) \leq \frac{n^2+5n}{14}$ *if* $n \equiv 0(\text{mod } 7)$,

$\lceil \frac{n(n+1)}{14} \rceil \leq \gamma(H_nG) \leq \lfloor \frac{n^2+7n}{14} \rfloor$ *if* $n \equiv 1, 2, 3, 4, 6(\text{mod } 7), n > 7$,

$\lceil \frac{n(n+1)}{14} \rceil \leq \gamma(H_nG) \leq \lfloor \frac{n^2+7n}{14} \rfloor + 1$ *if* $n \equiv 5(\text{mod } 7), n > 7$,

$\gamma(H_1G) = \gamma(H_2G) = 1$, $\gamma(H_3G) = 2$, $\gamma(H_4G) = \gamma(H_5G) = 3$, *and* $\gamma(H_6G) = 5$.

Proof A Grid dominates at most seven spaces of H_n giving the lower bounds in all five cases.

Define the set $S = \{H_n(i, j) : i \equiv 1(\text{ mod } 7) \text{ and } j \equiv 3(\text{mod } 7)\} \cup \{H_n(i, j) : i \equiv 2(\text{ mod } 7) \text{ and } j \equiv 1(\text{mod } 7)\} \cup \{H_n(i, j) : i \equiv 3(\text{ mod } 7) \text{ and } j \equiv 6(\text{mod } 7)\} \cup \{H_n(i, j) : i \equiv 4(\text{ mod } 7) \text{ and } j \equiv 4(\text{mod } 7)\} \cup \{H_n(i, j) : i \equiv 5(\text{ mod } 7) \text{ and } j \equiv 2(\text{mod } 7)\} \cup \{H_n(i, j) : i \equiv 6(\text{ mod } 7) \text{ and } j \equiv 0(\text{mod } 7)\} \cup \{H_n(i, j) : i \equiv 0(\text{mod } 7) \text{ and } j \equiv 5(\text{mod } 7)\} \cup \{H_n(i, 1), H_n(i, i) : i \equiv 0(\text{mod } 7)\}$. S uses one space out of every seven in each row except for those rows which are $\equiv 0(\text{mod } 7)$. In these rows which are multiples of seven, S uses the two additional spaces that are the leftmost and the rightmost. This seems like an efficient construction of S since a single Grid dominates at most seven spaces. A picture of S for $H_{14}G$ is in Figure 6.

If $n \equiv 0(\text{mod } 7)$, then S is a dominating set of size $\frac{n^2+5n}{14}$.

If $n \equiv 1(\text{mod } 7)$, then $S \cup \{H_n(n, j) : j \equiv 0(\text{mod } 7) \text{ and } j \neq n-1\}$ is a dominating set of size $\frac{n^2+7n-9}{14} = \lfloor \frac{n^2+7n}{14} \rfloor$.

If $n \equiv 2(\text{mod } 7)$, then $S \cup \{H_n(n, j) : j \equiv 5(\text{mod } 7)\}$ is a dominating set of size $\frac{n^2+7n-4}{14} = \lfloor \frac{n^2+7n}{14} \rfloor$.

If $n \equiv 3(\text{mod } 7)$, then $S \cup \{H_n(n, j) : j \equiv 4(\text{mod } 7)\} \cup \{H_n(n, n)\}$ is a dominating set of size $\frac{n^2+7n-2}{14} = \lfloor \frac{n^2+7n}{14} \rfloor$.

If $n \equiv 4(\text{mod } 7)$, then $S \cup \{H_n(n, j) : j \equiv 1(\text{mod } 7)\}$ is a dominating set of size $\frac{n^2+7n-2}{14} = \lfloor \frac{n^2+7n}{14} \rfloor$.

If $n \equiv 5(\text{mod } 7)$, then $S \cup \{H_n(n, j) : j \equiv 6(\text{mod } 7)\}$ is a dominating set of size $\frac{n^2+7n-18}{14} = \lfloor \frac{n^2+7n}{14} \rfloor + 1$.

If $n \equiv 6(\text{mod } 7)$, then $S \cup \{H_n(n, j) : j \equiv 4(\text{mod } 7)\} \cup \{H_n(n, 1), H_n(n, n-1)\}$ is a dominating set of size $\frac{n^2+7n-8}{14} = \lfloor \frac{n^2+7n}{14} \rfloor$.

If $1 \leq n \leq 6$, then finding $\gamma(H_nG)$ is a simple exercise. □

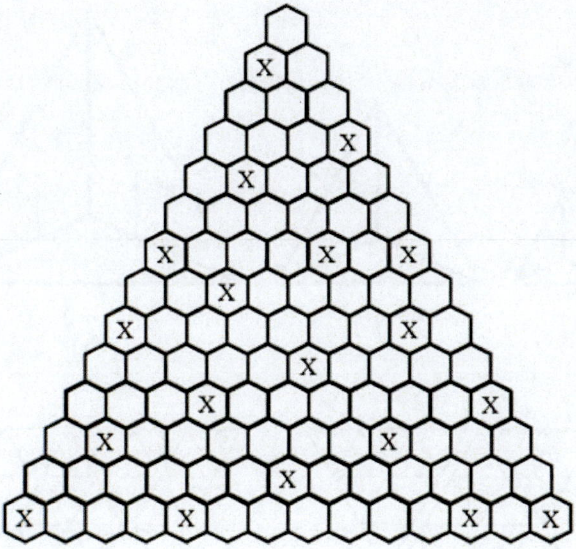

Fig. 6 S defined on $H_{14}G$

4 Triangular Boards on the Surface of a Tetrahedron

Questions of independence and domination can also be asked on 3-dimensional triangular boards that tile the surface of a tetrahedron.

4.1 Independence and Domination on T^n

The board T^n is defined to be the four-sided tetrahedron where each of the sides is tiled with one copy of T_n. See Fig. 7 for the boards T^1, T^2, and T^3.

Since T_n has n^2 spaces, T^n has $4n^2$ spaces. Rather than defining the moves of all six chess pieces on these new boards, this paper will focus only on one piece, the Grid, and leave the others for future work.

Grid

Just as in the two-dimensional case, the Grid on T^n can move to any space for which it shares a boarder of more than one point. It is often easier to think of T^n as a two-dimensional map as in Fig. 8. It shows the moves of Grids on T^3 in two different situations.

Define $T^n(i, j)$ to be the triangular space (or vertex) on T^n that is in row i and j spaces from the left using the two-dimensional map of T^n show in Fig. 8. For

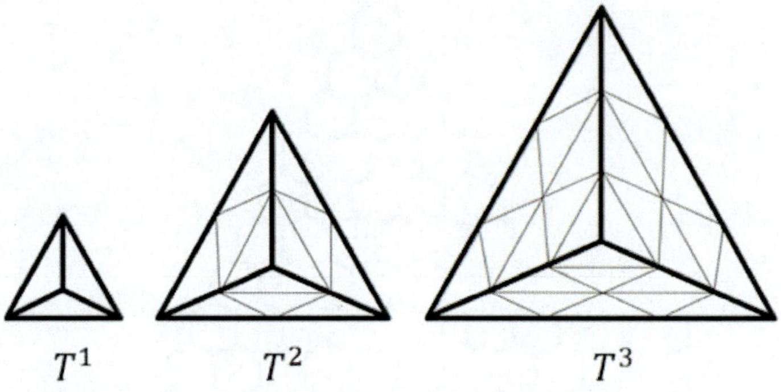

Fig. 7 T^n for $n = 1, 2, 3$

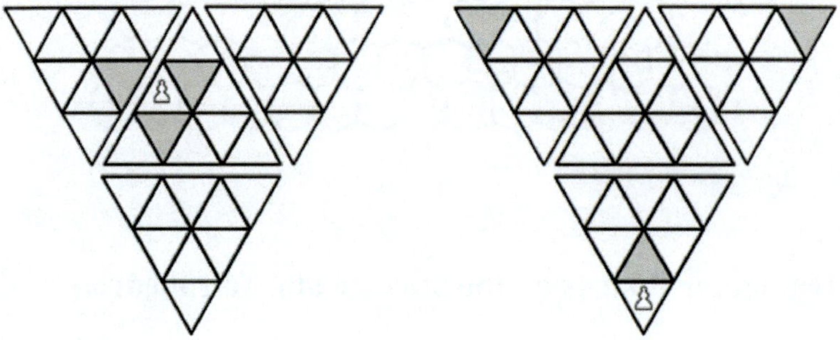

Fig. 8 The Grid G on T^3

example, the Grid on the left board is in space $T^3(2, 4)$ while the Grid on the right board is in space $T^3(6, 1)$.

Theorem 13 *$2n^2 - n \leq \beta(T^n G) \leq 2n^2$, and $\gamma(T^n G) = n^2$.*

Proof $\beta(T_{2n}G) \geq \beta(T^n G)$ because $T_{2n}G$ can be thought of as a subgraph of $T^n G$ on the same vertex set with fewer edges. For the same reason, $\gamma(T_{2n}G) \geq \gamma(T^n G)$. From [1], therefore $\beta(T^n G) \leq \beta(T_{2n}G) = \binom{2n+1}{2} = 2n^2 + n$. However, T^n can be decomposed into $2n^2$ disjoint copies of two adjacent triangular spaces, each which has an independence number of one. Therefore, $\beta(T^n G) \leq \frac{4n^2}{2} = 2n^2$. The set $I = \{T^n(i, j) : j \equiv 0 (\bmod 2)\}$ is an independent set of size $2n^2 - n$, so $\beta(T^n G) \geq 2n^2 - n$.

In Section 2, it was shown that if n is even, then $\gamma(T_n) = \frac{n^2}{4}$. Therefore $\gamma(T^n G) \leq \gamma(T_{2n}G) = \frac{(2n)^2}{4} = n^2$. However, it is still the case that a single Grid dominates at most 4 spaces. Since T^n has $4n^2$ total spaces, $\gamma(T^n G) \geq \frac{4n^2}{4} = n^2$. Therefore $\gamma(T^n G) = n^2$. □

4.2 Independence and Domination on H^n

The board H^n is defined to be the four-sided tetrahedron where each of the sides is tiled with a single copy of H_n. H^n would then have $4\binom{n+1}{2} = 2(n+1)(n+2)$ hexagonal spaces. Rather than defining all six chess pieces on H^n, this paper will consider only the Rook, R, and leave others for future work.

Rook

In the two-dimensional H_n, the Rook is able to travel in either direction along three straight lines starting with a space with which it shared an edge. For the three-dimensional H^n, this same rule applies recognizing that "lines" on the two-dimensional board will equate to "latitudinal lines" on H^n. Figure 9 shows the possible moves for a Rook on a map of H^6. The three different latitudinal lines are differentiated in this figure.

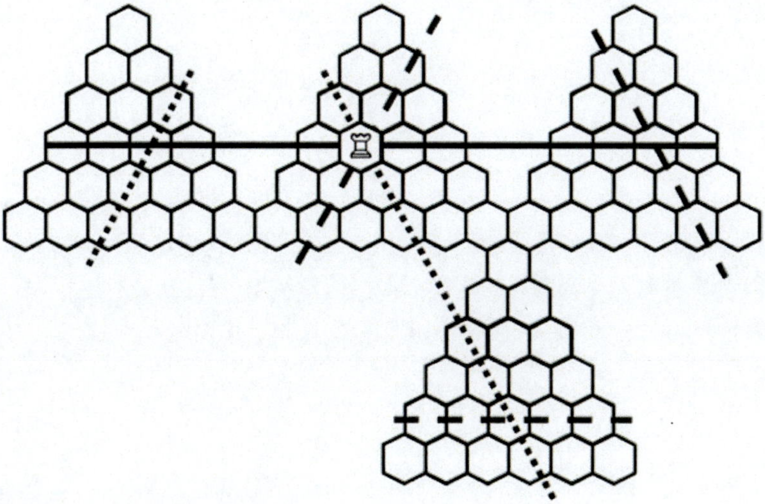

Fig. 9 The Rook, R, on the map of H^6

Theorem 14 $\lfloor \frac{2n+1}{3} \rfloor \leq \beta(H^n R)$*, and* $\gamma(H^n R) \leq n$.

Proof In [3], another proof attributed to Harborth is given to show that $\beta(H_n R) = \lfloor \frac{2n+1}{3} \rfloor$. The maximal independent sets given on H_n are still independent in H^n when placed on a single copy of H_n on one of the sides of the tetrahedron. Therefore $\beta(H^n) \geq \lfloor \frac{2n+1}{3} \rfloor$.

It is clear that if a Rook were placed in each of the n spaces along a single edge of H^n, then this would form a dominating set. Therefore $\gamma(H^n) \leq n$. This bound is not sharp. For example it true that $\gamma(H^3 R) = 2$, and $\gamma(H^4 R) = 3$. It is also known that $\gamma(H^1 R) = 1$ and $\gamma(H^2 R) = 2$. □

References

1. Heiko Dietrich and Heiko Harborth, Independence on Triangular Triangle Boards, *Abhandlungen der Braunschweigischen Wissenschaftlichen Gesellschaft* **54** (2004), 73-84.
2. Heiko Harborth and Vincent Kultan, Independence on Triangular Hexagon Boards, *Congressus Numerantium* **160** (2003), 215-222.
3. Joseph D. E. Konhauser, Dan Velleman, Stan Wagon: *Which Way Did the Bicycle Go?* Mathematical Association of America (MAA), 1996.

Efficient and Non-efficient Domination of $\mathbb{Z}$-stacked Archimedean Lattices

Lyle Paskowitz, Nathan Vallapureddy, and John Wierman

Abstract On a graph, a vertex v dominates vertex v' if $v = v'$ or v is adjacent to v'. A graph has an efficient dominating set if there exists a subset of vertices D such that every vertex in the graph is dominated by exactly one vertex in D. We investigate efficient domination on the stacked versions of each of the eleven Archimedean Lattices, and determine the existence or non-existence of efficient dominating sets on each lattice through integer programming. The proofs of existence are constructive, and the proofs of non-existence are generated by integer programs. We find efficient dominating sets on seven of the stacked lattices and prove that no such sets exist on the other four stacked lattices.

Keywords Efficient domination · Archimedean lattices · Integer programming

MSC Classification: 05C69, 05B35, 90C10

1 Introduction

1.1 Efficient Domination

Consider a simple undirected graph $G = (V_G, E_G)$, where V_G is the set of vertices and E_G the set of undirected edges on V_G. We define the closed neighborhood $N : V_G \to 2^{V_G}$ as $N[v] = \{v' : (v, v') \in E_G\} \cup \{v\}$, that is, the set of vertices either adjacent to v or v itself. A vertex v dominates vertex v' if and only if $v' \in N[v]$. Note of course that on a simple undirected graph, this relation is symmetric, so $v' \in N[v]$ if and only if $v \in N[v']$.

An *efficient dominating set* $D \subseteq V_G$ of G is a set such that $|N[v] \cap D| = 1$ for all $v \in V_G$, that is, every vertex in V_G is dominated by exactly one vertex in D. More

L. Paskowitz · N. Vallapureddy · J. Wierman (✉)
Department of Applied Mathematics and Statistics, Johns Hopkins University, Baltimore, MD 21218, USA
e-mail: jwierma1@jhu.edu

F. Hoffman (ed.), *Combinatorics, Graph Theory and Computing*,
Springer Proceedings in Mathematics & Statistics 388,
https://doi.org/10.1007/978-3-031-05375-7_15

generally, D is a *dominating set* of G if every vertex in V_G is dominated by at least one vertex in D.

1.2 $\mathbb{Z}$-stacked Archimedean Lattices

We consider efficient domination on a set of graphs related to the *Archimedean Lattices*. An Archimedean lattice is a two-dimensional lattice that is vertex-transitive and has regular polygons as faces. There are eleven such lattices, which include the recognizable square, triangular, hexagonal, and bathroom tile tessellations of the plane. In the typical naming convention, the numbers of edges of the polygons incident to a vertex are listed in the order they appear around the vertex, with exponents indicating the number of successive polygons of a given size. In this convention, the four previously mentioned lattices are denoted as the (4^4), (3^6), (6^3), and $(4, 8^2)$ lattices, respectively. Refer to Grünbaum and Shephard [4] for a discussion on the lattices.

Marge et al. [6] found efficient dominating sets on seven of the Archimedean lattices and proved that no efficient dominating sets can exist on the other four. Inspired by them, we consider efficient domination on the $\mathbb{Z}$*-stacked Archimedean lattices* [7].

A $\mathbb{Z}$-stacked Archimedean lattice is constructed as follows: take an embedding of the lattice L in the plane $z = 0$. Construct a copy of L in each plane $k \in \mathbb{Z}$, such that for each vertex $(i, j, 0)$ and every edge $\{(i, j, 0), (i', j', 0)\}$ in the original embedding, there is a corresponding vertex (i, j, k) and edge $\{(i, j, k), (i', j', k)\}$. Lastly, for each vertex (i, j, k) and layer $k \in \mathbb{Z}$, add a vertical edge $\{(i, j, k), (i, j, k + 1)\}$.

1.3 Domination Ratio and Periodic Graph

Define a *periodic graph* G as a locally-finite connected simple graph with a countably-infinite vertex set, which can be embedded in $\mathbb{R}^d$ for some $d < \infty$ such that G is invariant under translation by the unit vector in each coordinate axis direction in $\mathbb{R}^d$ and each compact set of $\mathbb{R}^d$ intersects only finitely many vertices of G. Note that it is actually the embedding which is periodic. Each of the eleven Archimedean lattices is a periodic graph in $\mathbb{R}^2$. Suding and Ziff [8] provide figures showing periodic embeddings, which we will call grid representations, of the Archimedean lattices. The stacked Archimedean lattices are periodic graphs in $\mathbb{R}^3$. Figures of the Archimedean lattices in both the original and grid representations are shown throughout this article.

For a periodic embedding in three dimensions of a periodic graph G, denote the subgraph of G induced by the vertices in the rectangle $[i_1, i_2) \times [j_1, j_2) \times [k_1, k_2) \subset \mathbb{R}^3$ by $R_G(i_1, i_2; j_1, j_2; k_1, k_2)$, where $i_1 < i_2$, $j_1 < j_2$, $k_1 < k_2$, and $i_1, i_2, j_1, j_2, k_1,$

$k_2 \in \mathbb{Z}$. We will refer to $R_G(i_1, i_2; j_1, j_2; k_1, k_2)$ as an $(i_2 - i_1) \times (j_2 - j_1) \times (k_2 - k_1)$ *block* of G.

Denote the minimum size of a dominating set for an $i \times j \times k$ block of G, known as its *domination number*, by $\gamma_{i,j,k}(G)$, and its number of vertices by $N_{i,j,k}(G)$. The *domination ratio* of G is defined by

$$\lim_{i,j,k\to\infty} \frac{\gamma_{i,j,k}(G)}{N_{i,j,k}(G)} = \inf_{i,j,k} \frac{\gamma_{i,j,k}(G)}{N_{i,j,k}(G)}.$$

For two-dimensional periodic graphs, Zhao [6] proved that the corresponding limit exists and is equal to the infimum, relying on *subadditivity* of the function $\gamma_{i,j,k}(G)$, and proved that it does not depend on the choice of periodic embedding of the periodic graph G. The proof is easily generalized to three or more dimensions.

1.4 Overview of Results

For each $\mathbb{Z}$-stacked Archimedean lattice, a construction of the efficient dominating set is shown in Sect. 3 if it exists. Archimedean lattices without efficient dominating sets are discussed in Sect. 4 and listed in the table below.

Efficiently dominated	Not efficiently dominated
$(4^4) \times \mathbb{Z}, \mathbb{Z}^n$	$(3, 12^2) \times \mathbb{Z}$
$(3^6) \times \mathbb{Z}$	$(3, 4, 6, 4) \times \mathbb{Z}$
$(6^3) \times \mathbb{Z}$	$(3^4, 6) \times \mathbb{Z}$
$(4, 8^2) \times \mathbb{Z}$	$(3, 6, 3, 6) \times \mathbb{Z}$
$(4, 6, 12) \times \mathbb{Z}$	
$(3^2, 4, 3, 4) \times \mathbb{Z}$	
$(3^3, 4^2) \times \mathbb{Z}$	

2 Proving Efficient Domination

2.1 Simplification of Criteria for Efficient Domination

First, we introduce a technique to simplify the proof of efficient domination, which we will use extensively later.

Let G be a d_G-regular vertex-transitive graph. Note that if there exists D efficiently dominating G, then the domination ratio $\frac{|D|}{|V_G|}$ must be $\frac{1}{d_G+1}$ since each vertex in D dominates exactly $d_G + 1$ vertices.

Proposition 1 *Let G be a finite vertex-transitive graph with degree d_G. For any $D \subseteq V_G$, consider the following three criteria:*

1. $N[v] \cap D \geq 1$ *for all* $v \in V_G$
2. $N[v] \cap D \leq 1$ *for all* $v \in V_G$
3. $\frac{|D|}{|V_G|} = \frac{1}{d_G+1}$

If any two of these criteria are satisfied, the third must also be satisfied.

We prove the Proposition in the following three lemmas. We discuss the application of the Proposition to blocks in the Archimedean lattices at the beginning of Sect. 3.

Lemma 1 *If criteria **1** and **2** are satisfied, then criterion **3** is also satisfied.*

Proof We have $N[v] \cap D = 1$ for all $v \in V_G$, so D is an efficient dominating set. Therefore, we must have $\frac{|D|}{|V_G|} = \frac{1}{d_G+1}$ as discussed earlier.

Lemma 2 *If criteria **1** and **3** are satisfied, then criterion **2** is also satisfied.*

Proof Let $\mathbf{1}$ be an indicator function, i.e. $\mathbf{1}_x = 1$ if x is true and $\mathbf{1}_x = 0$ otherwise. Since $\frac{|D|}{|V_G|} = \frac{1}{d_G+1}$, we have $|V_G| = |D|(d_G + 1)$. Now, suppose that there were some $v \in V_G$ such that $|N[v] \cap D| > 1$. Then

$$|V_G| = \sum_{v \in V_G} 1 < \sum_{v \in V_G} |N[v] \cap D| = \sum_{v \in V_G} \sum_{d \in D} \mathbf{1}_{d \in N[v]}$$
$$= \sum_{d \in D} \sum_{v \in V_G} \mathbf{1}_{v \in N[d]} = \sum_{d \in D} |N[d] \cap V_G| = \sum_{d \in D} d_G + 1 = |D|(d_G + 1)$$

forming a contradiction. Therefore $N[v] \cap D = 1$ for all $v \in V_G$. □

Lemma 3 *If criteria **2** and **3** are satisfied, then criterion **1** is also satisfied.*

Proof We again have $|V_G| = |D|(d_G + 1)$. Now, suppose that there were some $v \in V_G$ such that $|N[v] \cap D| < 1$. However,

$$|V_G| = \sum_{v \in V_G} 1 > \sum_{v \in V_G} |N[v] \cap D| = \sum_{v \in V_G} \sum_{d \in D} \mathbf{1}_{d \in N[v]}$$
$$= \sum_{d \in D} \sum_{v \in V_G} \mathbf{1}_{v \in N[d]} = \sum_{d \in D} |N[d] \cap V_G| = \sum_{d \in D} d_G + 1 = |D|(d_G + 1)$$

which is a contradiction. Therefore $N[v] \cap D = 1$ for all $v \in V_G$. □

In each case, we ended up with $|N[v] \cap D| = 1$ for all $v \in V_G$, which is exactly the criteria for efficient domination. So one method of proving efficient domination is to prove two of the above criteria.

2.2 Additional Conditions

Consider criterion 2 above, that is, $|N[v] \cap D| \leq 1$ for all $v \in V_G$. Define the distance metric $d : V_G \times V_G \rightarrow \mathbb{Z}$ by letting $d(v_1, v_2)$ be the minimum length of a path connecting vertex v_1 to v_2. Since we focus only on undirected graphs, this is also the minimum number of edges needed in a path connecting v_2 to v_1. Then we have the following lemma:

Lemma 4

$$\max_{v \in V_G} |N[v] \cap D| \leq 1 \iff \min_{v_i \neq v_j \in D} d(v_i, v_j) \geq 3.$$

Proof First, consider the forward direction, which we prove by contrapositive. Let's suppose that

$$\min_{v_i \neq v_j \in D} d(v_i, v_j) \leq 2,$$

i.e. there exist vertices, say v_1 and v_2, with either $d(v_1, v_2) = 1$ or $d(v_1, v_2) = 2$. In the first case, we have $|N[v_1] \cap D| \geq |N[v_1] \cap \{v_1, v_2\}| = 2$. In the second case, there exists a vertex v_{12} such that v_{12} lies on the length-2 path between v_1 and v_2, so v_{12} is adjacent to both v_1 and v_2, so $|N[v_{12}] \cap D| \geq |N[v_{12}] \cap \{v_1, v_2\}| = 2$.

Now, consider the reverse direction, which we prove via contradiction. Suppose $\min_{v_i \neq v_j \in D} d(v_i, v_j) \geqslant 3$ and there exists some v such that $|N[v] \cap D| > 1$. Then v is simultaneously adjacent to at least two vertices from the dominating set; so there exists a length-2 path between those two vertices (taken through v), a contradiction. □

The following theorem is a useful tool that enables us to extend proofs of efficient domination over a subset with a number of $\mathbb{Z}$-layers to efficient domination over the infinite lattice. For a set $A \subseteq \mathbb{Z}^3$, denote the nth $\mathbb{Z}$-layer of A as

$$A_n = \{(i, j, k) \in A : k = n\} \tag{1}$$

Let $t : \mathbb{Z}^3 \times \mathbb{Z} \rightarrow \mathbb{Z}^3$ be a simple translation function, where

$$t(A, z) = \{(i, j, k + z) : (i, j, k) \in A\} \tag{2}$$

We say two $\mathbb{Z}$-layers A_m and A_n are *equivalent*, or $A_m \equiv A_n$, if $t(A_m, n - m) = A_n$ (which also implies that $t(A_n, m - n) = A_m$).

Theorem 1 (Repeatibility of $\mathbb{Z}$-layers) *Let D be a subset of vertices in a $\mathbb{Z}$-stacked Archimedean lattice L. Consider three consecutive $\mathbb{Z}$-layers $L^{(3)} := L_1 \cup L_2 \cup L_3$ of L. Suppose each vertex in $L^{(3)}$ is efficiently dominated by the set $D_{rep} = t(D_3, -3) \cup D_1 \cup D_2 \cup D_3 \cup t(D_1, 3)$. Then the repetition set*

$$D' = \bigcup_{z=-\infty}^{\infty} t(D_1, 3z) \cup t(D_2, 3z) \cup t(D_3, 3z) \tag{3}$$

is an efficient dominating set for L.

Proof Note that by construction of a $\mathbb{Z}$-stacked Archimedean lattice, $t(L_z, z') = L_{z+z'}$. Consider $L_z \in L$ for any $z \in \mathbb{Z}$. By the first statement, $L_{z \pmod 3} = t\big(L_z, [z \pmod 3)] - z\big)$. Moreover, also by construction, $D'_z = t(D'_{z \pmod 3})$ for any $z \in \mathbb{Z}$. Then for all $v \in L_z$,

$$\begin{aligned} \left|N[v] \cap D'\right| &= \left|N[v] \cap \left(D'_{z-1} \cup D'_z \cup D'_{z+1}\right)\right| \\ = &\left|t(N[v], [z \pmod 3)] - z) \cap t\Big(\left(D'_{z-1} \cup D'_z \cup D'_{z+1}\right), [z \pmod 3)] - z\Big)\right| \\ &= 1 \end{aligned}$$

since $t\Big(N[v], [z \pmod 3)] - z\Big) \subset L^{(3)}$ and $t\Big(\left(D'_{z-1} \cup D'_z \cup D'_{z+1}\right), [z \pmod 3)] - z\Big) \subset D_{rep}$. The equality in the last line follows by the assumption of the theorem. Since this holds for all $L_z \subset L$, the entire lattice L is efficiently dominated by the repeated set D'. □

3 Efficiently Dominated Lattices

In this section, we exhibit dominating sets for seven of the stacked Archimedean lattices and prove that they are efficient dominating sets. An important tool in the proofs is Proposition 1 from Sect. 2, which is valid for finite graphs. We wish to apply Proposition 1 to blocks of the stacked Archimedean lattices. However, a block in an Archimedean lattice is not a vertex-transitive graph, because vertices on the boundary have smaller degrees than vertices in the interior of the block. To obtain a vertex-transitive finite graph, in the proofs in this section, we consider a block with *periodic boundary conditions*: Each lattice has a periodic dominating set in a block of size $a \times b \times c$ for some positive integers a, b, and c. A block of size $ai \times bj \times ck$ for positive integers i, j, and k with periodic boundary conditions is constructed by considering the edges leaving any face to be connected to the vertices on the opposite face, instead of connecting to the next layer outside the block. Since the number of vertices on the boundary is of smaller order than the volume of the block, the difference in the domination ratios of the original block and the block with periodic boundary conditions is negligible in the limit as i, j, and k tend to infinity.

3.1 $(4^4) \times \mathbb{Z}$ *Lattice and* $\mathbb{Z}^n$

Denote by $\mathbb{Z}^n$ the space of integer n-tuples, and consider that the $\mathbb{Z}$-stacked $(4^4) \times \mathbb{Z}$ lattice is simply the $n = 3$ subcase of $\mathbb{Z}^n$.

Definition 1 Consider the n-dimensional lattice $L_{\mathbb{Z}^n}$ defined by

$$L_{\mathbb{Z}^n} = (V_{L_{\mathbb{Z}^n}}, E_{L_{\mathbb{Z}^n}}) = \left(\mathbb{Z}^n, \{(v_1, v_2) : v_1 \in \mathbb{Z}^n, v_2 \in \mathbb{Z}^n, d(v_1, v_2) = 1\}\right)$$

where d is the Manhattan distance, so $d(v_1, v_2) = 1$ implies that v_1 and v_2 differ by exactly 1 in only one coordinate. Define the helper function $f : \mathbb{Z}^n \to \mathbb{Z}$ by

$$f(z_1, z_2, \ldots, z_n) = \sum_{i=1}^{n} i z_i \tag{4}$$

Let | denote *divides*, i.e. $k \mid n$ if and only if $n \equiv 0 \pmod{k}$.

Theorem 2 *$L_{\mathbb{Z}^n}$ has efficient dominating set*

$$D = \left\{(z_1, \ldots, z_n) : (2n+1) \mid f(z_1, \ldots, z_n)\right\} \tag{5}$$

Proof First, find the domination ratio of D. If an efficient dominating set of $L_{\mathbb{Z}^n}$ exists, it must have domination ratio $\frac{1}{2n+1}$ since each vertex is adjacent to $2n$ other vertices. Then, consider a chain of $2n + 1$ vertices

$$\left\{(z_1, z_2, \ldots z_n), (z_1 + 1, z_2, \ldots z_n), \ldots, (z_1 + 2n, z_2, \ldots, z_n)\right\}$$

Suppose $f(z_1, z_2, \ldots, z_n) = k$, so that

$$f(z_1, z_2, \ldots z_n) = k, f(z_1 + 1, z_2, \ldots z_n) = k + 1, \ldots, f(z_1 + 2n, z_2, \ldots, z_n) = k + 2n$$

so exactly one vertex in the chain $(z_1 + a, z_2, \ldots, z_n)$ will have $(2n + 1)$ dividing $f(z_1 + a, z_2, \ldots, z_n)$, so exactly one of the $2n + 1$ vertices will lie in the dominating set. Partition $\mathbb{Z}^n$ into disjoint chains of length $2n + 1$, where each chain has exactly one vertex in D. Then

$$\frac{|D|}{|V_{L_{\mathbb{Z}^n}}|} = \frac{1}{2n+1} \tag{6}$$

as desired. □

Next, we show that $|N[v] \cap D| \geq 1$ for all $v \in V_{L_{\mathbb{Z}^n}}$. Consider an arbitrary point $v = (z_1, z_2, \ldots, z_n) \in V_{L_{\mathbb{Z}^n}}$, and let $m = f(v) \pmod{2n+1}$. There are three cases to consider:

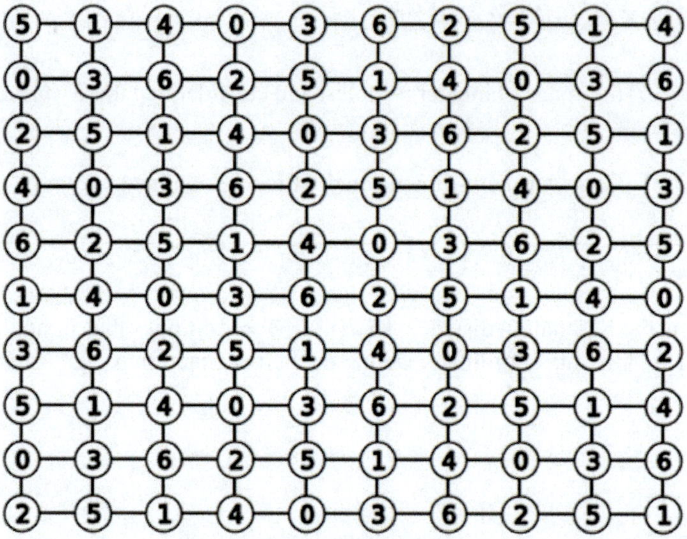

Fig. 1 Dominating set and stacked (4^4) lattice projected into a single $\mathbb{Z}$-layer. Numbered vertices are in the dominating set, with $\mathbb{Z}$-coordinate given by their number

1. If $m = 0$, then $v \in D$, so v is dominated.
2. If $1 \leq m \leq n$, the point $(z_1, \ldots, z_m - 1, \ldots, z_n)$ is in D and adjacent to v, so v is dominated.
3. If $n + 1 \leq m \leq 2n$, the point $(z_1, \ldots, z_{2n+1-m} + 1, \ldots, z_n)$ is in D and adjacent to v, so v is dominated.

In every case, v is dominated by some vertex in D, so $|N[v] \cap D| \geq 1$ for all $v \in V_{L_{\mathbb{Z}^n}}$ as desired. Apply Lemma 2 to conclude that D is an efficient dominating set. □

In particular, this pattern repeats every $2n + 1$ $\mathbb{Z}$-stacked layers. Figure 1 is a graphical representation of the dominating set on $n = 3$ after a translation. The number associated with each vertex is the layers (mod 7) for which the vertex is present in the dominating set, i.e. vertices numbered 0 are in the dominating set in layers $z_3 \equiv 0 \pmod 7$, vertices numbered 1 are in the dominating set in layers $z_3 \equiv 1 \pmod 7$, and so on.

3.2 $(3^6) \times \mathbb{Z}$ *Lattice*

Definition 2 *(Grid representation of the $\mathbb{Z}$-stacked (3^6) lattice)* Let $L_{(3^6)} = (V, E)$ be the $\mathbb{Z}^2 \times \mathbb{Z}$ lattice with additional undirected edges

$$E = \{((u_1, v_1, z), (u_2, v_2, z)) : (u_1 - v_1, u_2, -v_2) = (\pm 1, \mp 1)\} \tag{7}$$

The stacked triangular lattice $L_{(3^6)}$ can be considered as a cubic graph augmented by diagonal edges between the bottom-left and upper-right corner of each square within each $\mathbb{Z}$-layer.

Theorem 3 *The* $(3^6) \times \mathbb{Z}$ *lattice has an efficient dominating set*

$$D = \Big\{(x, y, z) : (x, y) \equiv (z, z) \pmod 3\Big\} \tag{8}$$

Proof Consider an arbitrary $v = (x, y, z) \in L_{(3^6)}$. Without loss of generality show that $u \equiv v \pmod 3$ is efficiently dominated by vertices in the proposed efficient dominating set.

Let $u = (i, j, 0) = v \pmod 3$. Note that $(0, 0) \preceq (i, j) \preceq (2, 2)$. We case-wise show $|N[u] \cap D| \geqslant 1$.

If $u = (0, 0, 0)$ then $u \in D$. If $u = (1, 0, 0)$ or $u = (0, 1, 0)$ then it is dominated by $(0, 0, 0) \in D$. If $u = (1, 1, 0)$ then it is dominated by $(1, 1, 1) \in D$. If $u = (2, 2, 0)$ then it is dominated by $(2, 2, -1) \in D$. If $u = (1, 2, 0)$ or $u = (0, 2, 0)$ then it is dominated by $(0, 3, 0) \in D$. If $u = (2, 1, 0)$ or $u = (2, 0, 0)$ then it is dominated by $(3, 0, 0) \in D$.

By definition of $L_{(3^6)}$, in each case, the difference between the vertex and its dominator is a valid edge connection. For example, $(1, 2) - (0, 3) = (1, -1)$. Thus, $|N[u] \cap D| \geqslant 1$ if and only if $|N[v] \cap D| \geqslant 1$.

Consider a $3 \times 3 \times 3$ block B of vertices with $|B| = 27$. For any $B_z \subset B$ by definition $(x, y, z) \in D$ if and only if $(x + 3i, y + 3j, z + 3k) \in D$ for all $i \in \mathbb{N}$, $j \in \mathbb{N}$ and $(x, y, z + 3k) \in D$ for all $k \in \mathbb{N}$. Thus in each z-layer there exists exactly one vertex in D. The domination ratio of B is $\frac{|B \cap D|}{|B|} = \frac{3}{27} = \frac{1}{9}$.

The domination ratio of $\frac{1}{9}$ combined with $|N[v] \cap D| \geqslant 1$ fulfill criteria 1 and 3, and thus by Lemma 2, D is an efficient dominating set. □

3.3 $(6^3) \times \mathbb{Z}$ Lattice

The (6^3), or hexagonal lattice, is a subgraph of the the (3^6) lattice.

Definition 3 *(Grid representation of the* $\mathbb{Z}$*-stacked* (6^3) *lattice)* Let the lattice be the $(3^6) \times \mathbb{Z}$ lattice with the following vertices and their edges removed (Fig. 2):

$$V' = \left\{ (i, j, k) \in L_{(3^6)} : (i, j) \equiv \begin{cases} (0, 1) \\ (1, 2) & \pmod{3, 3} \\ (2, 0) \end{cases} \right\} \tag{9}$$

Theorem 4 *The efficient dominating set* $D_{(6^3)}$ *is the same as the dominating set for* $D_{(3^6)}$.

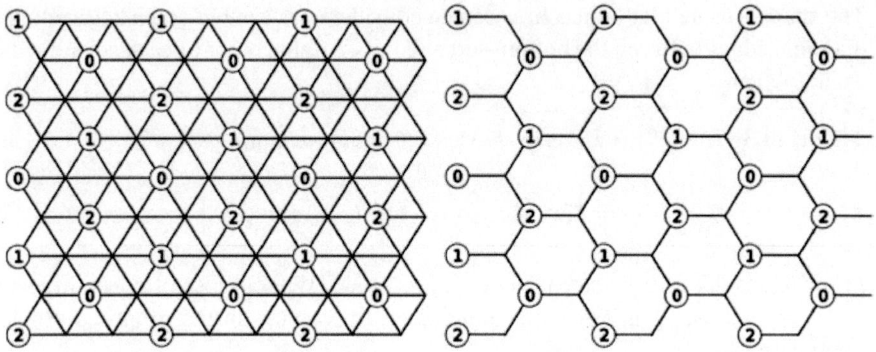

(a) The triangular (3^6) lattice. (b) The hexagonal (6^3) lattice.

Fig. 2 A shared efficient dominating set for the stacked triangular and hexagonal lattices

Proof The removed vertices have empty intersection with the dominating set. For any (i, j, k) with $(i, j) \pmod{3, 3} \in \{(0, 1) \cup (1, 2) \cup (2, 0)\}$, by definition $(i, j) \not\equiv (k, k) \pmod 3$. Thus $(i, j, k) \notin D_{(3^6)}$. Because the vertices in the $(3^6) \times \mathbb{Z}$ lattice are dominated by $D_{(3^6)}$, the remaining vertices in the $(6^3) \times \mathbb{Z}$ lattice are also dominated by $D_{(3^6)}$. □

3.4 $(4, 8^2) \times \mathbb{Z}$ *Lattice*

Definition 4 *(Grid representation of the* $\mathbb{Z}$*-stacked* $(4, 8^2)$ *lattice)*
The $\mathbb{Z}$-stacked $(4, 8^2) \times \mathbb{Z}$ lattice $L_{(4,8^2)}$ has vertex set $\mathbb{Z}^3$ and edge set which is the union of the following three sets:

$$\left\{((x, y, z), (x, y, z+1)) : x, y, z \in \mathbb{Z}\right\}$$

$$\left\{((x, y, z), (x+1, y, z)) : x, y, z \in \mathbb{Z}\right\}$$

$$\left\{(4x + 2a + b, 2y + a, z), (4x + 2a + b, 2y + a + 1, z) : x, y, z \in \mathbb{Z}; a, b \in \{0, 1\}\right\}$$

Then $L_{(4,8^2)} = (\mathbb{Z}^3, E_{L_{(4,8^2)}})$.

The first set comprises of edges that provide the $\mathbb{Z}$-stacking. The second set comprises of horizontal connections within a single layer. The third set contains vertical connections within a layer, initialized with a minimal set of vertices.

Theorem 5 *The following set* $D \subset \mathbb{Z}^3$ *efficiently dominates* $L_{(4,8^2)}$*:*

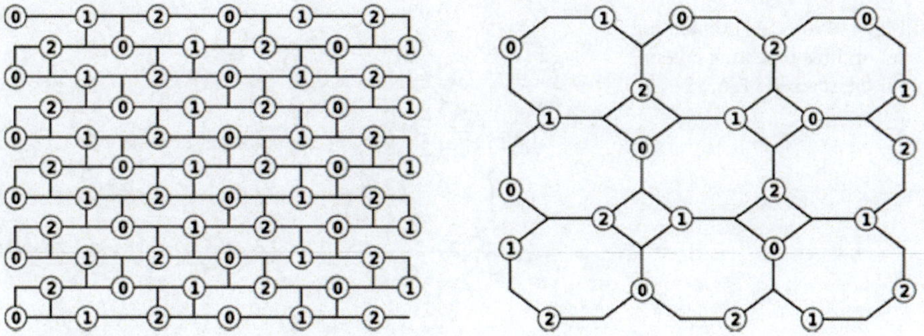

(a) The grid representation. (b) The original lattice.

Fig. 3 The dominating set of the stacked $(4, 8^2)$ lattice

$$D = \left\{(6, 0, 0)k + (3, 1, 0)m + (2, 0, 1)n : k, m, n \in \mathbb{Z}\right\} \tag{10}$$

Proof First, note that if a point $(x, y, z) \in D$ then

$$(x, y, z) = (6, 0, 0)k + (3, 1, 0)y + (2, 0, 1)z \implies k = \frac{x - 3y - 2z}{6},$$

so x must be congruent to $3y + 2z \pmod 6$. Thus, in every chain of points $\{(x, y, z), (x + 1, y, z), \ldots, (x + 5, y, z)\}$, exactly one point will be in D. Since we can decompose $\mathbb{Z}^3$ into disjoint chains of 6 vertices, each with one vertex in D, the domination ratio of D is $\frac{1}{6}$ as desired.

Next, we will show $|N[v] \cap D| \leq 1$ for all $v \in \mathbb{Z}^3$ by using Lemma 4. For a fixed Manhattan distance b between two points in the dominating set, with vertical distance of 0, the minimum distance is 4, achieved both between points in adjacent rows ($\Delta x = \pm 3$, $\Delta y = \pm 1$) and points in the same column ($\Delta x = 0$, $\Delta y = \pm 2$). The first case is distance 4 because no edge covers a Manhattan distance of more than 1, while the second case is not distance 2 since there are no two incident vertical edges. Note that between two layers with the same value of b but absolute value of z differing by 1, the vertical distance is 3, so the total distance is at least 3.

For vertices in different $\mathbb{Z}$-layers with vertical distance 1, the intra-layer distance is at least 2 for a total distance of 3. Therefore, $|N[v] \cap D| \leq 1$ for all $v \in \mathbb{Z}$, so D is efficiently dominating by Lemma 1. □

In particular, this pattern repeats every 3 $\mathbb{Z}$-stacked layers. Figure 3 is a graphical representation of the dominating set. The numbering of the vertices again represents the layers for which the vertices are present in the dominating set.

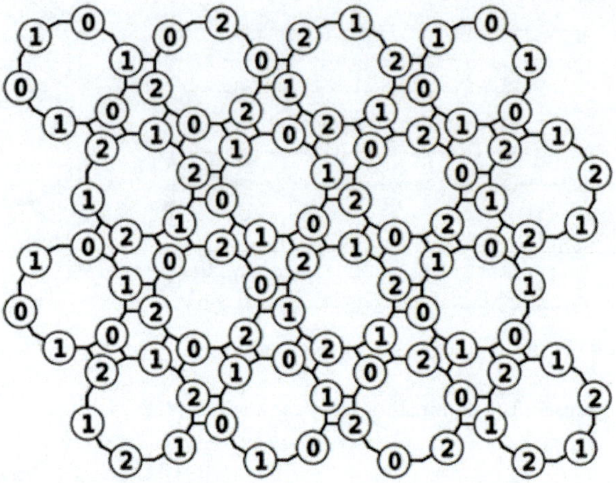

Fig. 4 Projected dominating set of three repeating layers for the stacked (4, 6, 12) lattice

3.5 (4, 6, 12) × $\mathbb{Z}$ *Lattice*

We present a 3-repeatable construction on the (4, 6, 12) lattice, again numbered to represent the layers for which the vertices are present in the dominating set. Consider the projection of the dominating set in these three layers onto one layer. Along each dodecagon, representing 36 vertices, there are six vertices in the dominating set. The domination ratio of the set on a dodecagon is thus $\frac{1}{6}$, and since the dodecagons provide a periodic subset of the (4, 6, 12) lattice, the domination ratio of the set on the graph is also $\frac{1}{6}$ (Fig. 4).

Moreover $|N[v] \cap D| \leqslant 1$ for all $v \in L_{(4,6,12)}$. Note that the distance between any two vertices on the projected dominating set is at least 2 (or at least 4 if they're in the same layer), and since they have a vertical distance of at least 1, the distance between any two points in the dominating set is at least 3, which is equivalent to $|N[v] \cap D| \leqslant 1$. We can then conclude that our construction is an efficient dominating set by Lemma 3. □

3.6 $(3^2, 4, 3, 4)$ × $\mathbb{Z}$ *Lattice*

Definition 5 *(Grid representation of the* $\mathbb{Z}$*-stacked* $(3^2, 4, 3, 4)$ *lattice)*
The $\mathbb{Z}$-stacked $(3^2, 4, 3, 4) \times \mathbb{Z}$ lattice $L_{(3^2,4,3,4)}$ is isomorphic to $\mathbb{Z}^3$ with additional edges connecting $(2k-1, 2l, m)$ to $(2k, 2l-1, m)$ for all $k, l, m \in \mathbb{Z}$, and additional edges connecting $(2k, 2l, m)$ to $(2k+1, 2l+1, m)$ for all $k, l, m \in \mathbb{Z}$.

Note: The additional edges of the $(3^2, 4, 3, 4)$ lattice, compared to those of the cubic lattice, comprise of alternating diagonal connections in every other square, within each layer (Fig. 5).

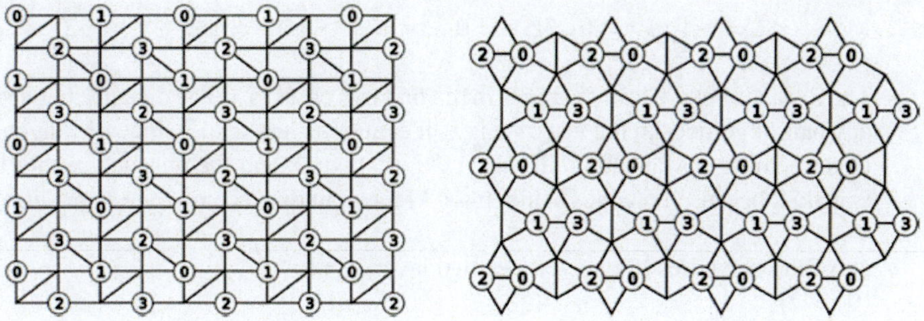

(a) The grid representation. (b) The original lattice.

Fig. 5 The dominating set of the $(3^2, 4, 3, 4)$ lattice

Theorem 6 *The following set $D \subset \mathbb{Z}^3$ efficiently dominates $L_{(3^2,4,3,4)}$:*

$$\begin{aligned} D = &\{(2k+1, 2k+1+4l, 4m) : k, l, m \in \mathbb{Z}\} \\ &\cup \{(2k+1, 2k-1+4l, 4m+1) : k, l, m \in \mathbb{Z}\} \\ &\cup \{(2k, 2k+4l, 4m+2) : k, l, m \in \mathbb{Z}\} \\ &\cup \{2k, 2k-2+4l, 4m+3) : k, l, m \in \mathbb{Z}\} \end{aligned}$$

Observe that this pattern repeats every 4 layers. A graphical representation of the dominating set is portrayed in Fig. 5. Again, numbering represents the layers for which the vertices are present in the dominating set.

Proof First consider the domination ratio $\frac{|D|}{|B|}$ over a $2 \times 4 \times 4$ block $B \subset L_{(3^2,4,3,4)}$ induced by vertices in the rectangle $[0, 2) \times [0, 4) \times [0, 4) \subseteq \mathbb{R}^3$. Note that there are 32 vertices in this block, of which 8 are in the dominating set. Since blocks are regular by definition, the domination ratio is $\frac{1}{8}$ as desired.

Moreover, no vertex is dominated more than once, i.e. $|N[v] \cap D| \leq 1$ for all $v \in V_{L_{(3^2,4,3,4)}}$, or equivalently, no two vertices in D are less than distance 3 away. Unfortunately, there are ten cases, four for vertices in the same layer and $6 = \binom{4}{2}$ for vertices in different layers. Each refers to the value of the layer taken mod 4 from which they come.

First, the four cases of two vertices in the same layer:

1. $(0 \to 0)$
2. $(1 \to 1)$
3. $(2 \to 2)$
4. $(3 \to 3)$

These cases may be solved simultaneously. Note that the Manhattan distance between any two points in the same layer is at least 4, which is achieved in one of three cases:

$$(\Delta x = \pm 4, \Delta y = 0), (\Delta x = 0, \Delta y = \pm 4), (\Delta x = \pm 2, \Delta y = \pm 2)$$

The distance between any vertices in the first two cases is 4. The distance between any pair of vertices in the last case is 3. It cannot be less than 2 since no diagonal edges achieve a Manhattan distance of more than 2, and the distance cannot be exactly 2 because no edges achieving a Manhattan distance of 2 are incident.

Next, the two cases of two vertices two layers apart:

5. $(0 \to 2)$
6. $(1 \to 3)$
These two cases can also be solved at the same time. For either, the vertical distance is 2, and travelling between $\mathbb{Z}$-layers can only be done with an edge preserving the x and y coordinates. Note that no points in those layers ($0 \to 2$ and $1 \to 3$) are repeated between vertical layers, so it would require at least 1 intra-layer edge in addition to the 2 inter-layer edges to create such a path.

Finally, the four adjacent layer cases:

7. $(0 \to 1)$
8. $(2 \to 3)$
Note that in cases (7) and (8), vertices have a vertical distance of 1 and an intra-layer gap of $(\Delta x = \pm 2, \Delta y = 0)$ or $(\Delta x = 0, \Delta y = \pm 2)$. These require at least 2 edges to achieve the same (x, y) values, for a total minimum distance of 3.

9. $(0 \to 3)$
10. $(1 \to 2)$
Note that in cases (9) and (10), vertices have a vertical distance of 1 and an intra-layer gap of $(\Delta x = 1, \Delta y = 1)$ or $(\Delta x = -1, \Delta y = -1)$. No single edges span this gap, so the minimum intra-layer distance is 2, and the shortest path between any vertices is 3.

Thus the minimum distance between any two points in the dominating set is 3, that is, $|N[v] \cap D| \leq 1$ for all $v \in V_{L_{(3^2,4,3,4)}}$. Lemma 3 concludes that D is an efficient dominating set. □

3.7 $(3^3, 4^2)$ x $\mathbb{Z}$ Lattice

Definition 6 *(Grid representation of the $(3^3, 4^2)$ lattice)* Let $L_{(3^3,4^2)}$ be the the $\mathbb{Z}$-stacked $(3^3, 4^2)$ lattice. $L_{(3^3,4^2)}$ is isomorphic to the cubic lattice $\mathbb{Z}^3$ with additional edges $(x, 4l + 1, z)$ to $(x + 1, 4l, z)$ for all $x, l, z \in \mathbb{Z}$ as well as $(x, 4l + 2, z)$ to $(x + 1, 4l + 3, z)$ for all $x, l, z \in \mathbb{Z}$.

Theorem 7 *The following set $D \subset \mathbb{Z}^3$ efficiently dominates $L_{(3^3,4^2)}$:*

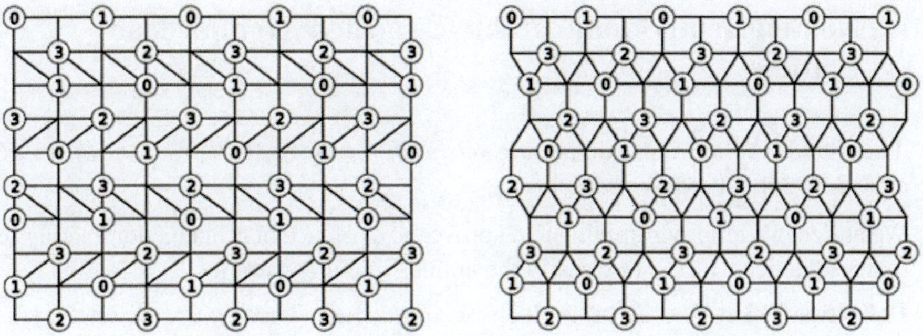

(a) The grid representation. (b) The original lattice.

Fig. 6 The dominating set of the $(3^3, 4^2)$ lattice

$$\begin{aligned} D = &\{(4k + l + 2, 4l + 2, 4m) : k, l, m \in \mathbb{Z}\} \\ &\cup \{(4k + l, 4l + 2, 4m + 1) : k, l, m \in \mathbb{Z}\} \\ &\cup \{(4k + l + 2, 4l + 3, 4m + 2) : k, l, m \in \mathbb{Z}\} \\ &\cup \{(4k + l, 4l + 3, 4m + 3) : k, l, m \in \mathbb{Z}\} \\ &\cup \{(4k + l, 4l, 4m) : k, l, m \in \mathbb{Z}\} \\ &\cup \{(4k + l + 2, 4l, 4m + 1) : k, l, m \in \mathbb{Z}\} \\ &\cup \{(4k + l + 3, 4l + 1, 4m + 2) : k, l, m \in \mathbb{Z}\} \\ &\cup \{(4k + l + 1, 4l + 1, 4m + 3) : k, l, m \in \mathbb{Z}\} \end{aligned}$$

Remark Observe that this pattern repeats every four $\mathbb{Z}$-layers. Figure 6 shows a projection of the dominating set from the $\mathbb{Z}$-axis. Again, the numbering indicates the layers for which vertices are present in the dominating set (Fig. 6).

Proof Note that in the $4 \times 4 \times 4$ block induced by the rectangle $[0, 4) \times [0, 4) \times [0, 4) \subset \mathbb{R}^3$, there are 8 vertices in the dominating set. By regularity of the blocks, the domination ratio of D to the lattice is $\frac{8}{4 \cdot 4 \cdot 4} = \frac{1}{8}$ as desired. We show that $|N[v] \cap D| \leq 1$ for all $v \in \mathbb{Z}^3$. Observe that any points in the same $\mathbb{Z}$-layer have a minimum Manhattan distance of 4 with either $\Delta x = 4$, $\Delta y = 0$ or $\Delta x = \pm 2$, $\Delta y = \pm 2$. Neither yields a distance of 2 since there are no two incident edges in the lattice both with a distance of 2. No points of vertical distance 2 from each other have the same x and y-coordinates, so the minimum distance between any two is 3. Any points in adjacent $\mathbb{Z}$-layers either have a distance of 3, or are distance 2 away from each other with no edge in between. Thus the minimum distance between any two points of D is 3. Therefore, $|N[v] \cap D| \leq 1$ for all $v \in \mathbb{Z}$. Then D is an efficient dominating set by Lemma 3. □

4 Non-efficiently-dominatable $\mathbb{Z}$-stacked Archimedean Lattices

Theorem 8 *No efficient dominating sets exist for the* $(3, 12^2)$, $(3, 4, 6, 4)$, $(3^4, 6)$, *and* $(3, 6, 3, 6)$ *$\mathbb{Z}$-stacked lattices.*

We utilize integer programming to disprove the existence of efficient dominating sets on these lattices. First, we classify the entities in our constraints:

Definition 7 Let $B \subset L$ be a block of a lattice L. For any $v \in B$, let $N_B[v] = N[v] \cap B$ denote the neighborhood of v in the block. We say $v \in B \subset L$ is *interior* with respect to B if $N_B[v] = N[v]$, and denote the set of such interior vertices as B^0.

We construct a binary integer linear program (IP) whose variables correspond to the vertices of a block $B \subset L$ in the lattice.

Definition 8 *(Integer Program for Efficient Domination on a Block)* Let L be a $\mathbb{Z}$-stacked Archimedean lattice with block $B \subset L$. For each v interior with respect to B, let $x_v \in \{0, 1\}$ be a corresponding binary variable. An integer program for efficient domination of L over B is thus defined as

$$\begin{aligned} \min \sum_{v \in B} x_v & & (11) \\ s.t. \sum_{x_{v'}: v' \in N_B[v]} x_{v'} &= 1 & \forall v \in B^0 \\ x_v &\in \{0, 1\} & \forall v \in B \end{aligned}$$

For exterior vertices, i.e. v for which $N[v] \not\subset B$, we do not place any domination constraints in the IP. The above integer program has a useful property for showing non-existence of dominating set. Since the domination constraint is removed for non-interior vertices, a solution to the integer program is not bijective with (a subset of) the true efficient dominating set D_L for the lattice L. However, infeasibility of a solution suffices to prove non-existence of the dominating set.

Proposition 2 *If there does not exist a feasible solution to the integer linear program (11) for lattice L over any block $B \subset L$, then L does not have an efficient dominating set.*

Proof The IP constraint $\sum_{x_{v'}: v' \in N_B[v]} x_{v'} = 1$ occurs if and only if $|N[v] \cap D| = 1$ for each x_v corresponding to an interior vertex $v \in B^0$.

For sake of contradiction, assume D is an efficient dominating set for L. Then $D \cap B$ would give the feasible solution $\left\{x_v = \mathbb{1}\{v \in D \cap B\}\right\}$ to (11), which is impossible. □

Note that an instance of the integer program is specified by a lattice and a block. Some integer programs may have feasible solutions, but the existence of any infeasible integer program for a single lattice suffices to show non-existence. Thus, to eliminate such false positives, integer programs for different block sizes were checked for each of the lattices. Intuitively, as blocks grow larger, both the solution space and number of constraints increase, making the integer programs less likely to have a feasible solution. The following table shows that infeasible solutions were found at block sizes $6 \times 7 \times 7$ or larger.

$\mathbb{Z}$-Stacked lattice	Infeasible block size
$(3, 6, 3, 6)$	$8 \times 8 \times 7$
$(3, 4, 6, 4)$	$8 \times 8 \times 8$
$(3^4, 6)$	$8 \times 8 \times 8$
$(3, 12^2)$	$6 \times 7 \times 7$

Interestingly, despite significantly different grid representations, each of the lattices listed above had a similar minimal block size for which solutions to the IP became infeasible. The $\mathbb{Z}$-stacked lattice $L_{(3,12^2)}$ with the smallest domination ratio of $1/6$ also had the smallest infeasible block size.

The integer program for each of the lattices was written in MATLAB.

5 Future Research

While it has been shown that no efficient domination set can exist on the $(3, 12^2)$, $(3, 4, 6, 4)$, $(3, 6, 3, 6)$, and $(3^4, 6)$ $\mathbb{Z}$-stacked lattices, the question of the minimum domination ratio for each of these lattices remains undecided.

Efficient domination could be considered over other vertex-transitive lattices in $3D$. The Bravais lattices, in particular, are closely related to such lattices. Of their four main categories (primitive, base-centered, body-centered, and face-centered), only primitive has been analyzed. However, the other three categories are not necessarily vertex-transitive, which would significantly increase the difficulty of determining the efficient domination set.

Finally, many other forms of domination exist—perfect domination [5], power domination [9], exponential domination [1], Roman domination [2], eternal domination [3], and many more. However, to our knowledge, no other forms of domination have been studied on the $\mathbb{Z}$-stacked Archimedean lattices.

References

1. Dankelmann, Peter; Day, David; Erwin, David; Mukwembi, Simon; and Swart, Henda, Domination with exponential decay, *Discrete Mathematics*, 309, 5877–5883, 2009, https://doi.org/10.1016/j.disc.2008.06.040".
2. Dreyer, Paul Andrew, *Applications and Variations of Domination in Graphs*, thesis, Rutgers University, 2000.
3. Goldwasser, John L. and Klostermeyer, William F. Tight bounds for eternal dominating sets in graphs, *Discrete Mathematics*, 308, 2589–2593, 2008, https://doi.org/10.1016/j.disc.2007.06.005.

4. Grünbaum, Branko and Shephard, G C, *Tilings and Patterns*,W. H. Freeman & Co., 1986, isbn = 0716711931.
5. Livingston, Marilynn and Stout, Quentin F. Perfect dominating sets. *Congressus Numerantium*, 79, 187–203, 1990.
6. Marge, Thomas; Zhao, Yunfan;and Wierman, John C. Efficient and non-efficient domination of Archimedean lattices, *Congressus Numerantium*, 231, 39–61, 2018.
7. Schrenk, K. J. and Araújo, N. A. M. and Herrmann, H. J. Stacked triangular lattice: Percolation properties, *Physical Review E*, 87, https://doi.org/10.1103/physreve.87.032123, 2013.
8. Suding, Paul N. and Ziff, Robert M. Site percolation thresholds for Archimedean lattices, *Phys. Rev. E*, 60, 275–283, 1999, https://doi.org/10.1103/PhysRevE.60.275.
9. Zhao, Min; Kang, Liying; and Chang, Gerard J. Power domination in graphs, *Discrete Mathematics*, 306, 1812–1816, 2006, https://doi.org/10.1016/j.disc.2006.03.037.

On Subdivision Graphs Which Are 2-steps Hamiltonian Graphs and Hereditary Non 2-steps Hamiltonian Graphs

Sin-Min Lee, Hsin-hao Su, and Yung-Chin Wang

Abstract Let G be a graph with vertex set $V(G)$ and edge set $E(G)$. A (p, q)-graph $G = (V, E)$ is said to be AL(k)-traversal if there exist a sequence of vertices $\{v_1, v_2, \ldots, v_p\}$ such that for each $i = 1, 2, \ldots, p-1$, the distance for v_i and v_{i+1} is equal to k. We call a graph G a k-steps Hamiltonian graph if it has a AL(k)-traversal in G and the distance between v_p and v_1 is k. A graph G is said to be hereditary k-steps hyperhamiltonian if it is k-steps Hamiltonian and for any v in G, the vertex-deleted subgraph $G - \{v\}$ is also k-steps Hamiltonian. Dually, a graph G is said to be hereditary non k-steps Hamiltonian if it is not k-steps Hamiltonian and for any v in G, the vertex-deleted subgraph $G - \{v\}$ is also not k-steps Hamiltonian. In this paper, we investigate subdivision graphs of a wheel graph and $C_4 \times K_2$ to see which are 2-steps Hamiltonian and hereditary non 2-steps Hamiltonian.

Keywords k-step traversal · AL(k)-traversal · k-steps Hamiltonian · k-steps hyperhamiltonian · Hereditary non 2-steps Hamiltonian · Subdivision

1 Introduction

In this paper we consider graphs with no loops.

The Hamiltonicity of a graph is the problem of determining for a given graph whether it contains a path or a cycle that visits every vertex exactly once. Hamiltonian graphs are related to the traveling salesman problem. Thus, it has been a well-studied

S.-M. Lee
1786 Plan Tree Drive, Upland 91784, USA

H. Su (✉)
Department of Mathematics, Stonehill College, Easton 02357, USA
e-mail: hsu@stonehill.edu

Y.-C. Wang
Department of Digital Media Design, Tzu-Hui Institute of Technology, Nanzhou, Pingtung, Taiwan

F. Hoffman (ed.), *Combinatorics, Graph Theory and Computing*, Springer Proceedings in Mathematics & Statistics 388,
https://doi.org/10.1007/978-3-031-05375-7_16

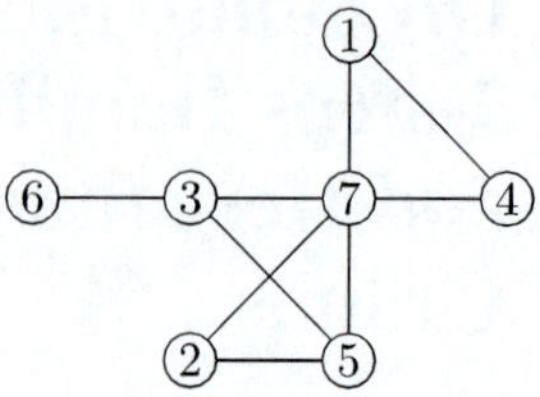

Fig. 1 An AL(2)-traversal graph

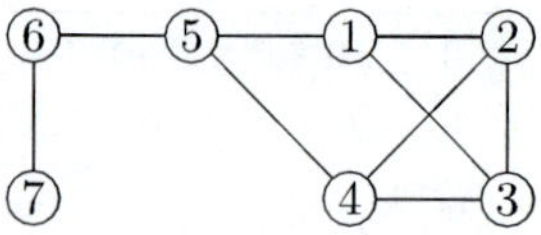

Fig. 2 An $D_2(G)$ graph from the graph in Fig. 1

topic in graph theory. However, we know very little about Hamiltonian graphs. A good reference for recent development and open problems is [3].

Inspired by Wallis's Magic Graph [13], Lee in [9] initiated the study of AL(k)-traversal graphs and 2-steps Hamiltonian graphs defined as follows:

Definition 1 For $k > 2$, a (p, q)-graph $G = (V, E)$ is said to have k-steps traversal if there exist a sequence of vertices, $v_1, v_2, \ldots, v_p$, such that, for each $i = 1, 2, \ldots, p-1$, the distance between v_i and v_{i+1} is equal to k. A graph admits a k-steps traversal is called the AL(k)-traversal graph.

Example 1 The graph showed in the Fig. 1 is AL(2)-traversal, but not AL(k)-traversal for all $k \geq 3$.

We can construct a new graph to study AL(k)-traversal graphs:

Definition 2 For integer $k \geq 2$, and a graph G, we construct a new graph $D_k(G)$ as follows: $V(D_k(G)) = V(G)$ and $(u, v) \in E(D_k(G))$ if and only if $d(u, v) = k$ in G. We call $D_k(G)$ as the distance k graph of G.

Example 2 The graph showed in the Fig. 2 is a $D_2(G)$ graph from the graph in Fig. 1.

Definition 3 We name a AL(k)-traversal in a graph G with the distance between vertices v_p and v_1 is k a k-steps Hamiltonian cycle.

Definition 4 We call a graph G a k-steps Hamiltonian graph if it has a k-steps Hamiltonian cycle.

Note here that in Fig. 1 the distance between the vertices labeled 1 and 7 is not 2. Thus, it is not a 2-steps Hamiltonion cycle. Moreover, after an exhaustive search, there is no labeling to make this graph 2-steps Hamiltonian.

Example 3 Figure 3 demonstrates a 2-steps Hamiltonian cubic graph.

The following observation which would be useful in this paper was recorded in [8].

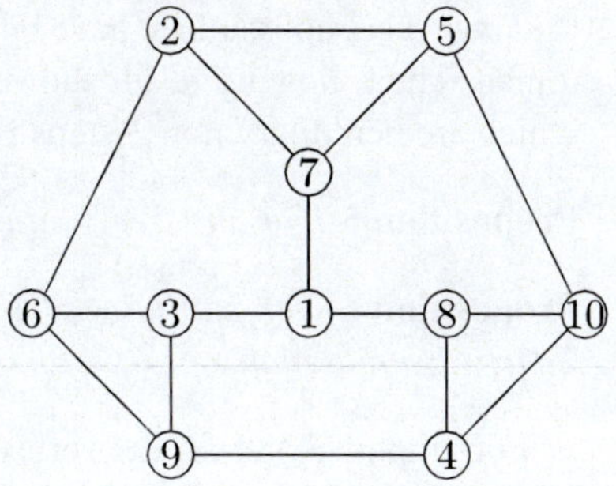

Fig. 3 A 2-steps Hamiltonian cubic graph

Proposition 1 *The cycle C_n is k-steps Hamiltonian if and only if* $\gcd(n, k) = 1$.

Proposition 2 *The graph G is k-steps Hamiltonian if and only if its distance k-graph is Hamiltonian.*

Proposition 3 *A bipartite graph is not* AL(2)*-traversal, thus, not 2-steps Hamiltonian.*

A Hamiltonian graph need not be k-steps Hamiltonian. One example is a cycle C_n with $n = 0 \pmod{k}$ is Hamiltonian but not AL(k)-traversal, hence cannot be k-steps Hamiltonian.

A graph property is called hereditary if it is closed with respect to deleting vertices. We define

Definition 5 A graph G is said to be k-steps hyperhamiltonian if it is k-steps Hamiltonian and for any v in G, the vertex-deleted subgraph $G - \{v\}$ is also k-steps Hamiltonian.

Example 4 The Möbius ladder M_8 is 2-steps hyperhamiltonian (Fig. 4).

The generalized Petersen graphs which are hyperhamiltonian had been studied in [11].

A graph is bipartite, or two-colorable, if it can be decomposed into two independent sets. It was shown in [5] that

Proposition 4 *If G is bipartite then G is not k-steps Hamiltonian for any even k.*

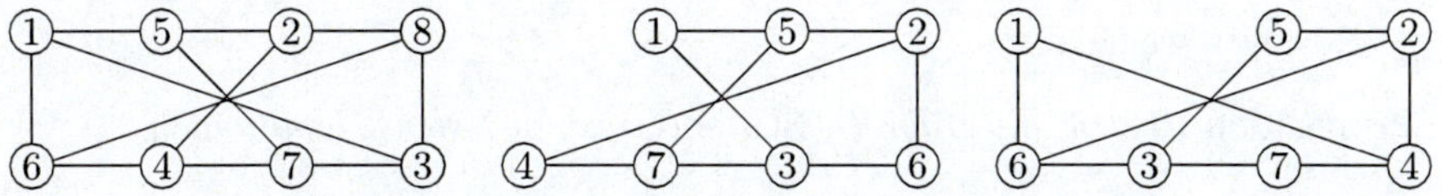

Fig. 4 M_8, $M_8 - \{v_1\}$ and $M_8 - \{v_2\}$ are all 2-steps Hamiltonian

Thus, it is impossible to have bipartite graphs which are hereditary 2-steps Hamiltonian. The following result showed that there exists an abundance bipartite graphs which are hereditary non 2-steps Hamiltonian.

Proposition 5 *For any tree T, it is hereditary non 2-steps Hamiltonian.*

Proposition 6 *For any integer $n > 2$, the 2-regular graph C_{2n} is hereditary non 2-steps Hamiltonian.*

For graphs G and H, the vertex gluing of G and H is the identifying of a vertex of G and H. It was shown in [4] that

Proposition 7 *The vertex-gluing of two cycles is not k-step Hamiltonian for all $k > 2$.*

We have the following obvious result

Proposition 8 *For any vertex-gluing of two cycles$T$, it is hereditary non 2-steps Hamiltonian.*

We also have the following for bipartite cubic graphs

Proposition 9 *For any integer $n > 2$, the prism graph $C_{2n} \times K_2$ is hereditary non 2-steps Hamiltonian.*

Dually, we can define

Definition 6 A graph G is said to be hereditary non k-steps Hamiltonian if it is not k-steps Hamiltonian and for any v in G, the vertex-deleted subgraph $G - \{v\}$ is also not k-steps Hamiltonian.

Definition 7 Let G be a graph, and $S \subseteq E(G)$, and $f : S \to N$. The subdivision graph Sub(G, S, f) is the graph obtained by for any e in S, if $f(e) = m$, then we insert m new m vertices along in e.

In literature, if $f : E(G) \to N$ with $f = 1$ for each $e \in E(G)$, then the subdivision graph Sub(G, E(G), f) is called the barycentric subdivision graph. We denote barycentric subdivision graph of G by BCSub(G).

We observe that C_{2k+1} is 2-steps Hamiltonian, however, BCSub(G) is isomorphic to C_{4k+2} which is not 2-steps Hamiltonian. It is natural to ask for what G in Gph, BCSub(G) is 2-steps Hamiltonian.

However, we have

Proposition 10 *For any graph G,* BCSub(G) *is not 2-steps Hamiltonian.*

Proof It is easy to see that we can group all original vertices as a group and all inserted vertices into another group to see that BCSub(G) is a bipartite graph. By Proposition 4, it is not 2-steps Hamiltonian. □

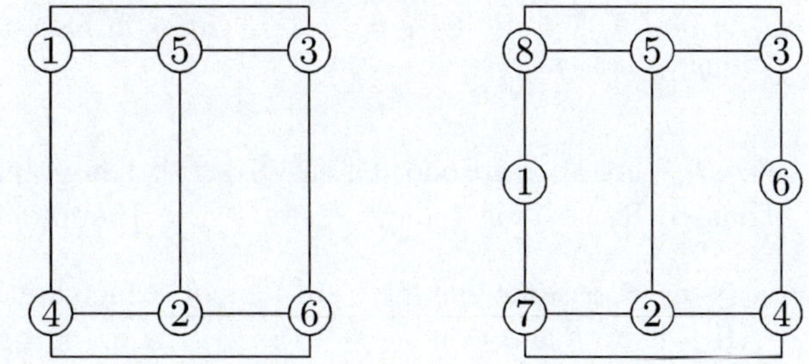

Fig. 5 $\mathrm{Sub}(C_3 \times K_2, S, f)$ is AL(2)-traversal but not 2-steps Hamiltonian

Example 5 Let $G = C_3 \times K_2$, and $S = \{(x_0, y_0), (x_2, y_2)\}$ and $f : S \to \mathbb{N}$ be defined by $f((x_0, y_0)) = f((x_2, y_2)) = 1$. We see in Fig. 5 that $Sub(G, S, f)$ is AL(2)-traversal but not 2-steps Hamiltonian while the original graph $G = C_3 \times K_2$ is 2-steps Hamiltonian.

For a cycle of order n, we denote its vertices by $\{v_1, v_2, \ldots, v_n\}$. If it has a chord between $\{v_1, v_c\}$, we denote this graph by $C_n(c)$. In [10], we investigated the subdivision graph $\mathrm{Sub}((C_n(c), \{v_1, v_c\}, f(\{v_1, v_c\}) = h)$ to see under what conditions the graph is 2-steps Hamiltonian. The reason we were interested in the Subdivision graph of a cycle with a chord is that some of them they are non-Hamiltonian.

In general, it is easy to see that when inserting too many vertices on an edge, it does not change it's Hamiltonicity,

Lemma 1 *A graph G with a subgraph P of a path of length* 5 *or more is* 2*-steps Hamiltonian if and only if the induced graph H from G by removing two middle vertices from the path P is* 2*-steps Hamiltonian.*

In this paper, we investigate subdivision graphs of a wheel graph and $C_4 \times K_2$ to see which are 2-steps Hamiltonian and hereditary non 2-steps Hamiltonian.

2 Subdivision Graphs of a Wheel Graph

A wheel graph $W(n)$ is a graph with n vertices where $n \geq 4$, formed by connecting a single vertex c to all vertices of an $(n-1)$-cycle $\{v_1, v_2, \ldots, v_{n-1}\}$. In a wheel graph, the hub c has degree $n-1$, and other vertices have degree 3.

Let us denote $X = \{(v_1, v_2), (v_2, v_3), \ldots, (v_{n-2}, v_{n-1}), (v_{n-1}, v_1)\}$ be the set of the cycle edges and be given a function $f : X \to N$ with $f((v_i, v_{i+1})) = h_i$. We can construct the graph $\mathrm{Sub(W(n), X, f)}$ and name the vertices between v_i and v_{i+1} by $v_{i,1}, v_{i,2}, \ldots, v_{i,h_i}$ for all $i = 1, 2, \ldots, n-1$.

Theorem 1 *The graph* $\mathrm{Sub(W(n), X, f)}$ *is* 2*-steps Hamiltonian if and only if one of the following conditions is satisfied:*

1. *n is even and $h_1 + h_2 + \cdots + h_{n-1}$ is even;*
2. *n is odd and $h_1 + h_2 + \cdots + h_{n-1}$ is odd;*

3. *n and $h_1 + h_2 + \cdots + h_{n-1}$ have different parity and there is a h_i for some i which equals to* 2.

Proof When all h_i are odd, it is easy to see that the graph Sub(W(n), X, f) is bipartite. Thus, by Proposition 4, they are not 2-steps Hamiltonian.

Now, we assume that there are some h_i which are even.

If n is even and $h_1 + h_2 + \cdots + h_{n-1}$ is even, then the cycle part of the graph Sub(W(n), X, f) has $h_1 + h_2 + \cdots + h_{n-1} + (n - 1)$ vertices. In this case, we have $h_1 + h_2 + \cdots + h_{n-1} + (n - 1)$ is odd. Thus, when start from any vertex and travel 2-steps on the cycle, it will go through all vertices and back to the first vertex. So, the labeling starting with c and then goes to $v_{1,1}$ following by every vertices in the cycle counterclockwise (can be done because of odd number vertices on the cycle) ending at $v_{n-1,h_{n-1}}$ to go back to c is a 2-steps Hamiltonian cycle. Thus, the graph Sub(W(n), X, f) is 2-steps Hamiltonian. (See Fig. 7 for an example.)

Similarly, when n is odd and $h_1 + h_2 + \cdots + h_{n-1}$ is odd, the cycle part of the graph Sub(W(n), X, f) has $h_1 + h_2 + \cdots + h_{n-1} + (n - 1)$, which is odd, vertices. The same labeling applys and it is 2-steps Hamiltonian.

Thus, we only need to consider the case where one of the n and $h_1 + h_2 + \cdots + h_{n-1}$ is odd and another is even. Note that in this case, there are even number of vertices in the cycle part of the graph Sub(W(n), X, f).

Next, we assume that there is an h_i which is equal to 2, w.l.o.g., we can assume that $h_1 = 2$. We would start labeling from c and then $v_{1,1}$ following by $v_{n-1,h_{n-1}}$ and travel clockwise through the cycle by every other vertices until reaching v_2. It is possible because the number of vertices in the cycle part is even and $h_1 = 2$. After that, we continue the labeling by jumping to v_1 following by $v_{n-1,h_{n-1}-1}$ (or v_{n-1} if $h_{n-1} = 1$) and travel clockwise again through the cycle by every other vertices until reaching $v_{1,2}$. Again, it is possible because the number of vertices in the cycle part is even and $h_1 = 2$. Since we have labeled every vertex and the distance between $v_{1,2}$ and c is 2, the graph Sub(W(n), X, f) is 2-steps Hamiltonian. (See Fig. 6 for an example.)

Finally, there is only one case left, that is, all even h_i are greater or equal to 4. By the Lemma 1, we know that if we insert 6 vertices or more on an edge then we can remove even vertices to keep the 2-steps Hamiltoniancy. Thus, we only need to consider then all even h_i are equal to 4.

If n is even and $h_1 + h_2 + \cdots + h_{n-1}$ is odd, then since $n - 1$ is odd, there must be at least two adjacent h_i and h_{i+1} (subscripts are module $n - 1$) to be 4, otherwise, there would be half of h_i are even and the other half are odd which makes the sum to be odd. Similarly, if n is odd and $h_1 + h_2 + \cdots + h_{n-1}$ is even, there must also be at least two adjacent h_i and h_{i+1} to be 4. With a pair of adjacent h_i and h_{i+1} to be 4, its D_2 graph would have a cycle of length 6 that have 4 consective order 2 vertices following with two order 4 vertices from the vertices v_i and v_{i+2}. By the Proposition 11 from [10] (for the completeness, we copy the Lemma we used from [10] right after the end of the proof), it cannot be Hamiltonian. Thus, the graph Sub(W(n), X, f) is not 2-steps Hamiltonian.

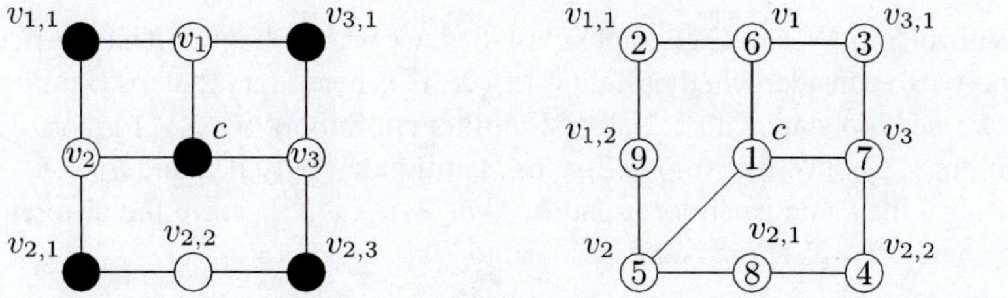

$h_1 + h_2 + h_3 = 1 + 3 + 1 = 5$ $\qquad$ $h_1 + h_2 + h_3 = 2 + 2 + 1 = 5$

Fig. 6 Two subdivision graphs of $W(4)$ with $\sum_{i=1}^{n-1} h_i$ is odd

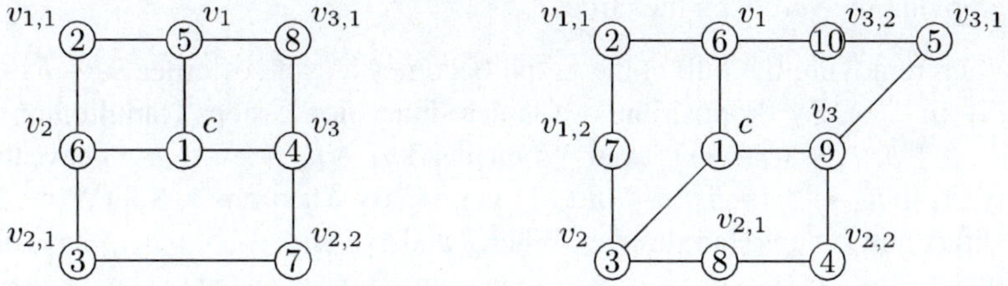

$h_1 + h_2 + h_3 = 1 + 2 + 1 = 4$ $\qquad$ $h_1 + h_2 + h_3 = 2 + 2 + 2 = 5$

Fig. 7 Two subdivision graphs of $W(4)$ with $\sum_{i=1}^{n-1} h_i$ is even

With all possible cases classified, the proof is complete. □

Proposition 11 (Lemma 1.4. in [10]) *If a distance* 2*-graph contains a subgraph* H *consisted with all order* 2 *vertices and two order* 3 *vertices where the distance between two order* 3 *vertices is greater than* 1*, then it is not Hamiltonian. Moreover, if* H *consists with* 3 *or more order* 3 *vertices and those order* 3 *vertices are adjacent to each other in two paths, then it is not Hamiltonian as well.*

Proof For a labeling cycle, it must enter the subgraph H through one of the two order 3 vertices. But, since the distance between two order 3 vertices is greater than 1, it is obvious that this cycle cannot be Hamiltonian.

Similarly, a path of adjacent order 3 vertices can be considered as one order 3 vertex in the purpose of our proof. □

Example 6 The following subdivision graphs of $W(4)$ with two different f and sum h_i are odd, one is bipartite another is not. However, they are not 2-steps hamiltonian.

Example 7 The following subdivision graphs of $W(4)$ with two different f and sum of h_i are even. They are 2-steps hamiltonian (Fig. 7).

With all Sub(W(n), X, f) graphs classified, we want to know if it has the hereditary peoperty. To consider whether Sub(W(n), X, f) is hereditary 2-steps Hamiltonian or not, we need to start with a 2-steps Hamiltonian Sub(W(n), X, f) graph. Thus, by Theorem 1, Sub(W(n), X, f) is 2-steps Hamiltonian only if n and $h_1 + h_2 + \cdots + h_{n-1}$ have the same parity or n and $h_1 + h_2 + \cdots + h_{n-1}$ have the different parity with some $h_i = 2$. These two cases lead to

Theorem 2 *The graph* Sub(W(n), X, f) *is hereditary non 2-steps Hamiltonian.*

Proof There are three kinds of vertices to remove in Sub(W(n), X, f):

1. removing the hub c;
2. removing the inserted vertex $v_{i,j}$;
3. removing the vertex on the circle v_i.

When removing the hub c, the graph becomes a cycle of order $h_1 + h_2 + \cdots + h_{n-1} + (n-1)$. By Proposition 1, it is hereditary non 2-steps Hamiltonian if $h_1 + h_2 + \cdots + h_{n-1} + (n-1)$ is even. When n and $h_1 + h_2 + \cdots + h_{n-1}$ have the same parity, $h_1 + h_2 + \cdots + h_{n-1} + (n-1)$ is odd. By Theorem 1, Sub(W(n), X, f) is hereditary non 2-steps Hamiltonian. When n and $h_1 + h_2 + \cdots + h_{n-1}$ have different parity, $h_1 + h_2 + \cdots + h_{n-1} + (n-1)$ is even. Thus, we need to check other vertex deleting subgrpahs.

Now, we know that we only need to focus on the case when n and $h_1 + h_2 + \cdots + h_{n-1}$ have different parity and there is a h_i for some i which equals to 2 by removing a $v_{i,j}$ vertex or a v_i vertex. By Theorem 1, for Sub(W(n), X, f) to be 2-steps Hamiltonian when n and $h_1 + h_2 + \cdots + h_{n-1}$ have different parity, it requires that there is a h_i for some i which equals to 2. But, no mater you remove a vertex $v_{i,1}$ or $v_{i,2}$, you have a subgraph with an order 1 vertex. Obviously, any grpah with an order 1 vertex cannot be k-steps Hamiltonian. Therefore, even when n and $h_1 + h_2 + \cdots + h_{n-1}$ have different parity, Sub(W(n), X, f) is hereditary non 2-steps Hamiltonian. □

Note that even though we can determine whether a Sub(W(n), X, f) graph is hereditary 2-steps Hamiltonian or not, we still have no clue what happen if we remove a vertex on the circle v_i. As far as we know, if we remove a vertex v_i where $h_i \geq 1$, then it will create an order 1 vertex $v_{i,1}$ which tells that it is not 2-steps Hamiltonian. But, when you remove a vertex v_i where $h_{i-1} = h_i = h_{i+1} \geq 0$ (subscripts are module $n-1$), then it becomes a broken fan graph, which is stil open to determine whether a broken fan is a 2-steps Hamiltonian or not.

3 Subdivision Graphs of $C_4 \times K_2$ on Its Perfect Matching

After looking at the wheel graphs, it is natural to extend the study to a graph with a circle on the outside. So, we turn our attention to $C_n \times K_2$ where $n \geq 4$. (The condition $n \geq 4$ comes from wheel graphs. The $C_3 \times K_2$ is studied in our next

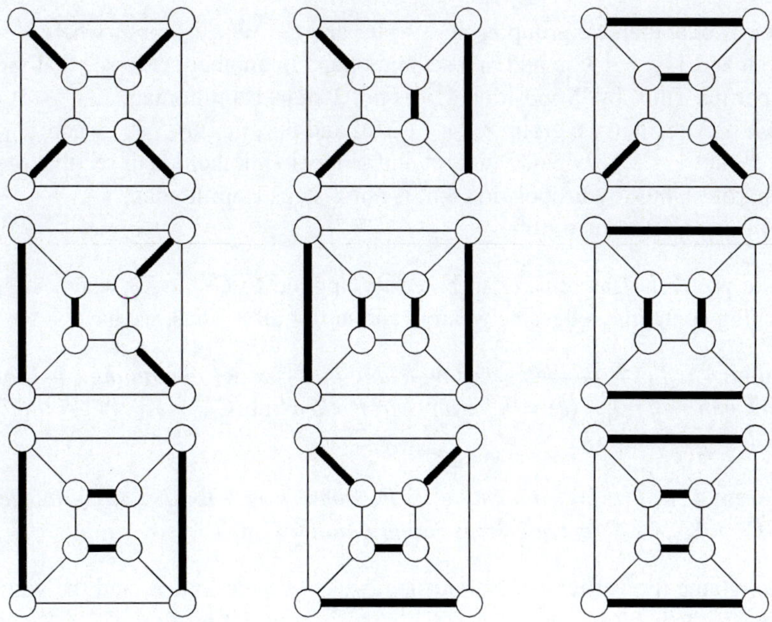

Fig. 8 Nine perfect matchings of the $C_4 \times K_2$

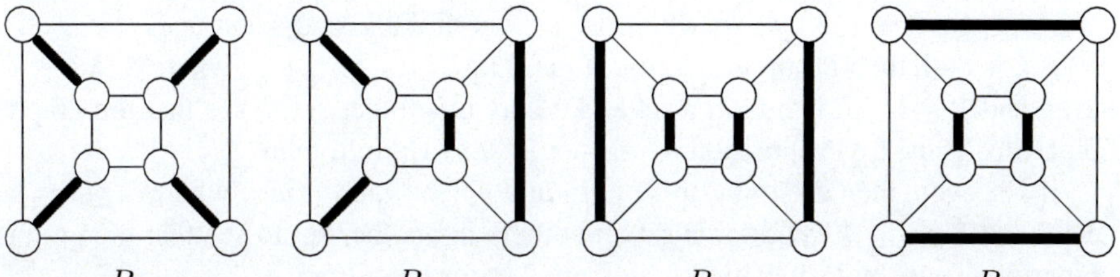

Fig. 9 Four up-to-isomorphic perfect matchings of the $C_4 \times K_2$

project.) Due to many edges to choose from and a AL(2)-traversal path needs to visit all vertices, we decide to start with $C_4 \times K_2$ with its perfect matchings (see Fig. 8).

Up to isomorphism there are four types of perfect matchings. (See Fig. 9 with the name for each perfect matching we are using in this paper.)

Theorem 3 *If $f(e) = k$ for any e in P_1 where k is a fixed positive integer, then* $\mathrm{Sub(C_4 \times K_2, P_1, f)}$ *is not 2-steps Hamiltonian for any k.*

Proof Name the vertices on the outside cycle by v_1, v_2, v_3, v_4 and the vertices on the inside cycle by w_1, w_2, w_3, w_4 where v_1 is adjacent to w_1 in $C_4 \times K_2$. We also name the inserted vertices in P_1 are on the edge between v_i and w_i for $i = 1, 2, 3, 4$ by $v_{i,1}, v_{i,2}, \ldots, v_{i,k}$ where v_i is adjacent $v_{i,1}$.

If k is odd, then we group v_1, w_1, v_3, w_3 and $v_{1,t}, v_{3,t}, v_{2,s}, v_{4,s}$ where $2 \leq t \leq k$ is even and $1 \leq s \leq k$ is odd in a set and others in another set to see that the graph is bipartite. Thus, by Proposition 4, it is not 2-steps Hamiltonian.

If k is even, then we group v_1, w_2, v_3, w_4 and $v_{1,t}, v_{3,t}, v_{2,s}, v_{4,s}$ where $2 \leq t \leq k$ is even and $1 \leq s \leq k$ is odd in a set and others in another set to see that the graph is bipartite. Thus, by Proposition 4, it is not 2-steps Hamiltonian.

This completes the proof. □

The proof in Theorem 3 can be easily applied to $C_{2n} \times K_2$ where the perfect matching containing edges between v_i and w_i for all i. Thus, we have

Corollary 1 *Let P be the perfect matching in $C_{2n} \times K_2$ containing edges between v_i and w_i for all i. If $f(e) = k$ for any e in P, then* $\mathrm{Sub}(\mathrm{C_{2n}} \times \mathrm{K_2}, \mathrm{P}, \mathrm{f})$ *is not 2-steps Hamiltonian for any k.*

Theorem 4 *If $f(e) = k$ for any e in P_3 where k is a fixed positive integer, then* $\mathrm{Sub}(\mathrm{C_4} \times \mathrm{K_2}, \mathrm{P_3}, \mathrm{f})$ *is not 2-steps Hamiltonian for any k.*

Proof Name the vertices on the outside cycle by v_1, v_2, v_3, v_4 and the vertices on the inside cycle by w_1, w_2, w_3, w_4 where v_1 is adjacent w_1 to in $C_4 \times K_2$. We also name the inserted vertices in the edge between v_i and v_{i+1} for $i = 1, 3$ by $v_{i,j}$ where $1 \leq j \leq k$ and v_i is adjacent $v_{i,1}$ and the inserted vertices in the edge between w_i and w_{i+1} for $i = 1, 3$ by $w_{i,j}$ where $1 \leq j \leq k$ and w_i is adjacent $w_{i,1}$.

If k is odd, then group v_1, v_2, w_3, w_4 and $v_{1,t}, w_{3,t}, v_{3,s}, w_{1,s}$ where $2 \leq t \leq k$ is even and $1 \leq s \leq k$ is odd in a set and others in another set to see that the graph is bipartite. Thus, by Proposition 4, it is not 2-steps Hamiltonian.

If k is even, then group v_1, w_2, v_3, w_4 and $v_{1,t}, v_{3,t}, w_{1,s}, w_{3,s}$ where $2 \leq t \leq k$ is even and $1 \leq s \leq k$ is odd in a set and others in another set to see that the graph is bipartite. Thus, by Proposition 4, it is not 2-steps Hamiltonian.

This completes the proof. □

The proof in Theorem 4 can be easily applied to $C_{2n} \times K_2$ where the perfect matching containing edges between v_{2i-1} and v_{2i} and between w_{2i-1} and w_{2i} for all $1 \leq i \leq n$. Thus, we have

Corollary 2 *Let P be the perfect matching in $C_{2n} \times K_2$ containing edges between v_{2i-1} and v_{2i} and between w_{2i-1} and w_{2i} for all $1 \leq i \leq n$. If $f(e) = k$ for any e in P, then* $\mathrm{Sub}(\mathrm{C_{2n}} \times \mathrm{K_2}, \mathrm{P}, \mathrm{f} : \mathrm{P} \to \mathrm{N})$ *is not 2-steps Hamiltonian for any k.*

While we investigate the subdivision graphs of $C_n \times K_2$ with perfectly matchings, we realize that if you insert too many vertices, then it would be impossible to be 2-steps Hamiltonian.

Theorem 5 *Let P be a perfect matching in $C_n \times K_2$. For any $k \geq 3$, if $f(e) = k$ for any e in P where k is a fixed positive integer, then* $\mathrm{Sub}(\mathrm{C_n} \times \mathrm{K_2}, \mathrm{P}, \mathrm{f})$ *is not 2-steps Hamiltonian.*

Proof Name the vertices on the outside cycle by $v_1, v_2, \ldots, v_n$ and the vertices on the inside cycle by $w_1, w_2, \ldots, w_n$ where v_1 is adjacent w_1 to in $C_n \times K_2$. Since there are n edges in P, we name the inserted vertices in an edge by $u_{i,j}$ where $1 \leq i \leq n$ and $1 \leq j \leq k$.

For any edge in the perfect matching, to reach the vertices in the middle of the path, i.e., $u_{i,j}$ where $2 \leq j \leq k-1$, we need to travel from and through two end vertices on the cycle part. Therefore, all v_i and w_i where $1 \leq i \leq n$ would be visited in any 2-steps Hamiltonian labeling. But, at the same time, since there are only three distance 2 vertices to any $u_{i,1}$ are $u_{i,3}$, v_{i+1} and v_{i-1}, to label $u_{i,1}$, we need to come from or go through v_{i+1} or v_{i-1}. Similarly, since there are only three distance 2 vertices to any $u_{i,k}$ are $u_{i,k-2}$, w_{i+1} and w_{i-1}, to label $u_{i,k}$, we need to come from or go through w_{i+1} or w_{i-1}. Thus, since there are $2n$ vertices in this kind of position, we would visit all v_i and w_i $2n$ times. Totally, we would visit all v_i and w_i $4n$ times. It is impossible. This completes the proof. □

Theorem 5 reduces the amount of the subdivision graphs, $\text{Sub}(\text{C}_\text{n} \times \text{K}_2, \text{P}, \text{f})$ for any n, we need to check from infinity to finite. Thus, from now on, we can only focus on $k = 1$ or 2.

Theorem 6 *If $f(e) = k$ for any e in P_2 where k is a fixed positive integer, then* $\text{Sub}(\text{C}_4 \times \text{K}_2, \text{P}_2, \text{f})$ *is 2-steps Hamiltonian if and only if k is* 1.

Proof By Theorem 5, $\text{Sub}(\text{C}_4 \times \text{K}_2, \text{P}_2, \text{f})$ is not 2-steps Hamiltonian when $k \geq 3$.

When $k = 2$, we name the vertices on the outside cycle by v_1, v_2, v_3 and v_4 and the vertices on the inside cycle by w_1, w_2, w_3 and w_4 where v_1 is adjacent w_1 to in $C_4 \times K_2$. We also name the inserted vertices in the edge between v_i and w_i for $i = 1, 2$ by $u_{i,j}$ where $1 \leq j \leq 2$ and v_i is adjacent $u_{i,1}$, the inserted vertices in the edge between w_3 and w_4 by $u_{3,j}$ where $1 \leq j \leq 2$ and w_3 is adjacent $u_{3,1}$, and the inserted vertices in the edge between v_3 and v_4 by $u_{4,j}$ where $1 \leq j \leq 2$ and v_3 is adjacent $u_{4,1}$. If we group $v_1, u_{1,2}, u_{2,1}, w_4, w_2, u_{4,2}, u_{3,1}, v_3$ and others in another set to see that the graph is bipartite. Thus, by Proposition 4, it is not 2-steps Hamiltonian.

When $k = 1$, the following graph shows that it is 2-steps Hamiltonian.

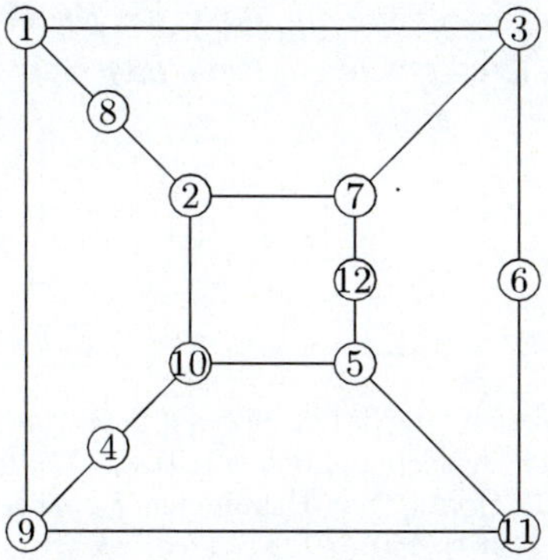

This completes the proof. □

Theorem 7 *If $f(e) = k$ for any e in P_4 where k is a fixed positive integer, then* $\mathrm{Sub}(C_4 \times K_2, P_4, f)$ *is 2-steps Hamiltonian if and only if k is* 1.

Proof By Theorem 5, $\mathrm{Sub}(C_4 \times K_2, P_2, f)$ is not 2-steps Hamiltonian when $k \geq 3$.

When $k = 2$, we name the vertices on the outside cycle by v_1, v_2, v_3 and v_4 and the vertices on the inside cycle by w_1, w_2, w_3 and w_4 where v_1 is adjacent w_1 to in $C_4 \times K_2$. We also name the inserted vertices in the edge between v_i and v_{5-i} for $i = 1, 2$ by $v_{i,j}$ where $1 \leq j \leq 2$ and v_i is adjacent $v_{i,1}$ and the inserted vertices in the edge between w_i and w_{i+1} for $i = 1, 3$ by $w_{i,j}$ where $1 \leq j \leq 2$ and w_i is adjacent $w_{i,1}$. If we group v_1, $v_{1,2}$, $v_{2,1}$, w_4, v_3, $w_{3,1}$, $w_{1,1}$, w_2 and others in another set to see that the graph is bipartite. Thus, by Proposition 4, it is not 2-steps Hamiltonian.

When $k = 1$, the following graph shows that it is 2-steps Hamiltonian.

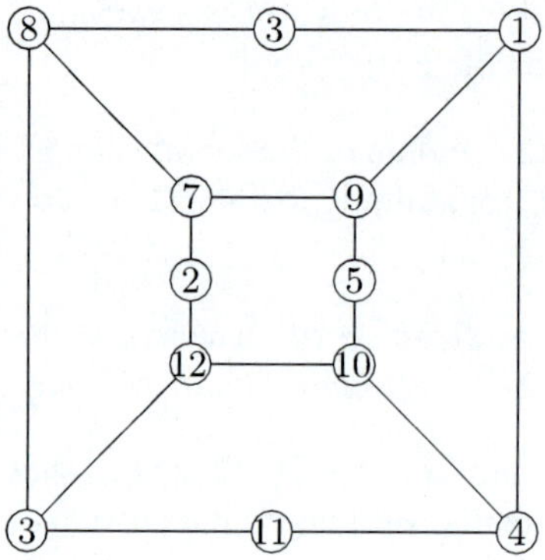

This completes the proof. □

Since we only insert vertices on the perfect matching of $C_4 \times K_2$, when we remove a vertex, it removes an edge on one of the perfect matching edge. It creates an order 1 vertex in the vertex deleting graph. Obviously, any grpah with an order 1 vertex cannot be k-steps Hamiltonian.

Theorem 8 *Let P be a perfect matching in $C_n \times K_2$. For any $k \geq 3$, if $f(e) = k$ for any e in P where k is a fixed positive integer, then* $\mathrm{Sub}(C_n \times K_2, P, f)$ *is hereditary non 2-steps Hamiltonian.*

References

1. J.C. Bermond, Hamiltonian Graphs. *In Selected Topics in Graph Theory. Edited by L. W. Beineke and R. J. Wilson*. Academic, London (1978), 127–167.
2. M.N. Ellingham and J.D. Horton, Non-Hamiltonian 3-connected Cubic Bipartite Graphs, *J. of Comb. Theory B* **34(3)** (1983), 350–353.
3. R. Gould, Advances on the Hamiltonian Problem, A Survey, *Graphs and Combinatorics*, **19** (2003), 7–52.
4. R.K. Guy and F. Harary, On the Möbius Ladders. *Canad. Math. Bulletin*, **10** (1967), 493–496.

5. Y.S. Ho, S.M. Lee and S.P.B. Lo, On 2-steps Hamiltonian Cubic Graphs, *J. Combin. Math. Combin. Comput.*, **98** (2016), 185–199.
6. D. Holton and R.E.L. Aldred, Planar Graphs, Regular Graphs, Bipartite Graphs and Hamiltonicity, *Austral. J. Combin.*, **20** (1999), 111–131.
7. G.C. Lau, S.M. Lee, K. Schaffer and S.M. Tong, On k-step Hamiltonian Bipartite and Tripartite Graphs, *Malaya J. Mat.*, **2(3)** (2014), 180–187.
8. G.C. Lau, S.M. Lee, K. Schaffer, S.M. Tong and S. Lui, On k-step Hamiltonian Graphs, *J. Combin. Math. Combin. Comput.*, **90** (2014), 145–158.
9. A.N.T. Lee and S.M. Lee, On super edge-magic graphs with many odd cycles, *Congr. Numer.*, **163** (2003), 65–80.
10. S.M. Lee and H.H. Su, The 2-steps Hamiltonian Subdivision Graphs of Cycles with a Chord, *J. Combin. Math. Combin. Comput.*, **98** (2016), 109–123.
11. T.C. Mai, J.J. Wang and L.H. Hsu, Hyper-Hamiltonian generalized Petersen graphs, *Computers & Mathematics with Applications*, **55** (2008), 2076–2085.
12. J. Moon and L. Moser, On Hamiltonian Bipartite Graphs, *Israel J. Math.*, **1** (1963),163–165.
13. W.D. Wallis, Magic Graphs. *Birkhäuser*, 2001, Boston, MA.

On the Erdős-Sós Conjecture for Graphs with Circumference at Most $k + 1$

A. M. Heissan and Gary Tiner

Abstract Let G be a graph with average degree $\bar{d}(G)$ greater than $k - 2$. Erdős and Sós conjectured that G contains every tree on k vertices as a subgraph. The circumference of the graph G, $c(G)$, is the number of edges on a longest cycle. Gilbert and Tiner proved that if $c(G)$ is at most k, then G contains every tree on k vertices. In this paper, we improve this result and show that the Erdős-Sós conjecture holds for graphs whose circumference is at most $k + 1$.

1 Introduction

The average degree of the graph G is $\bar{d}(G)$, where $\bar{d}(G) = 2e(G)/|V(G)|$. Erdős and Sós conjectured the following:

Erdős-Sós Conjecture. *If G is a graph with $\bar{d}(G) > k - 2$, then G contains every tree on k vertices.*

Various special cases of the conjecture have been proven, some of which place restrictions on the graph G. The cases where the graph G has number of vertices k, $k + 1$, $k + 2$, or $k + 3$ were proved by Zhou [11], Slater, Teo, and Yap [6], Woźniak [9], and Tiner [8], respectively. The case where G has $k + 4$ vertices was proved by Yuan and Zhang [10]. We state these results in a single theorem.

Theorem 1 *If G is a graph with $\bar{d}(G) > k - 2$ on at most $k + 4$ vertices, then G contains every tree on k vertices.*

The number of edges on a longest path in a tree T is the $diameter$ of T, or simply $diam(T)$. McLennan [5] proved the following:

A. M. Heissan (✉)
Manhattanville College, Purchase, NY 10577, USA
e-mail: a.heissan@mville.edu

G. Tiner
Faulkner University, Montgomery, AL 36109, USA
e-mail: gtiner@faulkner.edu

F. Hoffman (ed.), *Combinatorics, Graph Theory and Computing*,
Springer Proceedings in Mathematics & Statistics 388,
https://doi.org/10.1007/978-3-031-05375-7_17

Theorem 2 *If G is a graph with $\bar{d}(G) > k - 2$ then G contains every tree on k vertices that has diameter at most 4.*

If a tree T is made up of a path $a_1, \ldots, a_r$, where $r \geq 2$, and all of the remaining vertices $V(T) - \{a_1, \ldots, a_r\}$ are neighbors of a_r, then the tree T is a $broom$. If each remaining vertex is either a neighbor of a_1 or a_r, then the tree T is a $double\text{-}broom$. Notice that a path is a broom, and a broom is a double-broom. Tiner [7] proved the following theorem:

Theorem 3 *If G is a graph with $\bar{d}(G) > k - 2$, then the graph G contains every double-broom on k vertices.*

Eaton and Tiner [1] showed that the Erdős-Sós conjecture holds for values of k at most 8. Tiner and Tomlin proved the conjecture holds for $k = 9$ (see acknowledgments). We state this as a theorem.

Theorem 4 *For k at most 9, if G is a graph with $\bar{d}(G) > k - 2$, then G contains every tree on k vertices.*

Eaton and Tiner [2] proved the following theorem:

Theorem 5 *If G is a graph with $\bar{d}(G) > k - 2$ that has no path on $k + 4$ vertices, then G contains every tree on k vertices.*

For a subgraph W of G, the subgraph $G - W$ is obtained from G by removing $V(W)$ and each edge with an endpoint in W. The set of natural numbers is $\mathbb{N}$, and for $m \in \mathbb{N}$, the set $[m]$ is $\{1, \ldots, m\}$. Let G be a graph, and let u and v be two vertices in $V(G)$. The set of neighbors of v is $N(v)$, where $N(v) = \{w \in V(G) : vw \in E(G)\}$. The degree of v is $d_G(v)$, or simply $d(v)$, and $d(v) = |N(v)|$. The minimum degree among all vertices in $V(G)$ is $\delta(G)$, and the maximum degree is $\Delta(G)$. If $U \subseteq V(G)$, then $N(U) = \{w : w \in N(u) \text{ for some } u \in U\}$.

For two disjoint subgraphs $C, D \subseteq G$, the set of edges with one end-point in $V(C)$ and one in $V(D)$ is $E(C, D)$; the number of edges in $E(C, D)$ is $e(C, D)$. The subgraph induced by $V(C)$ is $G[C]$. The edge-set of $G[C]$ is $E(C, C)$ or simply $E(C)$, and $e(C)$ is the number of edges in $E(C)$.

Choose $A, B \subseteq V(G)$ and let $a \in A$ and $b \in B$. If $ab \in E(G)$, then the vertex a $hits$ B and the subset B $hits$ A. If no vertex in A hits B, then A $misses$ B. If A and B are disjoint sets, then the bipartite subgraph of G with bipartition A, B is $G[A, B]$, and $e(A, B) = e(G[A, B])$.

The $circumference$ of the graph G, $c(G)$, is the number of edges on a longest cycle. Gilbert and Tiner [4] proved the following:

Theorem 6 *If G is a graph with $\bar{d}(G) > k - 2$ and circumference at most k, then G contains every tree on k vertices.*

Let P be an r-path in a graph G, where $P = v_1, \ldots, v_r$. A *path on the vertex set V(P)*, or simply a *path on V(P)*, is an r-path in G whose vertex set is $V(P)$. For distinct vertices v_i and v_j on the path P, if there is a path (in G) on $V(P)$ whose end-vertices are v_i and v_j, then it is a v_i,v_j*-path on* $V(P)$. For a vertex v_t on the path P,

$$\alpha(P, v_t) = \{v_s \in V(P) : \text{there is a } v_s, v_t\text{-path on } V(P)\}.$$

For each $v_i \in N_P(v_1)$, the path $v_{i-1}, \ldots, v_1, v_i, \ldots, v_r$ is a v_{i-1}, v_r-path on $V(P)$. It follows that $v_{i-1} \in \alpha(P, v_r)$, and $e(v_1, P) \leq |\alpha(P, v_r)|$. We state this more generally in the following lemma:

Lemma 7 *If P is a path in a graph G, where $P = v_1, \ldots, v_r$, then $e(v_i, P) \leq |\alpha(P, v_r)|$ for all $v_i \in \alpha(P, v_r)$.*

Gilbert and Tiner [3, 4] proved the following two lemmas, respectively:

Lemma 8 *Let G be a graph that is minimal with $\bar{d}(G) > k - 2$, and let P be a path in G, where $P = v_1, \ldots, v_r$. If $r \leq k - 2$, then a vertex in $\alpha(P, v_1)$ hits $\lfloor \frac{1}{2}(k - r) \rfloor$ vertices outside of $V(P)$.*

Lemma 9 *Let G be a graph that is minimal with $\bar{d}(G) > k - 2$, and let Q be a path in G, where $Q = v_1, \ldots, v_r$. For $W = \alpha(Q, v_r)$, assume $N(W) \subseteq V(Q)$. If $r \leq k$, then G contains every tree on k vertices.*

Notice that in Lemma 9, if Q is a longest path having v_r as one end-vertex, then it must be that $N(W) \subseteq V(Q)$. In this paper, we use Lemma 9 to prove our main theorem (Theorem 10), a special case of the Erdős-Sós Conjecture.

Theorem 10 *If G is a graph with $\bar{d}(G) > k - 2$ and $c(G) \leq k + 1$, then G contains every tree on k vertices.*

Notice that in Theorem 10, no upper bound is imposed on the number of vertices in G, or on the length of a longest path in G.

2 Supporting Lemmas

The number of edges with at least one endpoint in A is $e^*_G(A)$ or simply $e^*(A)$. Note that

$$e^*(A) = \sum_{v \in A} d(v) - e(A) = e(A) + e(A, G - A).$$

A proof of the following lemma is in [1]:

Lemma 11 *Let G be a graph with $\bar{d}(G) > k - 2$. Let $W \subsetneq V(G)$ and $G' = G - W$. If $e^*(W) \leq \frac{1}{2}(k - 2)|W|$, then $\bar{d}(G') > k - 2$.*

The following two corollaries follow from Lemma 11:

Corollary 12 *If a graph G is minimal with $\bar{d}(G) > k-2$ and $W \subsetneq V(G)$, then $e^*_G(W) > \frac{1}{2} \cdot |W|(k-2)$. In particular,*
i. $\delta(G) \geq \lfloor \frac{k}{2} \rfloor$, *and*
ii. *if k is odd and $uv \in E(G)$, then one of $\{u, v\}$ has degree at least $\lfloor \frac{k}{2} \rfloor + 1$, and*
iii. *if (w, x, y) is a 3-cycle in G, then one of $\{w, x, y\}$ has degree at least $\lfloor \frac{k}{2} \rfloor + 1$.*

Corollary 13 *Let G be a graph that is minimal with $e(G) > \bar{d}(G) > k-2$, and let W be a subset of $V(G)$. If $1 \leq |W| \leq k-1$,*

$$e(W, G-W) > \frac{1}{2}|W|(k-2) - \binom{|W|}{2} = \frac{1}{2}|W|(k-|W|-1), \qquad (1)$$

and a vertex v in W hits at least $\frac{1}{2}(k-|W|)$ vertices in $G-W$.

A proof of Lemma 14 is in [7].

Lemma 14 *Let G be a graph that is minimal with $\bar{d}(G) > k-2$. Let Q be a path in G, where $Q = v_1, \ldots, v_r$, and let $W = \alpha(Q, v_r)$. If $N(W) \subseteq V(Q)$, then W hits a vertex in $\{v_{k-1}, \ldots, v_r\}$ and $r \geq k-1$.*

Gilbert and Tiner [4] proved the following:

Lemma 15 *Let G be a graph on k vertices. If $e(G) = \binom{k-1}{2} + 1$, then G contains every tree on k vertices that is not a star.*

Let T be a tree and let t be a vertex in T. The set of leaf neighbors of t is $L(t)$, and $L[t] = L(t) \cup \{t\}$. An embedding f of a tree T into a graph G is an injective map $f : V(T) \to V(G)$ that preserves edges, that is, if $ab \in E(T)$, then $f(a)f(b) \in E(f(T))$. Let $T' \subseteq T$ be a tree. If an embedding of T' into a graph G can be extended to an embedding of T into G, then the graph G is *T-extensible*.

Lemma 16 *Let G be a graph G on n vertices with more than $\frac{1}{2}(n-1)(k-2)$ edges, where $k \leq n \leq k+3$, and let T be a tree on k vertices. If G has a vertex v of degree $n-1$, then G contains T.*

Proof If T is a star, then G contains T since $d(v) \geq k-1$. Otherwise T is not a star. Let $a_0, \ldots, a_r$ be a longest path in T. If a_1 has two or more leaf neighbors in T, then let T' be obtained from T by removing two of the leaf neighbors or a_1. Otherwise a_1 has exactly one leaf neighbor, and let $T' = T - L[a_1]$. Let $k' = k-2$ and notice that T' has exactly k' vertices. Let $G' = G - v$, and for $n' = n-1$, notice that G' has n' vertices, and $n' \leq k' + 4$. It follows that

$$e(G') > \frac{1}{2}(n-1)(k-2) - (n-1) = \frac{1}{2}(n-1)(k-4) = \frac{1}{2}(n')(k'-2),$$

and G' contains T' (by Theorem 1). Since v hits every vertex in G', the embedding of T' into G' is T-extensible.

If a vertex t in a tree T has at least one leaf neighbor, and exactly one non-leaf neighbor, then the vertex t is a *penultimate* vertex.

Theorem 17 *Let G be an n vertex graph, where $n = k + 1$, and $k \geq 4$. If*

$$e(G) > \frac{1}{2}k(k-2)$$

then G contains every non-star tree on k vertices.

Proof Let T be a non-star tree on k vertices, and assume the graph G has exactly $\lfloor \frac{1}{2}k(k-2)+1 \rfloor$ edges. If a vertex u in G has degree less than $\lfloor \frac{k}{2} \rfloor$, then $G - u$ contains T (by Lemma 15).

Otherwise $\delta(G) \geq \lfloor \frac{k}{2} \rfloor$. It is worth noting that the statement holds for $k = 1$ and $k = 2$, but not for $k = 3$. The case for $k = 3$ fails only when the graph G consists of two vertex disjoint edges.

If $k = 4$, then T is P_4, the graph G has five vertices and five edges. Since each vertex has degree at least 2, the graph G is C_5, and G contains T.

Otherwise $k \geq 5$. If $k = 5$, the graph G has six vertices, eight edges, minimum degree 2, and the tree T is either a path or a broom. By the ES-conjecture, the graph G contains P_4. Let $v_1, \ldots, v_4$ be P_4 in G, and let v_5 and v_6 be the other two edges. Since there are five additional edges, and both v_5 and v_6 have minimum degree 2, it is easy to see that G contains T.

Otherwise $k \geq 6$. Notice that $k - 2 \leq \Delta(G) \leq k$. Let $V(G) = \{v_1, \ldots, v_k\}$, where the vertices are listed in non-increasing order. If $d(v_1) = k$, then G contains T (by Lemma 16). Otherwise $d(v_1)$ is either $k - 1$ or $k - 2$.

Case 1 $d(v_1) = k - 1$

If T has a vertex t with at least two leaf neighbors, then let $T' = T - L(t)$ (and notice that the sub-tree T' has at most $k - 2$ vertices). Let $N_{T'}(t) = \{t_1, \ldots, t_s\}$. Let $\{H_1, \ldots, H_s\}$ be the components of $T' - t_i$, where t_i is a vertex in H_i for $1 \leq i \leq s$. Notice that each H_i is a tree and that the disjoint union of the trees in $\{H_1, \ldots, H_s\}$ is a forest. Let T'' be the tree on $V(T) - L[t]$ obtained from the s trees in forest by adding the $s - 1$ edges $\{t_1t_2, \ldots, t_{s-1}t_s\}$. Notice that the tree T'' has $|V(T')| - 1$ vertices (i.e., at most $k - 3$ vertices).

Let $G' = G - \{v_1, v_j\}$, where v_j is the single vertex in $V(G - v_1)$ that is missed by v_1. Notice that G' has $k - 1$ vertices, and

$$e(G') \geq e(G) - (2k-2) > \frac{1}{2}k(k-2) - (2k-2) > \frac{1}{2}(k-1)(k-5),$$

and G' contains T'' (by Lemma 16). Since the vertex v_1 hits each vertex in $\{t_1, \ldots, t_s\}$, and since $d(v_1)$ has degree $k - 1$, we see that G contains T.

If $\mathrm{diam}(T) = 3$, then let $v_j \in N(v_1)$, and consider the edge v_1v_j in G. Since v_1 and v_j have degrees $k - 1$ and at least m, respectively, the edge v_1v_j is T-extensible.

Otherwise $\mathrm{diam}(T) \geq 4$ (and $k \geq 5$). If T is a broom (which could be a path), then let $G' = G - v_1$. Notice that G' has k vertices, and

$$e(G') \geq e(G) - (k-1) > \frac{1}{2}k(k-2) - (k-1) > \frac{1}{2}k(k-4).$$

Thus G' contains the path $u_1, \ldots, u_{k-2}$ (by Theorem 1). Since v_1 hits one of $\{u_1, u_{k-2}\}$ (and since $d(v_1) = k-1$), it is easy to see that the graph G contains T.

Otherwise T is not a broom (or a path). If $k = 6$, then the only non-broom tree of diameter at least four is the 4-path $a_0, \ldots, a_4$ along with the edge a_2x. Define G' and the path $u_1, \ldots, u_{k-2}$ as in the previous paragraph, where $k-2 = 4$. If the vertex v_1 hits both u_1 and u_4, then the path u_2, u_1, v_1, u_4, u_3 it T-extensible. Otherwise the vertex v_1 misses one of u_1 and u_4, and thus hits both u_2 and u_3. It follows that the path u_1, u_2, v_1, u_3, u_4 is T-extensible in G.

Otherwise $k \geq 7$. If $d(v_j) \leq k-4$, then let $T' = T - L[a_1]$, and let $G' = G - \{v_1, v_j\}$. Notice that T' has $k-2$ vertices, the graph G has $k-1$ vertices, and

$$e(G') > \frac{1}{2}k(k-2) - (2k-5) \geq \frac{1}{2}(k-1)(k-4). \tag{2}$$

The latter inequality holds since $k \geq 5$. By Theorem 1, the graph G' contains T'. Since v_1 hits every vertex in G', we see that G contains T.

Otherwise $k-3 \leq d(v_j) \leq k-1$. If $d(v_{k-1}) \geq k-2$, then the degree sum S of G is such that

$$k(k-2) > S \geq 1(k-1) + (k-2)(k-2) + 2m \geq k(k-2),$$

a contradiction.

Otherwise $d(v_{k-1}) \leq k-3$. Since v_j misses v_1 and at most two other vertices in G, it must hit one of $\{v_{k-1}, v_k, v_{k+1}\}$, each of which has degree at most $k-3$. Assume $d(v_k) \leq k-3$ (and notice that v_1 hits v_k). Let $T' = T - L[a_1]$, let $G' = G - \{v_1, v_k\}$, and notice that $e^*(\{v_1, v_k\}) \leq (2k-5)$. Thus G' contains T' (see the paragraph containing Inequality (2) above). Since $N(\{v_1, v_k\}) = V(G)$, we see that G contains T.

Case 2 $d(v_1) = k-2$

Let $B \subseteq V(G)$ be the set of degree $k-2$ vertices. Let S be the sum of the degrees of the vertices in G, and notice that $S \geq k(k-2)+1$. If $|B| \leq 3$, then

$$S \leq 3(k-2) + (k+1-3)(k-3) \leq k(k-2) < S,$$

a contradiction.

Otherwise $|B| \geq 4$. If a vertex v in G misses B, then $d(v)$ vertices in G have degree at most $k-3$, and

$$S \leq d(v) + d(v)(k-3) + (k+1-(d(v)+1))(k-2) \leq k(k-2) < S,$$

a contradiction.

Otherwise every vertex in G hits a vertex in B. If $\text{diam}(T) = 3$, then T is a double-broom. Let $v \in B$, and let $\alpha \in \overline{N[v]}$. Let $w \in B$ be a neighbor of α. Since α has degree at least $\lfloor \frac{k}{2} \rfloor$, and since v has degree $k-2$ (and v misses w), the 3-path v, α, w is T-extensible.

Otherwise $\text{diam}(T) \geq 4$ (and $k \geq 6$). If two vertices $u, v \in B$ miss each other, then G contains T. Let $G' = G - \{u, v\}$ and let $T' = T - \{a_0, a_r\}$. Notice that T' has $k-2$ vertices, and the subgraph G' has $k-1$ vertices. Since

$$e(G') > \frac{1}{2}k(k-2) - 2(k-2) = \frac{1}{2}(k-2)(k-4), \tag{3}$$

the graph G' contains T' (by the induction hypothesis).

If $N(u) = N(v)$, then let q be the single vertex in $V(G) - \{u, v\}$ that misses $\{u, v\}$. Since one of $\{u, v\}$ hits $f(a_{r-1})$, suppose v hits it, and set $f(a_r) = v$. If u hits $f(a_1)$, then set $f(a_0) = u$, and f is an embedding of T into G. Otherwise u misses $f(a_1)$ and hits $f(a_2)$. Set $f(a_1) = u$ and f is T-extensible.

Otherwise $N(u) \neq N(v)$ (so $N[u] \cup N[v] = V(G)$). Since one of $\{u, v\}$ hits $f(a_{r-1})$, suppose v hits it, and set $f(a_r) = v$. If u hits $f(a_1)$, then set $f(a_0) = u$, and f is an embedding of T into G. Otherwise u misses $f(a_1)$ and hits $f(a_2)$. Set $f(a_1) = u$, and we see that f is T-extensible.

Otherwise, no two vertices $u, v \in B$ miss each other. Thus B is a clique in G. If $k = 6$, then $3 \leq d(v) \leq 4$ for each $v \in V(G)$, and the degree sequence is $(4, 4, 4, 4, 4, 3, 3)$. Since B is a 5-clique in G, no vertex in B hits a vertex in $G - B$, a contradiction (since the two vertices in $V(G) - B$ have degree 3).

Otherwise $k \geq 7$. Since every vertex has a neighbor in B, it is easy to see that all of the vertices in B cannot have the same closed neighborhoods. Let $u, v \in B$ have different closed neighborhoods.

If $N(u) \cup N(v) = V(G)$, then let $G' = G - \{u, v\}$ and let $T' = T - \{a_0, a_r\}$. By Inequality (3) above, G' contains T'.

Since one of $\{u, v\}$ hits $f(a_{r-1})$, suppose v hits it, and set $f(a_r) = v$. If u hits $f(a_1)$, then set $f(a_0) = u$, and f is an embedding of T into G. Otherwise u misses $f(a_1)$ and hits $f(a_2)$. Set $f(a_1) = u$ and f is T-extensible.

Otherwise, $N(u) \cup N(v) \neq V(G)$. Let q be the single vertex that both u and v miss. Let $G' = G - \{u, v, q\}$ and let $T' = T - L[a_{r-1}] - \{a_r\}$. Notice that T' has $k-3$ vertices, the subgraph G' has $k-2$ vertices, and

$$e(G') > \frac{1}{2}k(k-2) - (3k-6) = \frac{1}{2}(k-3)(k-4),$$

and G' contains T' (by the induction hypothesis).

Since one of $\{u, v\}$ hits $f(a_2)$, suppose v does. Set $f(a_r) = v$. If v hits $f(a_1)$, then set $f(a_0) = v$ and f is an embedding of T into G. Otherwise v misses $f(a_1)$, and it hits $f(a_2)$. Set $f(a_1) = v$ and the embedding f is T-extensible.

The set of bipartite graphs with bipartition sizes of m and n, respectively, is $\mathcal{B}_{m,n}$. The following is an implication of a lemma proved by Eaton and Tiner [2] (see Lemma 2.5 in [2]).

Lemma 18 *Let T be a tree on k vertices with $diam(T) \geq 5$, and let $m = \lfloor \frac{k}{2} \rfloor$. For non-negative integers m_1 and m_2, let $B \in \mathcal{B}_{m+m_1,(k-3)+m_2}$. If $\delta(B) \geq m$, and*

$$e(B) \geq (m + m_1)(k - 3 + m_2) - [(m - 1) + (m_1 + m_2)].$$

then G contains T.

The set of trees having bipartition sizes of m and n is $\mathcal{T}_{m,n}$; clearly each tree T in $\mathcal{T}_{m,n}$ is also in $\mathcal{B}_{m,n}$. We state the following folklore lemma without proof.

Lemma 19 *Let A, B be the bipartition of the bipartite graph $H \in \mathcal{B}_{m,n}$, where A and B have numbers of vertices m and n, respectively, and let $T \in \mathcal{T}_{r,s}$. If $d(a) \geq s$ for each $a \in A$, and $d(b) \geq r$ for each $b \in B$, then the graph H contains T.*

Lemma 20 *For $k = 2m + 1$, where $m \geq 5$, let T be a tree on k vertices with $diam(T) \geq 5$. Let G be a graph on n vertices and more than $\frac{1}{2}(n-1)(k-2)$ edges, and let Y, Z be a vertex bipartition of $V(G)$. If $|Y| = m + 1$ and $|Z| \geq |Y| + 2$, and $e(Z) = 0$, then G contains T.*

Proof Let $s = |Z|$ and notice that $n = m + 1 + s \geq 2m + 3$. Since $\delta(G) \geq m$, we see that $e(Y, Z) \geq ms$.

If $k = 11$, then $n \geq 13$, $y = 6$ and $z \geq 7$. If $z = 7$, then $n = 13$, and $e = 55$. Thus $d(y_1) = 12$, and G contains T (by Lemma 16). Similarly, for $k = 13$, if $z = 8$ or 9, then $d(y_1) = n - 1$, and G contains T (by Lemma 16).

Otherwise, for $k = 11$ or 13, we have $z \geq k - 3$. For $k \geq 15$, we see that $e(G) \leq \binom{|Y|}{2} + |Y||Z|$. It follows that

$$\begin{aligned}\binom{m+1}{2} + (m+1)z \geq e(G) > \frac{1}{2}(n-1)(k-2) &= \frac{1}{2}(m+z)(2m-1) \\ \Longrightarrow z > \frac{1}{3}m(m-2) > 2m - 3 &= k - 4, \qquad (4)\end{aligned}$$

and $z \geq k - 3$.

Let $Z = \{z_1, \ldots, z_s\}$, and let $Y = \{y_1, \ldots, y_{m+1}\}$, where the vertices in each set are listed in non-increasing order. If at least $m - 2$ vertices in Z have degree $m + 1$, then at most $s - m + 2$ have degree m, and

$$\begin{aligned}e(Y, Z) &\geq (m-2)(m+1) + (s - m + 2)(m) \\ &= (m+1)(s) - [(m-1) + (1 + (s - (k-3)))], \qquad (5)\end{aligned}$$

and G contains T (by Lemma 18).

Otherwise at most $m-3$ vertices in Z have degree $m+1$. Since $e(G) > \frac{1}{2}(m+s)(2m-1)$, it follows that

$$\frac{1}{2}(m+s)(2m-1) < \binom{m+1}{2} + (m-3)(m+1) + (s-m+3)(m)$$
$$\implies s \geq m^2 - 4m + 7 \geq 3m - 4. \tag{6}$$

Assume that $s = 3m-4$ and that $d(z) = m$ for each $z \in Z$. It follows that $e(Y, Z) = m(3m-4)$. If $d(y_{m+1}) \geq k-3 = 2m-2$, then G contains T (by Lemma 19). Otherwise $d(y_{m+1}) \leq k-4 = 2m-3$. Therefore,

$$e(Y - y_{m+1}, Z) \geq m(3m-4) - (2m-3)$$
$$= m(3m-4) - [(m-1) + (s-(k-3))] \tag{7}$$

and G contains T (by Lemma 18).

3 Proof of the Main Theorem

In this section, we prove the main theorem (Theorem 10). The following is an implication of a theorem by Gilbert and Tiner [3] (see Lemma 8 in [3]):

Lemma 21 *Let C be an r-cycle in a graph G, let Q be an s-cycle in $G - C$, and assume C is a longest cycle in G. If $e(C, Q) = js + 1$ for $j \in \mathbb{N}$, then*

$$r \geq (j+1)(s+1).$$

As evident in the proof of the above lemma, the result holds when the term s-cycle is replaced with edge, and s is replaced with 2. We state this as a lemma.

Lemma 22 *Let C be an r-cycle in a graph G, let uv be an edge in $G - C$, and assume C is a longest cycle in G. If $e(C, Q) \geq 2j + 1$ for $j \in \mathbb{N}$, then*

$$r \geq 3(j+1).$$

We now restate and prove Theorem 10.

Theorem 10. Let T be a tree on k vertices. If G is a graph with $\bar{d}(G) > k-2$ and $c(G) \leq k+1$, then G contains T.

Proof If a subgraph G' of G that is minimal with $\bar{d}(G') > k-2$ contains every tree on k vertices, then so does G. Since G has circumference at most $k+1$, so does G'. For these reasons, we will simply assume that G is minimal with $\bar{d}(G) > k-2$, and has circumference at most $k+1$.

Let $m = \lfloor \frac{k}{2} \rfloor$. By Corollary 12, we see that $\delta(G) \geq m$. Let T be at tree on k vertices. If $k \leq 9$, then G contains T (by Theorem 4).

Otherwise $k \geq 10$. If $\mathrm{diam}(T) \leq 4$, then the graph G contains T (by Theorem 2).

Otherwise $\mathrm{diam}(T) \geq 5$. Let Q be a longest path in G, where $Q = v_1, \ldots, v_t$. If $t \leq k+3$, then G contains every tree on k vertices (by Theorem 5).

Otherwise $t \geq k+4$. Let $v_r \in V(Q)$ be a neighbor of v_1 in G, and choose the path Q on t vertices so that r is as large as possible. If $r \leq k$, then G contains T (by Theorem 9). Otherwise $r \geq k+1$. If $r \geq k+2$, then the cycle $(v_1, \ldots, v_r)$ has more than $k+1$ edges, and $c(G) > k+1$, a contradiction.

Otherwise $r = k+1$. Let C be the cycle $(v_1, \ldots, v_{k+1})$. Let $\mathcal{K}$ be the collection of components of $G - C$. We partition $\mathcal{K}$ as follows.

$Z \in \mathcal{K}$ is the component that contains vertices v_{k+2} and v_{k+3}.

$\mathcal{X} \subseteq \mathcal{K} - Z$ is the set of components that hit a vertex on $C - x_{k+1}$.

$\mathcal{Y} \subseteq \mathcal{K} - Z$ is the set of components that hit only x_{k+1} on C.

Claim 3 *No vertex in* $Z - v_{k+2}$ *hits a vertex in* $C - v_{k+1}$.

To the contrary, suppose a vertex w in $V(Z - v_{k+2})$ hits a vertex v_s on $C - v_{k+1}$. For simplicity, since there is a w, v_{k+2}-path in component Z, assume $w = v_{k+3}$, and so v_{k+3} hits v_s. Assume that $s \leq \frac{1}{2}(k+1)$ (otherwise, we could simply reverse the labels on the vertices on $v_1, \ldots, v_k$). Let R be the path $v_1, \ldots, v_{k+1}$ chosen so that s is as small as possible. By our choice of R, note that v_1 may no longer hit v_{k+1}.

If $1 \leq s \leq 2$, then the cycle $(v_s, \ldots, v_{k+3})$ has more than $k+1$ vertices, a contradiction.

Otherwise $3 \leq s \leq \frac{1}{2}(k+1)$. If v_1 hits $v_i \in \{v_{s+1}, \ldots, v_{2s-1}\}$, then the path $v_s, \ldots, v_1, v_{s+1}, \ldots, v_{k+1}$ contradicts our choice of R (since v_{k+3} hits v_s).

Otherwise v_1 misses $\{v_{s+1}, \ldots, v_{2s-1}\}$. If v_1 hits two consecutive vertices v_i, v_{i+1} on the $(k+2s+2)$-path $v_{2s}, \ldots, v_{k+1}$, then the path $v_2, \ldots, v_i, v_1, v_{i+1}, \ldots, v_{k+1}$ contradicts our choice of R (since v_{k+3} hits v_s).

Otherwise v_1 does not hit two consecutive vertices v_i, v_{i+1} on the path $v_{2s}, \ldots, v_{k+1}$. Thus we see that the vertex v_1 hits at most $\lceil \frac{1}{2}(k-2s+2) \rceil$ vertices on $\{v_{s+1}, \ldots, v_{k+1}\}$. Let $S = v_1, \ldots, v_s$. For $W = \alpha(S, v_s)$, we see that $1 \leq |W| \leq s-1$.

Recall (by Lemma 7), that $e(w, S) \leq |W|$ for each $w \in W$. By Corollary 13, a vertex in W hits $\frac{1}{2}(k - |W|)$ outside of W; assume it is v_1. Since v_1 hits at least $\frac{1}{2}(k - |W|)$ vertices outside of W (and outside of $S - v_s$), it might hit v_s, and it must hit $\frac{1}{2}(k - |W|) - 1$ vertices on the path $v_{2s}, \ldots, v_{k+1}$, no two of which are consecutive. Since the subpath has $k - 2s + 2$ vertices, and since v_1 does not hit two consecutive vertices on it, the vertex v_1 hits at most $\lceil \frac{1}{2}(k-2s+2) \rceil$ vertices on the subpath. This implies

$$\lceil \frac{1}{2}(k - |W|) \rceil - 1 < \lceil \frac{1}{2}(k - 2s + 2) \rceil.$$

Since $|W|$ is at most $s-1$, this is a contradiction for $s \geq 4$ when k is even, and for $s \geq 5$ when k is odd.

Otherwise either $s = 3$, or $s = 4$ and k is odd. If $s = 3$, then for even k we see that v_1 must hit both v_2 and v_3, and $\frac{1}{2}(k-2) - 2$ vertices on $v_6, \ldots, v_{k+1}$. If k is odd, then one of v_1 and v_2 has degree $m + 1$ (by Corollary 12.*ii*). Thus if the vertex v_1 hits v_3, we will assume $d(v_1) \geq m + 1$. This implies that $N(v_1) = \{v_2, v_3, v_6, v_8, \ldots, v_{k+1}\}$. If v_1 misses v_3, then $N(v_1) = \{v_2, v_6, v_8, \ldots, v_{k+1}\}$. Whether even or odd k, every vertex v_j on the path $v_5, \ldots, v_{k+1}$ is distance either 1 or 2 (on the path) away from a neighbor v_i of v_1. Thus if the vertex v_2 hits v_j, then suppose that v_1 hits v_{j+2}. It follows that the cycle $(v_3, \ldots, v_j, v_1, v_{j+2}, \ldots, v_{k+3})$ has more than k_1 vertices, a contradiction.

Otherwise the vertex v_2 misses the path $v_5, \ldots, v_{k+1}$. Since k is at least 10, the vertex v_2 must hit a vertex x that is not on the cycle C. Since $x, v_2, \ldots, v_t$ is also a longest path, the neighborhood of x is similar to the neighborhood of v_1. In fact it is easy to see that the neighbors of x on the path $v_6, v_8, \ldots, v_{k+1}$ are identical to the neighbors on v_1 on the path. Suppose v_1 and x both hit v_i and v_{i+1} on the path $v_6, \ldots, v_{k+1}$. It follows that the cycle $(v_3, \ldots, v_i, v_1, v_2, v_{i+2}, \ldots, v_{k+3})$ has more than $k + 1$ vertices, a contradiction.

Otherwise $s = 4$ (and k is odd). Since v_1 hits at least $\frac{1}{2}(k-3)$ vertices in $v_8, \ldots, v_{k+1}$, no two of which are consecutive, we reach the same conclusions as in the previous paragraph for $s = 3$. Therefore, Claim 3 holds true.

If no component of $G - C$ hits $C - v_{k+1}$, then v_{k+1} is a cut-vertex. Since $G[C]$ has $k + 1$ vertices and more than $\frac{1}{2}k(k-2)$ edges, the graph G contains T (by Lemma 17).

Otherwise at least one component of $G - C$ hits a vertex on $C - v_{k+1}$. Let $H_1, \ldots, H_\ell$ be the components of $G - C$ that hit a vertex on $C - v_{k+1}$.

Claim 4 *Each component in $\mathcal{X}$ is a vertex.*

To the contrary, suppose the component $X \in \mathcal{X}$ has more than one vertex. If X is an edge uv, then $e(uv, C) \geq k - 2$ (by Corollary 12). Let $j = \lfloor \frac{1}{2}(k-1) \rfloor$. Since $e(uv, C) \geq 2j + 1$ we see that $|V(C)| \geq 3(j+1) > k + 1$ (by Lemma 22), a contradiction.

Otherwise X is not an edge, and X contains a path on at least three vertices. Let H be a longest path in X, where $H = h_1, \ldots, h_b$, and h_1 hits a vertex v_s on $C - v_{k+1}$. Assume that $s \leq \frac{1}{2}(k+1)$ (otherwise, we could simply reverse the labels on the vertices on $v_1, \ldots, v_k$). Among all $(k+1)$-paths on $V(C)$ having v_{k+1} as one endpoint, choose the path $v_1, \ldots, v_{k+1}$ so that s is as small as possible. (Note that v_1 may no longer hit v_{k+1}.)

Both h_1 and h_b might hit v_{k+1}, but if either hits a vertex in the $2b$-set $\{v_{k+1-b}, \ldots, v_b\} - v_{k+1}$, then we easily find a path on more than t vertices, a contradiction.

Otherwise each of h_1 and h_b hits at most one vertex (namely v_{k+1}) on the path $v_{k-b+1}, \ldots, v_b$. If either h_1 or h_b hits two consecutive vertices on the cycle C, then we easily find a cycle on $k + 2$ vertices, a contradiction.

Since both h_1 and h_b hit the cycle C, suppose that h_1 hits v_{b+1}. If the vertex h_b hits a vertex in $\{v_{b+2}, \ldots, v_{2b}\}$, then the cycle $(v_1, \ldots, v_{b+1}, h_1, \ldots, h_b, v_{2b}, \ldots, v_{k+1})$ has more than $k + 1$ vertices, a contradiction.

Otherwise, the vertex h_b misses $\{v_{b+2}, \ldots, v_{2b}\}$. Since the vertex h_1 hits at least $\lfloor\frac{1}{2}(k-b)\rfloor$ vertices on C, it hits at least $\lfloor\frac{1}{2}(k-b)\rfloor - 2$ on the $(k-3b-1)$-path $v_{2b+2}, \ldots, v_{k-b}$, no two of which are consecutive. Thus h_1 hits at most $\lfloor\frac{1}{2}(k-3b)\rfloor$ vertices on the path. Since $b > 2$, we see that $\lfloor\frac{1}{2}(k-b)\rfloor - 2 > \lfloor\frac{1}{2}(k-3b)\rfloor$, a contradiction. Therefore Claim 4 holds true.

Claim 5 *If $x_{k+2} \in Z$ hits a vertex v_i on $C - x_{k+1}$, then $\mathcal{Y} = \emptyset$.*

To the contrary, suppose the vertex v_{k+1} hits v_i on $C - x_{k+1}$, and Y is a component in $\mathcal{Y}$. Let P be a longest path in $G[Y + v_{x+1}]$ that has v_{k+1} as one endpoint, where $P = v_{k+1}, y_2, \ldots, y_r$. Let $W = \alpha(P, v_{k+1})$ and notice that $N(W) \subseteq V(P)$. If $r \leq k-2$, then a vertex in W hits a vertex on v_j on $C - v_{k+1}$ (by Lemma 8), a contradiction.

Otherwise $r \geq k-1$. This implies that one of the paths

$y_r, \ldots, v_1, v_{k+1}, v_k, \ldots, v_i, v_{k+2}, \ldots, v_t$, or

$y_r, \ldots, v_1, v_{k+1}, v_1, \ldots, v_i, v_{k+2}, \ldots, v_t$

has more than t vertices, a contradiction. Therefore Claim 5 holds true.

By Claims 1, 2, and 3, it is easy to see that either v_{k+1} or v_{k+2} is a cut-vertex of G. Let $G' \subseteq G$ be the (2-connected) block of G that contains the cycle C.

If v_{k+1} is cut-vertex, then notice that $V(G') = V(C) \cup \mathcal{X}$. Otherwise, v_{k+1} is not a cut-vertex, but v_{k+2} is a cut-vertex, and $V(G') = V(C) \cup \mathcal{X} \cup \{v_{k+2}\}$.

Let $n' = |v(G')|$, and notice that the subgraph G' has more than $\frac{1}{2}(n'-1)(k-2)$ edges. If $v_{k+2} \in V(G')$, and $d_{G'}(v_{k+2}) < m$, then $G' - v_{k_2}$ has more then $\frac{1}{2}(n'-2)(k-2)$ edges. For this reason, we will assume that $d_{G'}(v_{k+2}) \geq m$.

Let x be a vertex in $\mathcal{X}$. If the vertex x hits v_i and v_{i+2}, then the vertex v_i is a $Terminal\text{-}P_1$. If x hits v_i and v_{i+3}, then the 2-path v_{i+1}, v_{i+2} is a $Terminal\text{-}P_2$. Finally, if x hits v_i and v_{i+4} (and misses v_{i+2}), then the 3-path $v_{i+1}, v_{i+2}, v_{i+3}$ is a $Terminal\text{-}P_3$.

We now state and prove five claims that help characterize the subgraph G'.

Claim 6 *No vertex in $\mathcal{X}$ hits two consecutive vertices on the cycle C.*

If $x \in \mathcal{X}$ hits both v_i and v_{i+1}, then the cycle $(v_1, \ldots, v_i, x, v_{i+1}, \ldots, v_{k+1})$ has $k+2$ vertices, a contradiction.

The proof of the Claim 7 follows from Claim 6.

Claim 7 *Each vertex $x \in \mathcal{X}$ has degree m or $m+1$.*

Claim 8 *For $x \in \mathcal{X}$ hits the vertex v_i on C, then no vertex in $\mathcal{X}$ hits either v_{i-1} or v_{i+1}.*

By Claim 6, we know x misses $\{v_{i-1}, v_{i+1}\}$. Suppose $x' \in \mathcal{X}$ hits v_{i+1} (if x' hits v_{i-1}, the proof is similar). For $v_j \in N(x) - v_i$, if x' hits $v_\ell \in \{v_{j+1}, v_{j+2}\}$, then the cycle $(x', v_\ell, \ldots, v_i, x, v_j, \ldots, v_{i+1})$ has more than $k+1$ vertices, a contradiction. It follows that there are $2(m-1)$ vertices on the cycle C for which x' does not hit. Since $d(x') \geq m$, we reach a contradiction.

Claim 9 *If v_i and v_j on C are two neighbors of $x \in \mathcal{X}$, where $j \geq i+2$, then the vertex v_{i+1} misses v_{j+1}, and the vertex v_{i-1} hits v_{j-1}.*

If the vertex v_{i+1} hits v_{j+1}, then the cycle $(v_1, \ldots, v_i, x, v_j, \ldots, v_{i+1}, v_{j_1}, \ldots, v_{k+1})$ has $k+2$ vertices, a contradiction. The proof is similar if the vertex v_{i-1} misses v_{j-1}.

Claim 10 *The set of Terminal-P_1's on C are an independent set, neither vertex in a Terminal-P_2 hits a Terminal-P_1, and there is at most one edge connecting one Terminal-P_2 to another Terminal-P_2.*

The first two of the three statements follow from Claim 9. Consider two Terminal-P_2's v_i, v_{i+1} and v_j, v_{j+1} on C. By Claim 9, we know that v_i misses v_j, and the vertex v_{i+1} misses v_{j+1}. If $v_i v_{j+1}, v_{i+1} v_j \in E(G)$, then the cycle $(v_1, \ldots, v_j, v_{j+1}, v_j, v_{i+1}, \ldots, v_{j-1}, x, v_{j+2}, \ldots, v_{k+1})$ has $k+2$ vertices, a contradiction.

Let $X = \mathcal{X}$, let $Y = N(X)$, and let $Z = V(C) - Y$. Notice that the three sets form a disjoint union of $V(G')$.

Case 1 k is even.

Notice that $k = 2m$, and the cycle C has $2m+1$ vertices. Thus Z consists of $m-1$ Terminal-P_1's and one Terminal-P_2, and $e(Z) = 1$ (by Claim 9). For $x \in X$, we see that $N(X) = N(x)$ by claim 9. Let $Z' = Z \cup X$ and we see that $e(Z') = 1$. It follows that $|Y| = m$, $|Z| \geq m+2$, and $e(G) \leq \binom{|Y|}{2} + |Y||Z| + 1$. For $s = |Z|$, we see that

$$\frac{1}{2}(m+s-1)(2m-2) < e(G) \leq \binom{m}{2} + ms + 1$$
$$\implies s > \frac{1}{2}(m^2 - 3m - 2).$$

For $k = 10$, we have $s \geq 7 = k-3$. For $k \geq 12$, we see that $s \geq k-3$. Since $e(Y, Z) \geq ms - 2$, the graph G' contains T (by Lemma 18).

Case 2 k is odd.

Notice that $k = 2m+1$, and the cycle C has $2m+2$ vertices. We have three cases to cover:

(a) The cycle C has $m+1$ Terminal-P_1's (and $|Y| = m+1$),

(b) The cycle C has two Terminal-P_2's and $m-1$ Terminal-P_1's (and $|Y| = m$), and

(c) The cycle C has one Terminal-P_3 and $m-1$ Terminal-P_1's (and $|Y| = m$).

Case 21 *The cycle C has $m+1$ Terminal-P_1's (and $|Y| = m+1$)*

For $Z' = Z \cup X$, we see that $e(Z') = 0$. Since there is at least one vertex in X, we see that $|Z'| \geq m+2$, and the graph G contains T (by Lemma 20).

Case 22 *The cycle C has two Terminal-P_2's and $m-2$ Terminal-P_1's (and $|Y| = m$).*

Since $|Y| = m$ and Z has two Terminal-P_2's, we see that $|Z| = m + 2$, $e(Z) \leq 3$ (by Claim 9), and for $Z' = Z \cup X$, we have $e(Z') = e(Z) \leq 3$. Let $s = |Z'|$. It follows that $e(Y, Z') \geq ms - 3$. Since $e(G) \leq \binom{|Y|}{2} + |Y||Z| + 3$, and

$$\frac{1}{2}(m+s)(2m-1) < e(G) \leq \binom{m}{2} + ms + 3$$
$$\implies s > \frac{1}{2}(m^2 - 2m - 6) \geq 2m = k - 1.$$

Since $|Z'| \geq k - 1$ and $e(Y, Z') \geq ms - 6 \geq ms - [(m-1) + 2]$, the graph G' contains T (by Lemma 20).

Case 23 *The cycle C has one Terminal-P_3 and $m-1$ Terminal-P_1's (and $|Y| = m$).*

Notice that $Y = N(X)$ and $N(x) = Y$ for each $x \in X$. Let v_i, v_{i+1}, v_{i+2} be the Terminal-P_3 on the cycle C. For each $x \in X$, notice that $N(x) = Y = \{v_{i+3}, v_{i+5}, \ldots, v_{i-1}\}$.

If v_j is a Terminal-P_1, then we have $N(v_j) \subseteq Y \cup \{v_{i+1}\}$ (by Claims 9 and 10). Suppose v_j hits v_{i+1}. If the vertex v_i hits v_{i+1}, then the cycle $(v_1, \ldots, v_i, v_{i+2}, v_{i+1}, v_j, \ldots, v_{i+3}, x, v_{j+1}, \ldots, v_{k+3})$ has $k+2$ vertices, a contradiction.

Otherwise the vertex v_i misses v_{i+1}. Thus for $Y' = Y \cup v_{v_i}$ and $Z' = (V(C) - Y) \cup X$, we see that $e(Z') = 0$. This case was proven in Case 1.

Otherwise, no Terminal-P_1 hits v_{i+1}. Thus the neighborhood of each $x \in X$ and of each Terminal-P_1 is Y, and each vertex on the Terminal-P_3 misses X and misses each Terminal-P_1. For $Z' = Z \cup X$, we see that $e(Z') = e(\{v_i, v_{i+1}, v_{i+2}\}) \leq 3$. For $s = |Z'|$, it follows that $e(Y, Z') \geq ms - 3$. By the same argument in Case 2.2, the graph G contains T.

Acknowledgements The authors thank our undergraduate student Zachary Tomlin (ztomlin@faulkner.edu) for proving the Erdős-Sós Conjecture for $k = 9$. This required 13 individual proofs, one for each of the 13 trees on nine vertices for which the conjecture was not known to hold.

References

1. N. Eaton and G. Tiner, On the Erdős-Sós Conjecture and Graphs with large minimum degree, *Ars Combinatoria* **95** (2010), pp. 373–382.
2. N. Eaton and G. Tiner, On the Erdős-Sós Conjecture for graphs having no path with k+4 vertices, *Discrete Mathematics* **313(16)** (2013), pp. 1621–1629.
3. A. Gilbert and G. Tiner, On the Erdős-Sós Conjecture for spiders in graphs having no path with 2k+1 vertices, *Ars Combinatoria* **139** (2013), pp. 297–313.

4. A. Gilbert and G. Tiner, On the Erdős-Sós Conjecture for Graphs with Circumference at most k, *Congressus Numerantium* **223** (2015), pp. 127–137.
5. A. McLennan, The Erdős-Sós Conjecture for Trees of Diameter Four, *J. of Graph Theory* **49** (2005), pp. 291–301.
6. P. Slater, S. Teo and H. Yap, Packing a Tree with a Graph of the Same Size, *J. of Graph Theory* **9** (1985), pp. 213–216.
7. G. Tiner, On the Erdős-Sós Conjecture and double-brooms, *JCMCC* **93** (2015), pp. 291–296.
8. G. Tiner, On the Erdős-Sós Conjecture for Graphs on $n = k + 3$ Vertices, *Ars Combinatoria* **95** (2010), pp. 143–150.
9. M. Woźniak, On the Erdős-Sós Conjecture, *J. of Graph Theory* **21** (1996), pp. 229–234.
10. L. Yuan, X. Zhang, On the Erdős-Sós Conjecture for Graphs on n=k+4 Vertices, arXiv preprint arXiv:1403.5430, (2014).
11. B. Zhou, A note on the Erdős-Sós Conjecture, *Acta Math. Sci.* **4** (1984), pp. 287–289.

Regular Graph and Some Vertex-Deleted Subgraph

Yoshimi Egawa and Kenji Kimura

Abstract In this paper, we consider a relationship between a regular graph and a regular factor of its vertex-deleted subgraph. Katerinis proved that if r is even integer and k is integer with $1 \leq k \leq \frac{r}{2}$, and G is an r-regular, r-edge-connected graph of odd order, then $G - x$ has a k-factor for each $x \in V(G)$. When the result "for each $x \in V(G)$" of Katerinis is replaced "for some $x \in V(G)$", we consider what condition can be followed. One of our main results is that let r and k be an even integer such that $4 \leq k \leq \frac{r}{2}$, and ℓ be a minimum integer such that $\ell \geq \frac{r}{r-2k+4}$, and G be an r-regular, 2ℓ-edge-connected graph of odd order. Then, there is some $x \in V(G)$ such that $G - x$ has a k-factor. Moreover, if $r \geq 4k - 8$, then we can replace 2ℓ-edge-connected with 2-edge-connected.

Keywords Regular graph · Regular factor · Vertex-deleted subgraph

1 Introduction

We consider finite undirected graphs that may have *loops* and *multiple edges*. Let G be a graph. For $x \in V(G)$, we denote by $\deg_G(x)$ the *degree* of x in G. The set of *neighbours* of $x \in V(G)$ is denoted by $N_G(x)$ and let $N_G(X) = \bigcup_{x \in X} N_G(x)$ for $X \subseteq V(G)$. We denote by $G[X]$ the subgraph of G induced by X for a subset X of $V(G)$. The number of components of a graph G is denoted by $\omega(G)$. If $\deg_G(x) = r$ for any $x \in V(G)$, we call the graph *r-regular graph*. For subsets S and T of $V(G)$, we denote by $e_G(S, T)$ the number of the edges joining S and T. If S is a singleton $\{x\}$, we write $S = x$ instead of $S = \{x\}$. For example, we write $e_G(x, T)$ instead of

Y. Egawa
Department of Applied Mathematics, Tokyo University of Science, Kagurazaka, Shinjuku, Tokyo 162-8601, Japan

K. Kimura (✉)
Department of Information Technology and Electronics, Ishinomaki Senshu University, Shinmito 1, Minamizakai, Ishinomaki-Shi, Miyagi 986-8580, Japan
e-mail: kimura@isenshu-u.ac.jp

F. Hoffman (ed.), *Combinatorics, Graph Theory and Computing*,
Springer Proceedings in Mathematics & Statistics 388,
https://doi.org/10.1007/978-3-031-05375-7_18

$e_G(\{x\}, T)$. Let k be a constant. A spanning subgraph F of G such that $\deg_F(x) = k$ for each $x \in V(G)$ is called a *k-factor* of G. We denote by rG, r copies of G for constant r and graph G. When no fear of confusion arises, we often introduce the definition of $v[i, j] = v_{ij}$ for vertex v_{ij}. Furthermore, we define $X\left[i_j\right] = X_{i_j}$ and $X\left[i^j\right] = X_{i^j}$ for subscript i and subsubscript j of set X and for subscript i and subsuperscript j of set X, respectively.

Petersen proved the next theorem in 1891.

Theorem A (Petersen [1]) *Let r be an even integer. Then every r-regular graph can be decomposed into $\frac{r}{2}$ disjoint 2-factors.*

This theorem implies that if r and k are even integers, and G is an r-regular graph, then G has a k-factor for every k such that $2 \leq k \leq r$.

Katerinis showed the next theorem in 1985.

Theorem B (Katerinis [3]) *Let a, b, and c be odd integers such that $1 \leq a < b < c$, and let G be a connected graph of even order. If G has both a-factor and c-factor, then G has a b-factor.*

Assume r is even integer. If an r-regular graph G has a 1-factor, we can obtain an $(r-1)$-factor by excluding the 1-factor from G. By the 1-factor and the $(r-1)$-factor of G and by Theorem B, G has a k-factor for any odd integer k such that $1 \leq k \leq r-1$. Thus, by the above two theorems, if an r-regular graph G has a 1-factor, then G has a k-factor for every integer k such that $1 \leq k \leq r$. Note that the order of G is even. For the case that the order of G is odd, Katerinis proved the next theorem in 1994.

Theorem C (Katerinis [4]) *Let r be an even integer, and let k be an integer such that $1 \leq k \leq \frac{r}{2}$, and let G be an r-regular, r-edge-connected graph of odd order. Then for every $x \in V(G)$, $G - x$ has a k-factor.*

Lu, Wang and Bai generalized Theorem C.

Theorem D (Lu, Wang and Bai [5]) *Let r and ℓ be an even integer with $4 \leq \ell \leq r$, and let k be an integer such that $2 \leq k \leq \frac{r}{2}$, and let G be an r-regular, ℓ-edge-connected graph of odd order. Then for every $x \in V(G)$, $G - x$ has a k-factor in the following cases:*

1. *k is even, and $\ell \geq 2k$;*
2. *k is odd, and $\ell \geq 2k$ and $\ell > \frac{r}{2}$;*

Let r be an even integer and k be an integer. In [4], if $k = 1$ or $k = \frac{r}{2}$, then Katerinis showed that the condition of Theorem C is the best possible. In [5], Lu, Wang and Bai showed that the condition of Theorem D is the best possible. Now, we consider other cases. If k is odd and $1 < k < r$, then the condition of r-edge-connected can be substituted by $\max\{2k, \frac{r}{2} + 1\}$-edge-connected. Moreover, we consider the case that $|V(G)|$ is even. If k is odd, then $G - x$ clearly has no k-factor for every $x \in V(G)$. Thus, we consider the case that k is even. We now define the graph H_F as the following. First, we consider a bipartite graph F with bipartition (A, B). Let H_F be a graph obtained from F, adding edges subject to $|E(F[A])| \leq \frac{k}{2} - 1$ and $|E(F[B])| \leq \frac{k}{2} - 1$. Then, if r, k are even integers with $2 \leq k \leq \frac{r}{2}$ and $G \neq H_F$ is an r-regular, $2k$-edge-connected, then $G - x$ has a k-factor for every $x \in V(G)$. Furthermore, if r is odd and k is even with $2 \leq k \leq \frac{r}{2}$, and $G \neq H_F$ is an r-regular, $(2k - 1)$-edge-connected, then we can conclude that the graph $G - x$ has a k-factor for every $x \in V(G)$. We summarize above results as the remark following.

Remark 1 Let r, ℓ and k be integers with $2 \leq k \leq \frac{r}{2}$ and let G be an r-regular, ℓ-edge-connected graph. Then, for every $x \in V(G)$, $G - x$ has a k-factor in the following cases:

1. $|V(G)|$ is odd, and k and r are even, and $\ell = 2k$;
2. $|V(G)|$ and $k \neq \frac{r}{2}$ are odd, and r is even, and $\ell = \max\{2k, \frac{r}{2} + 1\}$;
3. $G \neq H_F$ and $|V(G)|$, k and r are even, and $\ell = 2k$;
4. $G \neq H_F$ and $|V(G)|$ and k are even, and r is odd, and $\ell = 2k - 1$.

We show that we cannot replace the edge-connectivity of Remark 1 with weaker condition. Let r, ℓ and k be as stated in the hypotheses of Remark 1. We consider r-regular and ℓ-edge-connected bipartite graph H_1 with bipartition (A, B) and $|A| = |B|$. Let G_1 be a graph obtained from H_1 after a deletion of $\lceil \frac{\ell}{2} - 1 \rceil$ independent edges such that G_1 remains ℓ-edge-connected. Suppose H_2 is an r-regular, ℓ-edge-connected graph. Let also G_2 be a graph obtained from H_2 after a deletion of $\lceil \frac{\ell}{2} - 1 \rceil$ independent edges such that G_2 remains ℓ-edge-connected. We form G as follows. We add $2\lceil \frac{\ell}{2} - 1 \rceil$ independent edges having one end-vertex in G_1 that have degree $r - 1$ and the another in G_2 that have degree $r - 1$. Such a graph G is r-regular, and $(\ell - 1)$ or $(\ell - 2)$-edge-connected. Now suppose that $x \in A$. Let $S = A$ and $T = B$. If $\ell = 2k$ or $2k - 1$, then we have $\delta_{G-x}(S - x, T; k) = -k + (\lceil \frac{\ell}{2} - 1 \rceil) - 1 \leq -2$ (see the following for the definition of $\delta_G(S, T; k)$). Thus, G has no k-factor.

Next, we consider the case with $\ell = \frac{r}{2} + 1$. Let also r, ℓ and k be as the above and we consider r-regular and ℓ-edge-connected graph H_3 of odd order. Assume that $\frac{r}{2}$ is even, and G_3 is a graph obtained from H_3 after a deletion of $\frac{r}{4}$ independent edges such that G_3 remains ℓ-edge-connected. We form G as follows. We start from $2G_3$. We add vertex x, and r edges joining x to each vertex $v \in V(2G_3)$ with $\deg_{2G_3}(v) = r - 1$. The resulting graph G is r-regular, $\frac{r}{2}$-edge-connected and has an odd number of vertices. On the other hand, when $\frac{r}{2}$ is odd, G_3' also is a graph obtained from H_3 after a deletion of $\lfloor \frac{r}{4} \rfloor$ independent edges, such that G_3' remains ℓ-edge-connected. We form G' as follows. We start from $2G_3'$. We add vertex x' with one loop and $r - 2$ edges joining x' to every vertex $v \in V(2G_3')$ with $\deg_{2G_3'}(v) = r - 1$. The resulting graph G' is r-regular, $(\frac{r}{2} - 1)$-edge-connected and has an odd number of vertices. First, we consider G. Let $S = \{x\}$ and $T = \emptyset$. Now $k|V(G_3)| + e_G(V(G_3), T)$ is an odd number since $|V(G_3)|$ and k are odd and $e_G(V(G_3), T) = 0$. Thus $h_{G-x}(S - x, T; k) = 2$ (see also the following for the definition of $h_G(S, T; k)$). Hence $\delta_{G-x}(S - x, T; k) = -2$. Therefore $G - x$ has no k-factor. Similarly, $G' - x'$ has no k-factor.

The above examples show that the assumption of edge-connectivity in Remark 1 is sharp. Let us focus our attention that the result "for each $x \in V(G)$" of statements is replaced by "for some $x \in V(G)$". What condition can be followed under the weakened result? Now we will present our theorems.

Theorem 1 *Let r be an integer such that $r \geq 4$, let G be an r-regular, 2-edge-connected graph. If G is not bipartite, then there is some $x \in V(G)$ such that $G - x$ has a 2-factor.*

Moreover, if $k \geq 4$, then following result holds.

Theorem 2 *Let r and k be even integers such that $4 \leq k \leq \frac{r}{2}$, and let ℓ be a minimum integer such that $\ell \geq \frac{r}{r-2k+4}$, and let G be an r-regular, 2ℓ-edge-connected graph of odd order. Then, there is some $x \in V(G)$ such that $G - x$ has a k-factor. In particular, if $r \geq 4k - 8$, then we can replace 2ℓ-edge-connected with 4-edge-connected.*

Furthermore, we shall prove next theorem.

Theorem 3 *Let r be an integer and k be an even integer such that $2 \leq k \leq \frac{r}{2}$, and let G be an r-regular, 2-edge-connected graph having a 2-edge cut. If either $|V(G)|$ is odd, or $k = 2$ and G is not bipartite, then there is some $x \in V(G)$ such that $G - x$ has a k-factor.*

2 Prepare for Proofs

In order to prove Theorems 1, 2 and Remark 1, we use the following Tutte's Theorem. Let G be a graph. For disjoint subsets S and T of $V(G)$, we define $\delta_G(S, T; k)$ by

$$\delta_G(S, T; k) = k|S| - \sum_{y \in T}(k - \deg_G(y)) - e_G(S, T) - h_G(S, T; k),$$

where $h_G(S, T; k)$ is the number of components C of $G - (S \cup T)$ such that $k|V(C)| + e_G(V(C), T)$ is odd. These components are called *odd* components.

Theorem E (Tutte [6]) *Let G be a graph, and let k be a positive integer. Then*

1. $\delta_G(S, T; k) \equiv k|V(G)| \pmod 2$ *for each pair of disjoint subsets S and T of $V(G)$, and*
2. *G has a k-factor if and only if $\delta_G(S, T; k) \geq 0$ for each pair of disjoint subsets S and T of $V(G)$.*

Proposition 1 *Let r and k be non-negative integers, let G be an r-regular graph and let S and T be disjoint subsets of $V(G)$. Define $U = V(G) - (S \cup T)$, $\theta = \frac{k}{r}$,*

$$m_S = 2e_G(S, S) + e_G(S, U) \quad \text{and}$$
$$m_T = 2e_G(T, T) + e_G(T, U).$$

Then,

$$\delta_G(S, T; k) = \theta m_S + (1-\theta)m_T - h_G(S, T; k).$$

Proof

$$\begin{aligned}\delta_G(S, T; k) &= k|S| - \theta e_G(S, T) + (r-k)|T| - (1-\theta)e_G(S, T) - h_G(S, T; k)\\ &= \theta(r|S| - e_G(S, T)) + (1-\theta)(r|T| - e_G(S, T)) - h_G(S, T; k)\\ &= \theta m_S + (1-\theta)m_T - h_G(S, T; k).\end{aligned}$$

□

Katerinis dealt with the idea of calculation of following Lemma 1 in [4]. Now, we recalculate for our proofs.

Lemma 1 *Let r, ℓ and k be integers and let G be an r-regular, ℓ-edge-connected graph. Suppose $G - x$ has no k-factor for some $x \in V(G)$. Then, we have following results such that $S, T \subseteq V(G)$ with $S \neq T$.*

$$(r-2k)(|S| - |T|) \geq \ell\omega(G[U]) - 2h_{G-x}(S-x, T; k) - 2k + 4 + 2e_G(S, S) + 2e_G(T, T).$$

Proof Since $G - x$ has no k-factor, by Theorem E, there are $S', T \subseteq V(G) - x$ with $S' \cap T = \emptyset$ such that $\delta_{G-x}(S', T) \leq -2$. Let $S = S' \cup \{x\}$ and $U = V(G) - (S \cup T)$. Then, we have

$$\delta_{G-x}(S-x, T; k) = k|S| - k - \sum_{y\in T}(k - \deg_G(y)) - e_G(S, T) - h_{G-x}(S-x, T; k) \leq -2.$$

Then, we have

$$k|S| - k|T| + \sum_{y\in T} \deg_{G-S}(y) - h_{G-x}(S-x, T; k) \leq k-2. \tag{1}$$

Thus,

$$\sum_{y\in T} \deg_{G-S}(y) \leq k - 2 + h_{G-x}(S-x, T; k) - k|S| + k|T|. \tag{2}$$

On the other hand, since G is r-regular,

$$r|S| = 2e_G(S, S) + e_G(S, T) + e_G(S, U). \tag{3}$$

Similarly,

$$r|T| = 2e_G(T, T) + e_G(S, T) + e_G(T, U). \tag{4}$$

By (3) and (4), we have

$$\begin{aligned} r|S| &= 2e_G(S,S) + r|T| - 2e_G(T,T) - e_G(T,U) + e_G(S,U) \\ &= r|T| - 2e_G(T,T) - 2e_G(T,U) + e_G(T,U) + e_G(S,U) + 2e_G(S,S). \quad (5) \end{aligned}$$

Since G is ℓ-edge-connected,

$$e_G(T,U) + e_G(S,U) \geq \ell\omega(G[U]). \quad (6)$$

Combining (5) with (6),

$$\begin{aligned} r|S| &\geq r|T| - 2e_G(T,T) - 2e_G(T,U) + \ell\omega(G[U]) + 2e_G(S,S) \\ &= r|T| - 4e_G(T,T) - 2e_G(T,U) + \ell\omega(G[U]) + 2e_G(S,S) + 2e_G(T,T) \\ &= r|T| - 2\sum_{y\in T}(\deg_{G-S}(x)) + \ell\omega(G[U]) + 2e_G(S,S) + 2e_G(T,T). \quad (7) \end{aligned}$$

Now using (2), (7) implies,

$$\begin{aligned} r|S| \geq\ & r|T| - 2(k - 2 + h_{G-x}(S-x,T;k) - k|S| + k|T|) + \ell\omega(G[U]) \\ & +2e_G(S,S) + 2e_G(T,T). \end{aligned}$$

Thus,

$$\begin{aligned} (r-2k)(|S|-|T|) \geq\ & \ell\omega(G[U]) - 2h_{G-x}(S-x,T;k) - 2k + 4 \\ & +2e_G(S,S) + 2e_G(T,T). \end{aligned}$$

□

We use the following Kano's theorem to prove Theorem 3. We remark that Kano actually proved a stronger statement than Theorem F.

Theorem F (Kano [2]) *Let r be an integer, and let k be an even integer such that $2 \leq k \leq \frac{r}{2}$, and let G be an r-regular, 2-edge-connected graph. Then, G has a k-factor containing e and another k-factor avoiding e for every edge $e \in E(G)$.*

2.1 *Proof of Theorem 3*

Let r and k be as stated in the hypotheses of Theorem 3. Suppose the Theorem is false and choose a counterexample G such that $|V(G)|$ is as small as possible. Since G has a 2-edge cut, there are edges $f_1, f_2 \in E(G)$ such that $G - f_1 - f_2$ has two components C_1, C_2. Define the graph $D_i = C_i \cup \{e_i\}$, where e_i is edge such that D_i becomes r-regular with $1 \leq i \leq 2$. Note that D_i is a 2-edge-connected graph. We consider two cases.

Case 1. $|V(G)|$ is odd.

Then since either $|V(D_1)|$ is odd or $|V(D_2)|$ is odd, without loss of generality we may assume that $|V(D_1)|$ is odd. By the induction hypothesis, there is a vertex $x \in V(D_1)$ such that $D_1 - x$ has a k-factor F_1 since $|V(D_1)|$ is odd. If $F_1 \cap \{e_1\} = \emptyset$, we obtain a k-factor F_2 from D_2 such that $F_2 \cap \{e_2\} = \emptyset$ by Theorem F. Thus, since $F_1 \cup F_2$ is a k-factor in G, this is a contradiction. If $F_1 \cap \{e_1\} \neq \emptyset$, we obtain a k-factor F_2 from D_2 such that $F_2 \cap \{e_2\} \neq \emptyset$ by Theorem F. Thus, since $F_1 \cup F_2 - \{e_1, e_2\} \cup \{f_1, f_2\}$ is a k-factor in G, this is a contradiction.

Case 2. $|V(G)|$ is even.

Then, we have $k = 2$. If $|V(D_i)| \equiv 1 \pmod 2$ with $i \in \{1, 2\}$, then we can conclude that $G - x$ has a 2-factor as in the case 1 since there is a vertex x such that $D_i - x$ has a 2-factor by the induction hypothesis. Therefore, this is a contradiction. Thus we may assume $|V(D_i)| \equiv 0 \pmod 2$ with $i \in \{1, 2\}$. If both of D_1 and D_2 are bipartite, G becomes a bipartite graph and this is a contradiction. Thus we may assume that D_1 is not bipartite. Hence, as above, we can conclude that $G - x$ has a 2-factor, contradicting our assumption that $G - x$ has no 2-factor. □

3 Proof of Theorem 2

Suppose G is a graph of counterexample. Let S and T be disjoint subsets of $V(G)$. Define $U = V(G) - (S \cup T)$, $\delta_G(S, T) = \delta_G(S, T; k)$ and $h_G(S, T) = h_G(S, T; k)$. First, we prove some basic properties of G.

Claim 1 *For every $x \in V(G)$, there are some disjoint subsets S and T of $V(G)$ with $S \ni x$ such that $\delta_G(S, T) \le k - 2$.*

Proof Since $G - x$ has no k-factor, by Theorem E, there are disjoint subsets $S', T \subset V(G) - \{x\}$ such that $\delta_{G-x}(S', T) \le -2$. Let $S = S' \cup \{x\}$. Then, $\delta_G(S, T) \le k - 2$. □

Claim 2 $|S| = |T|$.

Proof Assume $|S| > |T|$. Since $m_S - m_T = r(|S| - |T|)$ by Proposition 1, we have $m_S \ge m_T + r$. By the definition of the odd component and k is even, $m_T \ge h_G(S, T)$. Therefore,

$$k - 2 \ge \delta_G(S, T) \ge \theta(m_T + r) + (1 - \theta)m_T - h_G(S, T) \ge \theta r = k.$$

This is a contradiction.

Assume $|S| < |T|$. Then, $m_T \ge m_S + r$. For every component C of U, if $e_G(V(C), T)$ is odd, then $e_G(V(C), S)$ is odd since r is even. Thus, $m_S \ge h_G(S, T)$ by the definition of the odd component and k is even. Hence,

$$k - 2 \ge h_G(S, T) \ge \theta m_S + (1 - \theta)(m_S + r) - h_G(S, T) \ge r - \theta r = r - k.$$

Therefore, $k-1 \geq \frac{r}{2}$. This contradicts $\frac{r}{2} \geq k$. □

By Claim 2, we have $m_S = m_T$ and $\delta_G(S,T) = m_T - h_G(S,T)$. Since G is a 2ℓ-edge-connected, $m_S + m_T = 2m_T \geq 2\ell h_G(S,T)$. Thus, $\frac{m_T}{\ell} \geq h_G(S,T)$. Then by Claim 1,

$$k-2 \geq \delta_G(S,T) \geq m_T - \frac{m_T}{\ell} \geq m_T\left(1 - \frac{r-2k+4}{r}\right) = \frac{2(k-2)}{r} m_T.$$

Therefore, we have $m_T \leq \frac{r}{2}$.

Define a pair $\{S,T\}$ such that $S \cap T = \emptyset$, $|S| = |T|$ and $m_S < \frac{r}{2}$. We call these pairs *Tutte pair of 1st kind*. Define a pair $\{S,T\}$ such that $S \cap T = \emptyset$, $|S| = |T|$, $m_S = m_T = \frac{r}{2}$, and $G-(S \cup T)$ consists of h components and these h components are odd components. We call these pairs *Tutte pair of 2nd kind*. Then, $\ell = \frac{r}{r-2k+4}$, $h = \frac{r}{2} - k + 2$, $E(G[S]) = \emptyset$ and $E(G[T]) = \emptyset$ hold. By Claim 1, we can select Tutte pairs $\{S_1,T_1\},\ldots,\{S_p,T_p\}$ such that $V(G) = \bigcup_{1\leq i\leq p}(S_i \cup T_i)$ and $\sum_{i=1}^{p}|S_i \cup T_i|$ is as small as possible. Let $\{S_i,T_i\}$ be a Tutte pair of 2nd kind for $1 \leq i \leq q$ and $\{S_i,T_i\}$ be a Tutte pair of 1st kind for $q+1 \leq i \leq p$. For $1 \leq i, j \leq p$, suppose

$$\begin{aligned}
S_i' &= S_i \cap U_j,\\
T_i' &= T_i \cap U_j,\\
S_j' &= S_j \cap U_i,\\
T_j' &= T_j \cap U_i,\\
U_i' &= V(G) - (S_i' \cup T_i') \text{ and}\\
U_j' &= V(G) - (S_j' \cup T_j').
\end{aligned}$$

Now, we prove following claims.

Claim 3 *If $i \neq j$ and $(S_i \cup T_i) \cap (S_j \cup T_j) \neq \emptyset$, then $1 \leq i, j \leq q$.*

Proof Assume the contrary. Without loss of generality, we may assume $i \geq q+1$. Then, $m_{S_i} < \frac{r}{2}$. Then,

$$e_G(S_i' \cup T_i', U_i') + e_G(S_j' \cup T_j', U_j') \leq e_G(S_i \cup T_i, U_i) + e_G(S_j \cup T_j, U_j).$$

On the other hand,

$$\begin{aligned}
m_{S_i'} + m_{T_i'} + m_{S_j'} + m_{T_j'} &= e_G(S_i' \cup T_i', U_i') + e_G(S_j' \cup T_j', U_j')\\
&\quad +2\left(e_G(S_i', S_i') + e_G(T_i', T_i') + e_G(S_j', S_j') + e_G(T_j', T_j')\right)\\
&\leq e_G(S_i \cup T_i, U_i) + e_G(S_j \cup T_j, U_j)\\
&\quad +2\left(e_G(S_i, S_i) + e_G(T_i, T_i) + e_G(S_j, S_j) + e_G(T_j, T_j)\right)\\
&= m_{S_i} + m_{T_i} + m_{S_j} + m_{T_j}\\
&< 2r.
\end{aligned}$$

We may assume $m_{S_i'} + m_{T_i'} < r$. Then, $|S_i'| = |T_i'|$ by $m_{S_i'} - m_{T_i'} = r(|S_i'| - |T_i'|)$. Thus, $\{S_i', T_i'\}$ is a Tutte pair. This contradicts minimality of $\sum_{i=1}^{p} |S_i \cup T_i|$ if we replace $\{S_i, T_i\}$ with $\{S_i', T_i'\}$. □

Claim 4 $|(S_i \cup T_i) \cap (S_j \cup T_j)| \equiv 0 \pmod 2$ *for* $1 \leq i, j \leq p$.

Proof First we note that $r = 2\ell h$ since $\{S_i, T_i\}$ is a Tutte pair of 2nd kind with $1 \leq i \leq p$. Suppose $|(S_i \cup T_i) \cap (S_j \cup T_j)| \equiv 1 \pmod 2$ for $1 \leq i, j \leq p$ and $i \neq j$. Without loss of generality, we may assume $i = 1$ and $j = 2$. Suppose that $C_1, C_2, \ldots, C_h$ are odd components of $G - (S_1 \cup T_1)$ and that $D_1, D_2, \ldots, D_h$ are odd components of $G - (S_2 \cup T_2)$. Note that $h \geq 2$. Thus we may assume $|(S_1 \cup T_1) \cap V(D_1)| \equiv 1 \pmod 2$ and $|(S_2 \cup T_2) \cap V(C_1)| \equiv 1 \pmod 2$. Then, by the proof of Claim 3, we obtain

$$m_{S_1'} + m_{T_1'} = r \text{ and } \mathrm{m}_{\mathrm{S}_2'} + \mathrm{m}_{\mathrm{T}_2'} = \mathrm{r}.$$

On the other hand, since $|(S_1 \cup T_1) \cap V(D_1)|$ is odd, $m_{S_1 \cap V(D_1)} + m_{T_1 \cap V(D_1)} \geq r$. Then, for $2 \leq a \leq h$

$$e_G((S_1 \cup T_1) \cap V(D_a), V(G) - ((S_1 \cup T_1) \cap V(D_a))) = 0$$

since $e_G((S_1 \cup T_1) \cap V(D_1), (S_1 \cup T_1) \cap V(D_a)) = 0$. Thus, $(S_1 \cup T_1) \cap V(D_a) = \emptyset$ with $2 \leq a \leq h$. Similarly, $(S_2 \cup T_2) \cap V(C_a) = \emptyset$ with $2 \leq a \leq h$. Note that

$$e_G((S_1 \cup T_1) \cap (S_2 \cup T_2), V(C_a) \cap V(D_b)) = 0$$

for $1 \leq a, b \leq h$ since $m_{S_1} + m_{T_1} = r$, $m_{S_2} + m_{T_2} = r$, $m_{S_1 \cap V(D_1)} + m_{T_1 \cap V(D_1)} = r$ and $m_{S_1 \cap V(C_1)} + m_{T_1 \cap V(C_1)} = r$. Thus, for any component C_a with $2 \leq a \leq h$, $e_G(V(C_a), S_1 \cup T_1) = e_G(V(C_a), (S_1 \cup T_1) \cap V(D_1))=2\ell$. However C_a joins either S_1 or T_1 for any component C_a with $2 \leq a \leq h$ since $|(S_1 \cup T_1) \cap V(D_1)|$ is odd. This contradicts that C_a is a odd component with $2 \leq a \leq h$ since $2\ell h = r$. □

From now on, we use definition of $X[i_j] = X_{i_j}$ and $X[i^j] = X_{i^j}$. We consider $q' \leq q$ such that $i_1, i_2, \ldots, i_{q'}$ is as small as possible and subject to $\left|(S[i_1] \cup T[i_1]) \cap (S[i_2] \cup T[i_2]) \cap, \ldots, \cap \left(S[i_{q'}] \cup T[i_{q'}]\right)\right| \equiv 1 \pmod 2$.

Define $j_a \in \{0, 1\}$ and

$$X[j_{q'} j_{q'-1} \ldots j_1] = \bigcap_{j_a=1} (S[i_a] \cup T[i_a]) \bigcap_{j_a=0} U[i_a]$$

with $1 \leq a \leq q'$. For example, when $q' = 4$, $j_4 = 1$, $j_3 = 1$, $j_2 = 0$ and $j_1 = 1$, we have

$$X_{1101} = (S[i_4] \cup T[i_4]) \cap (S[i_3] \cup T[i_3]) \cap (S[i_1] \cup T[i_1]) \cap U[i_2].$$

Since $j_{q'} j_{q'-1} \dots j_1$ is a sequence of 0 and 1, we can consider this sequence as a binary number. Then, we reconstruct decimal b from binary number $j_{q'} j_{q'-1} \dots j_1$ and we define

$$Y_b = X\left[j_{q'} j_{q'-1} \dots j_1\right]$$

for $0 \le b \le 2^{q'} - 1$. Note that $\left|Y\left[2^{q'} - 1\right]\right|$ is odd since $\left|(S[i_1] \cup T[i_1]) \cap (S[i_2] \cup T[i_2]) \cap, \dots, \cap \left(S\left[i_{q'}\right] \cup T\left[i_{q'}\right]\right)\right| \equiv 1 \pmod 2$. Now we have $Y[2^a] = \left(S\left[i_{a+1}\right] \cup T\left[i_{a+1}\right]\right) \bigcap_{b \ne a} U\left[i_{b+1}\right]$ with $0 \le a, b \le q' - 1$. Then, since $|Y[2^a]|$ is odd by Claim 4, $\left|Y\left[2^{q'} - 1\right]\right| \equiv 1 \pmod 2$ and the minimality of q', we have

$$\sum_{0 \le a \le q'-1} e_G(Y\left[2^a\right], V(G) - Y\left[2^a\right]) = rq'. \tag{8}$$

On the other hand, as there are h odd components for $\{S[i_a], T[i_a]\}$ with $1 \le a \le q'$, and $\{S[i_a], T[i_a]\}$ is a Tutte pair of 2nd kind, and G is 2ℓ-edge-connected, then certainly

$$2\ell hq' = \sum_{1 \le a \le q'} e_G(S[i_a] \cup T[i_a], V(G) - (S[i_a] \cup T[i_a])).$$

Thus, by (8) and the definition of Y_b we have

$$\begin{aligned} 2\ell hq' &= \sum_{1 \le a \le q'} e_G(S[i_a] \cup T[i_a], V(G) - (S[i_a] \cup T[i_a])) \\ &\ge \sum_{0 \le a \le q'-1} e_G(Y\left[2^a\right], V(G) - Y\left[2^a\right]) \\ &= rq'. \end{aligned}$$

Now, since $2\ell h = r$, we have

$$\begin{aligned} &\sum_{1 \le a \le q'} e_G(S[i_a] \cup T[i_a], V(G) - (S[i_a] \cup T[i_a])) \\ &\qquad = \sum_{0 \le a \le q'-1} e_G(Y\left[2^a\right], V(G) - Y\left[2^a\right]). \end{aligned} \tag{9}$$

We rewrite both sides of (9) by using summation of $e_G(Y_i, Y_j)$. Then, we have

$$\sum_{1 \le a \le q'} e_G(S[i_a] \cup T[i_a], V(G) - (S[i_a] \cup T[i_a])) = \sum_{0 \le i < j \le 2^{q'}-1} b_{ij} e_G(Y_i, Y_j)$$

and

$$\sum_{0 \le a \le q'-1} e_G(Y\left[2^a\right], V(G) - Y\left[2^a\right]) = \sum_{0 \le i < j \le 2^{q'}-1} c_{ij} e_G(Y_i, Y_j),$$

Then, $b_{ij} \geq c_{ij}$ for $0 \leq i < j \leq 2^{q'}$. Moreover, $b_{0j} > c_{0j}$ for $1 \leq j \leq 2^{q'}$. Thus, $e_G\left(Y\left[2^{q'}-1\right], V(G) - Y\left[2^{q'}-1\right]\right) = 0$. Therefore, $Y\left[2^{q'}-1\right] = \emptyset$. However, this contradicts $\left|Y\left[2^{q'}-1\right]\right|$ is odd. □

4 Proof of Remark 1 and Theorem 1

Proof of Remark 1 Let r, k, ℓ and G be as stated in the hypotheses of Remark 1. Assume on the contrary that $G - x$ has no k-factor for some $x \in V(G)$. Then by Theorem E and Lemma 1, there are some disjoint subsets $S, T \subseteq V(G)$ such that

$$\begin{aligned}(r-2k)(|S|-|T|) \geq\ & \ell\omega(G[U]) - 2h_{G-x}(S-x, T; k) - 2k + 4 \\ & +2e_G(S,S) + 2e_G(T,T).\end{aligned} \tag{10}$$

We consider the cases $\omega(G[U]) \geq 1$ and $\omega(G[U]) = 0$ separately.

Case 1. $\omega(G[U]) \geq 1$.

Without loss of generality, we may assume $\ell \geq 2k - 1 \geq 1$. Then, by (10),

$$(r-2k)(|S|-|T|) \geq 1.$$

Thus, we have

$$|S| > |T|. \tag{11}$$

If $h_{G-x}(S-x, T; k) = 0$, by (1) and (11),

$$1 \leq |S| - |T| \leq 1 - \frac{2}{k}.$$

This is a contradiction. Thus, we may assume $h_{G-x}(S-x, T; k) \geq 1$. Now, we consider two cases.

Case 1–1. k is even.

Then by the definition of the odd component, $\sum_{x \in T} \deg_{G-S} \geq h_{G-x}(S-x, T; k)$. Hence (1) implies $|T| \geq |S|$. However, this contradict (11).

Case 1–2. k is odd.

We now consider the case of Remark 1 (2). Let $h = h_{G-x}(S-x, T; k)$. By (1) and (11),

$$1 \leq |S| - |T| \leq 1 + \frac{h-2}{k} < h \tag{12}$$

since if $1 + \frac{h-2}{k} \geq h$, then we have $1 > \frac{k-2}{k-1} \geq h$ and this contradicts $h \geq 1$. Note that (12) implies $h \geq 2$.

By (10) and (12),

$$
\begin{aligned}
(r-2k)\left(1+\frac{h-2}{k}\right) \geq (r-2k)(|S|-|T|) \geq (\ell-2)h-2k+4 \\
(r-2k)(k+h-2) \geq (\ell-2)kh-2k^2+4k \\
rk+rh-2r \geq \ell kh \\
r(k+h-2) \geq \ell kh
\end{aligned} \tag{13}
$$

Since $h \geq 2$, $k+h-2 \geq 0$. Assume $\ell = 2k \geq \frac{r}{2}+1$. Then, we have $4k-2 \geq r$. Thus, by (13), we have

$$
\begin{aligned}
(4k-2)(k+h-2)-2k^2h \geq 0 \\
4k^2+4kh-10k-2h+4-2k^2h \geq 0 \\
2h(2k-1-k^2)+4k^2-10k+4 \geq 0
\end{aligned}
$$

Now since $k \geq 3$, $2k-1-k^2 \leq 0$. Recall that $h \geq 2$. Hence, we have

$$
4(2k-1-k^2)+4k^2-10k+4 = -2k \geq 0.
$$

This is a contradiction.

Assume $\ell = \frac{r}{2}+1$. From (13),

$$
\begin{aligned}
r(k+h-2) \geq \left(\frac{r}{2}+1\right)kh \\
2r(k+h-2) \geq (r+2)kh \\
2rk+2rh-4r-rkh-2kh \geq 0 \\
k(2r-rh-2h)+2rh-4r \geq 0.
\end{aligned}
$$

Since $h \geq 2$, $2r-rh-2h < 0$. Recall that $k \geq 3$. Hence, we have

$$
\begin{aligned}
3(2r-rh-2h)+2rh-4r \geq 0 \\
6r-3rh-6h)+2rh-4r \geq 0 \\
2r-rh-6h \geq 0.
\end{aligned}
$$

However, since $h \geq 2$, $2r-rh-6h < 0$ and this is a contradiction.

Case 2. $\omega(G[U]) = 0$.

By the definition of $h_G(S,T;r)$, $\omega(G[U]) \geq h_G(S,T;r)$, i.e. $h_G(S,T;r) = 0$. Since G is r-regular, by Theorem E,

$$
\delta_G(S,T;r) = r|S| - \sum_{y \in T}(r-\deg_G(y)) - e_G(S,T) \geq 0. \tag{14}
$$

Subtracting (1) from (14), we have

$$(r - k)(|S| - |T|) + k \geq 2. \tag{15}$$

Thus, $|S| \geq |T|$. We now consider two cases.

Case 2–1. $|V(G)|$ is odd.

Then, we obtain $|S| > |T|$. However, by (1), we have

$$\sum_{y \in T} \deg_{G-S}(y) \leq -2.$$

This is a contradiction.

Case 2–2. $|V(G)|$ is even.

We may assume $|S| = |T|$. Then by (1),

$$\sum_{y \in T} \deg_{G-S}(y) = 2e_G(T, T) \leq k - 2.$$

Since $e_G(S, S) = e_G(T, T)$ by $|S| = |T|$ and $\omega(G[U]) = 0$, we have

$$2e_G(S, S) \leq k - 2.$$

Thus, G becomes H_F. This is a contradiction. □

Proof of Theorem 1 Choose a counterexample G. Let $h_{even} = h_G(S, T; r)$ if r is even and $h_{odd} = h_G(S, T; r)$ if r is odd. Note that $h_{even} = h_G(S, T; r) = h_{G-x}(S - x, T; 2)$. Now, since $k = 2$ and $\ell = 2$ in Lemma 1, we have

$$\begin{aligned}(r - 4)(|S| - |T|) &\geq 2\omega(G[U]) - 2h_{even} + 2e_G(S, S) + 2e_G(T, T) \\ &\geq 0.\end{aligned} \tag{16}$$

On the other hand, from (1), we have

$$2\, S| - 2\, T| + \sum_{y \in T} \deg_{G-S}(y) - h_{even} \leq 0. \tag{17}$$

Then by the definition of an odd component, $\sum_{y \in T} \deg_{G-S}(y) \geq h_{even}$. Thus, by (17), we have

$$|S| \leq |T|. \tag{18}$$

Assume $r = 4$. Now, since G is 4-regular, by Theorem E,

$$4\, S| - 4\, T| + \sum_{y \in T} \deg_{G-S}(y) - h_{even} \geq 0. \tag{19}$$

Subtracting (17) from (19),

$$2\,S| - 2\,T| \geq 0$$
$$|S| \geq |T|. \tag{20}$$

By (18) and (20),

$$|S| = |T|.$$

Assume $r \geq 5$. Then by (16) and (18),

$$|S| = |T|. \tag{21}$$

Thus, we may assume $|S| = |T|$ if $r \geq 4$.

By (16) and (21),

$$\omega(G[U]) = h_{even},\ e_G(S, S) = e_G(T, T) = 0, \text{ and hence}$$
$$\sum_{y \in T} \deg_{G-S}(y) = h_{even} = e_G(S, U) = e_G(T, U). \tag{22}$$

If $\omega(G[U]) = 0$, G becomes bipartite graph by (21) and (22). This contradicts that G is not bipartite. Thus we may assume $\omega(G[U]) \geq 1$. Furthermore, (22) implies $E_G(V(G) - V(C_i), V(C_i)) = \{e_{i1}, e_{i2}\}$ for every component C_i of $G[U]$ and $\{e_{i1}, e_{i2}\}$ becomes 2-edge cut. Then by Theorem 3, there is a vertex $x \in V(G)$ such that $G - x$ has a 2-factor. This is a contradiction. □

5 Sharpness

First, we show that we cannot replace the 2-edge-connected of Theorem 1 with edge-connected. We consider a graph H such that $\deg_G(x_1) = r - 1$ for some $x_1 \in V(H)$ and $\deg_G(x) = r$ for every $x \in V(H) - \{x_1\}$. Define $W = (r - 1)H$. We form G_1 as follows. We add vertex y and join y to $(r - 1)$ vertices of degree $r - 1$ in W. The resulting graph G_1 contains one vertex of degree $r - 1$ and others vertices of degree r. Let G be as follows. We consider $2G_1$ and join the vertex of degree $r - 1$ to the vertex of degree $r - 1$. Such a graph is r-regular and connected. It is easily checked that $G - x$ has no 2-factor for every $x \in G$.

Next, we show that the condition of Theorem 2, $|V(G)|$ is odd, cannot be dropped. Let m, r be integers and k be an even integer such that $4 \leq k \leq \frac{r}{2}$ and $m \geq 3$. We will describe a graph G such that G is r-regular and $(2k - 2)$-edge-connected of an even order. We form G as follows. We start from a complete bipartite graph $K^i_{r,r}$ with bipartition (A_i, B_i) where $A_i = \{a_{i1}, a_{i2}, \ldots, a_{ir}\}$ and $B_i = \{b_{i1}, b_{i2}, \ldots, b_{ir}\}$ with $1 \leq i \leq m$. Now we use the definition of $a[i, j] = a_{ij}$. Remove the edges $a[i, 1]b[i, 1]$, $a[i, 2]b[i, 2], \ldots, a[i, k - 1]b[i, k - 1]$ from $K^i_{r,r}$ with $1 \leq i \leq m$. We add the edges

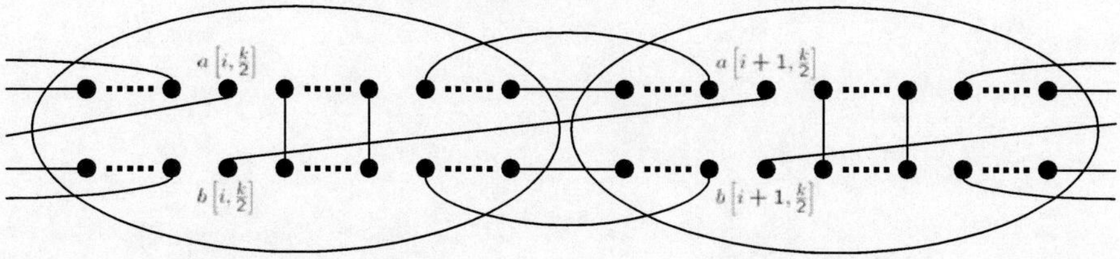

Fig. 1 Counterexample when $|V(G)|$ is even

$a[i, 1]a[i-1, r], a[i, 2]a[i-1, r-1], \ldots, a\left[i, \frac{k}{2}-1\right]a\left[i-1, r-\left(\frac{k}{2}-2\right)\right]$ with $2 \le i \le m$ and let $i-1=r$ if $i=1$. Similarly, we add the edges $b[i, 1]b[i-1, r], b[i, 2]b[i-1, r-1], \ldots, b\left[i, \frac{k}{2}-1\right]b\left[i-1, r-\left(\frac{k}{2}-2\right)\right]$ with $2 \le i \le m$ and let $i-1=r$ if $i=1$. Moreover we add the edges $a\left[i, \frac{k}{2}\right]b\left[i-1, \frac{k}{2}\right]$ and $b\left[i, \frac{k}{2}\right]a\left[i+1, \frac{k}{2}\right]$ with $2 \le i \le m-1$, and $a\left[1, \frac{k}{2}\right]b\left[m, \frac{k}{2}\right], b\left[1, \frac{k}{2}\right]a\left[2, \frac{k}{2}\right]$ and $a\left[m, \frac{k}{2}\right]b\left[m-1, \frac{k}{2}\right]$. The resulting graph G is r-regular, $(2k-2)$-edge-connected and has an even order. (Note that G is not also H_F, where we defined H_F in introduction.) We can easily check that $G-x$ has no k-factor for any $x \in G$. This graph G is shown in Fig. 1.

Finally, we present the following conjecture.

Conjecture 1 Let r and k be even integers such that $2 \le k \le \frac{r}{2}$, and G be an r-regular, 2-edge-connected graph of odd order. Then there is some $x \in V(G)$ such that $G-x$ has a k-factor.

References

1. Bondy, J.A., Murty, U.S.R.: Graph Theory with Applications, North Holland, Amsterdam, (1976).
2. Kano, M.: Sufficient conditions for a graph to have factors, Discrete Math. 80, 159–165 (1990).
3. Katerinis, P.: Some conditions for the existence of f-factors, J. Graph Theory. 9, 513–521 (1985).
4. Katerinis, P.: Regular factors in vertex-deleted subgraphs of regular graphs, Discrete Math. 131, 357–361 (1994).
5. Lu, H., Wang W., Bai, B.: Vertex-deleted subgraphs and regular factors from regular graph, Discrete Math. 311, 2044–2048 (2011).
6. Tutte, W.T.: The factors of graphs, Canad. J. Math. 4, 314–328 (1952).

Connectivity and Extendability in Digraphs

LeRoy B. Beasley

Abstract In this article we give several definitions of connectedness and extendability of paths and cycles in directed graphs. We define sets of digraphs by various types of connectedness or extendability and give some containments as well as give examples to show proper containment.

1 Introduction

The study of path and cycle extendability in graphs and digraphs began with articles by J. W. Moon in 1969 [13] and later by G. R. T. Hendry in 1989–1990 [8–10]. In [9] the question was asked: Is every Hamiltonian chordal graph cycle extendable? That is, given a Hamiltonian chordal graph and a cycle in that graph of length k, is there a cycle in the graph of length $k + 1$ with the same incident vertex set, plus one vertex? This question was studied by several researchers and was shown to be "yes" in several subsets of Hamiltonian chordal graphs, like interval graphs [7]. However, in 2013, Lafond and Seamone [11, Theorem 2.2] showed that not all Hamiltonian chordal graphs are cycle extendable. In Sect. 3 we shall return to chordal Hamiltonian graphs and extendability. For an excellent review of pancyclicity and cycle extendability in undirected graphs see Deborah Arangno's Ph.D. thesis [2].

The situation for connectivity and extendability of directed graphs is more complex than for undirected graphs and hence we shall address directed graph connectivity and extendability after relative definitions are presented. Few articles have appeared lately about path or cycle extendability. A few exceptions are [2–5, 7, 14].

For graph theoretical background see [6] and for background on tournaments see [12].

L. B. Beasley (✉)
Utah State University, Logan, UT 84322, USA
e-mail: leroy.b.beasley1@gmail.com

F. Hoffman (ed.), *Combinatorics, Graph Theory and Computing*,
Springer Proceedings in Mathematics & Statistics 388,
https://doi.org/10.1007/978-3-031-05375-7_19

2 Preliminaries

The length of a directed or undirected path or cycle in a digraph or graph is the number of edges/arcs in that path or cycle. Since our graphs are loopless, there are no cycles of length one, and a path of length one is an edge or arc. Throughout we shall reserve the symbol n to represent the number of vertices in a graph or digraph and let the vertex set be $V = \{v_1, v_2, \ldots, v_n\}$, so given a graph G, $G = (V, E(G))$ where $E(G)$ is the edge set of G.

Definition 1 Let G be an undirected graph. If u and v are two vertices in G, then $d(u, v)$ is the length of the shortest path connecting u to v. It is called the *distance from u to v*.

Definition 2 Let D be a directed graph. If u and v are two vertices in D, then $d(u, v)$ is the length of the shortest directed path from u to v or from v to u. It is also called the *distance from u to v*.

Definition 3 A path in an undirected graph is called a *Hamilton path* if it is of length $n - 1$, the longest possible path. In a directed graph, a Hamilton path is a directed path of length $n - 1$.

Definition 4 An undirected graph is *pan-connected* if every two distinct vertices is connected by a path of every possible length greater than or equal to the distance between them.

3 Connectivity

One of the basic properties that a graph may or may not possess is that of being connected. For undirected graphs, we say a graph is *connected* if given any two vertices, there is a path between them. For directed graphs, the situation is more complex. Let $\mathcal{D}_n$ denote the set of all simple, loopless digraphs on the n vertex set $V = \{v_1, v_2, \ldots, v_n\}$,

There are several basic concepts of connectedness for directed graphs:

Definition 5 A directed graph in $\mathcal{D}_n$ is:

1. *connected* if the underlying undirected graph is connected.
2. *(weakly) path connected* if given any two distinct vertices u and v there is a directed path from u to v, or from v to u.
3. *strongly (path) connected* if given any two distinct vertices u and v there are directed paths, one from u to v and one from v to u.
4. *weakly-Hamilton connected* if given any two distinct vertices u and v there is a Hamilton path connecting either u to v or v to u.
5. *(strongly)-Hamilton connected* if given any two distinct vertices u and v there is are Hamilton paths, one connecting u to v and one connecting v to u.

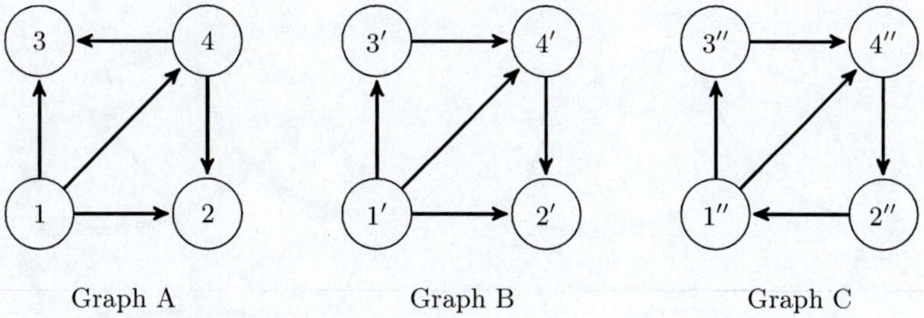

Fig. 1 Connected digraphs

6. *weakly-pan-connected* if given any two distinct vertices, u and v, and any $k \in \{d(u, v), \ldots, n\}$, there is a directed path of length k from vertex u to vertex v or from vertex v to vertex u.
7. *strongly-pan-connected* if given any two distinct vertices, u and v, and any $k \in \{d(u, v), \ldots, n\}$, there is a directed path of length k from vertex u to vertex v and one from vertex v to vertex u.

Definition 3 part 4 and Definition 3 part 5 above are the directed version of a Hamilton connected undirected graph which is an undirected graph for which there is a Hamilton path between any two distinct vertices.

Consider the three graphs on four vertices in Fig. 1. Graph A is connected but not path connected nor strongly connected, there is no path between vertex 2 and vertex 3. Graph B is weakly path connected but not strongly connected, there is no path from vertex $2'$ to vertex $3'$. Graph C is strongly connected.

Let $\mathcal{CD}_n$ denote the set of all connected digraphs in $\mathcal{D}_n$, let $\mathcal{PC}_n$ denote the set of all (weakly) path connected digraphs in $\mathcal{D}_n$, and let $\mathcal{SC}_n$ denote the set of all strongly connected digraphs in $\mathcal{D}_n$. Then clearly, $\mathcal{SC}_n \subseteq \mathcal{PC}_n \subseteq \mathcal{CD}_n \subseteq \mathcal{D}_n$. The digraphs in Fig. 1 together with any digraph with an isolated vertex show that these containments are all strict. The subscript n is usually omitted if the order of the graph is obvious from the context.

Let $\mathcal{HC}_w$ denote the set of digraphs in $\mathcal{D}_n$ that are weakly-Hamilton connected and let $\mathcal{HC}_s$ denote the set of digraphs in $\mathcal{D}_n$ that are strongly-Hamilton connected.

Let the set of all weakly-pan-connected digraphs in $\mathcal{D}_n$ be denoted $\mathcal{P}an_w$ and the set of all strongly-pan-connected digraphs in $\mathcal{D}_n$ be denoted $\mathcal{P}an_s$.

As noted above, all our graphs have the same vertex set, so the union of two graphs G and H is the graph $G \cup H = (V, E(G) \cup E(H))$.

3.1 Examples and Containment

Note that any two vertices in a strongly-pan-connected digraph are either not adjacent or are connected by arcs in both directions, forming a digon.

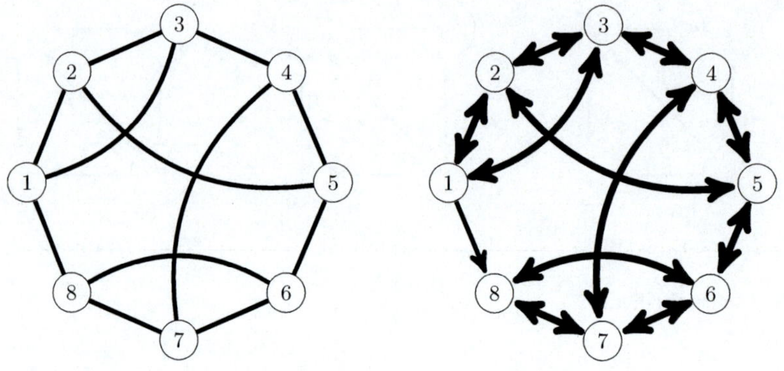

Fig. 2 Graphs Ch_8 (left) and $Ch_{8,d}$ (right)

Example 1 See [1]. Let G be an undirected graph. Define G^2 to be the graph whose edge set is the set of all edges uv such that there is a path of length 2 from u to v. It was shown in [1, Theorem 2] that if C_n is an undirected Hamilton cycle, then $C^{[2]} = C_n \cup C_n^2$ is pan-connected, so replacing all the edges in $C^{[2]}$ with arcs in both directions to get $\overrightarrow{C^{[2]}}$, we have a strongly-pan-connected digraph. By deleting one arc from this digraph, we have a digraph, $C^{[2,d]} = \overrightarrow{C^{[2]}} \setminus \{(u, v)\}$, that is weakly-pan-connected. It is not strongly-pan-connected since $d(u, v) = 1$ but there is no arc from u to v.

In [6, p. 191] an example of a Hamilton connected graph was presented. For $n = 8$ it is the graph Ch_8 on the left in Fig. 2. If all the edges of Ch_8 are replaced by two arcs, one in each direction we get the digraph $\overrightarrow{Ch_8}$ which is an example of a strongly-Hamilton connected digraph. If all the edges except the one connecting vertex 1 and vertex 8 are replaced by two arcs, one in each direction, and the edge between vertex 1 and vertex 8 replaced by a single arc from vertex 1 to vertex 8, we have a weakly-Hamilton connected digraph since there is no directed Hamilton path from vertex 6 to vertex 2. This digraph is the digraph $Ch_{8,d}$ on the right in Fig. 2. Note that the digraph $Ch_{8,d}$ is not weakly-pan-connected.

The following containments are easily established:

Proposition 1 *For* $n \geq 3$,

- $\mathcal{P}an_s \subsetneqq \mathcal{P}an_w$, *not equal by* $C^{[2,d]}$;
- $\mathcal{P}an_w \subsetneqq \mathcal{HC}_w$, *not equal by* $Ch_{8,d}$;
- $\mathcal{P}an_s \subsetneqq \mathcal{HC}_s$, *not equal by* $\overrightarrow{Ch_8}$;
- $\mathcal{HC}_s \subsetneqq \mathcal{HC}_w$, *not equal by* $Ch_{8,d}$.

Motivated by the above discussion of connectedness, we shall proceed to investigate concepts of path and cycle extendability in digraphs. In the next section we shall define several concepts of extendability and end the section with a table showing relations between the various sets defined above and in Sect. 4.2.

4 Extendability

We begin by giving several definitions of extendability in digraphs. We divide this section into three subsections. In the first subsection we give the traditional definitions of extendability; in the second subsection we give notation for sets of graphs defined by extendability conditions. We end with a subsection summarizing the results of this section and Sect. 2.

4.1 Definitions—Path- and Cycle-Extendability

1. A digraph $D \in \mathcal{D}_n$ is said to be *path dense* if given any set of vertices $V' \subseteq V$ with $|V'| \in \{2, 3, \ldots, n\}$ there is a directed path in D whose set of incident vertices is V'.
2. A digraph $D \in \mathcal{D}_n$ is said to be *k-path dense* if given any set of vertices $V' \subseteq V$ with $|V'| \in \{k, k+1, \ldots, n\}$ there is a directed path in D whose set of incident vertices is V'.
   The digraph $Ch_{8,d}$ is 6-path dense, but not 5-path dense, there is no path with incident vertex set $\{v_1, v_2, v_3, v_4, v_6\}$.
3. A path in D whose incidence vertex set is V', $(|V'| \leq n-1)$, is *weakly extendable* if there is some vertex w in $V \setminus V'$ such that D contains a path whose set of incident vertices is $V' \cup \{w\}$.
4. A digraph in $\mathcal{D}_n$ is said to be *weakly-path-extendable* if every path of length 1 through $n-2$ is weakly extendable.
5. A path in D whose incidence vertex set is V', $(|V'| \leq n-1)$, with initial vertex v_I and terminal vertex v_T is *strongly extendable* if there is some vertex w in $V \setminus V'$ such that D contains a path whose incidence vertices is $V' \cup \{w\}$ and whose initial vertex is v_I and whose terminal vertex is v_T.
6. A digraph in $\mathcal{D}_n$ is said to be *strongly-path-extendable* if every path of length 1 through $n-2$ is strongly extendable.
   A slight generalization of the concept of strongly extendable requires that the endpoints of the path be the same but not necessarily that the initial vertex be the initial vertex, etc.
7. A path in D whose incidence vertex set is V', $(|V'| \leq n-1)$, with initial vertex v_I and terminal vertex v_T is *almost strongly extendable* if there is some vertex w in $V \setminus V'$ such that D contains a path whose incidence vertices is $V' \cup \{w\}$ and whose initial vertex is either v_I or v_T and whose terminal vertex is either v_T or v_I, respectively.
8. A digraph in $\mathcal{D}_n$ is said to be *almost-strongly-path-extendable* if every path of length 1 through $n-2$ is almost strongly extendable.
9. A directed cycle in D whose incidence vertex set is V', $(|V'| \leq n-1)$, is *extendable* if there is some vertex w in $V \setminus V'$ such that D contains a directed cycle whose set of incident vertices is $V' \cup \{w\}$.

10. A digraph in $\mathcal{D}_n$ is said to be *cycle-extendable* if every cycle of length 2 through $n-1$ is extendable. Note that a 2-cycle is called a *digon*.

4.2 Definitions—Sets of Graphs Defined by Extendability

1. Let $\mathcal{PD}_n$ denote the set of all digraphs in $\mathcal{D}_n$ that are path dense.
2. Let $\mathcal{PE}_{w,n}$ denote the set of all digraphs in $\mathcal{D}_n$ that are weakly path extendable.
3. Let $\mathcal{PE}_{s,n}$ denote the set of all digraphs in $\mathcal{D}_n$ that are strongly path extendable.
4. Let $\mathcal{PE}_{as,n}$ denote the set of all digraphs in $\mathcal{D}_n$ that are almost strongly path extendable.
5. Let $\mathcal{CE}_n$ denote the set of all digraphs in $\mathcal{D}_n$ that are cycle extendable.
   Any acyclic digraph is cycle extendable and any digraph that dominates a cycle dominates a Hamilton cycle. Thus we define:
6. Let $\mathcal{CE}_{H,n}$ denote the set of all digraphs in $\mathcal{D}_n$ that are cycle extendable and are not acyclic.

Note that the empty graph $\overline{K_n}$ is (vacuously) a member of all of the above sets except $\mathcal{PD}_n$. Further, we will omit the subscript n if the order is obvious from the context.

4.3 Examples and Containment

Some obvious containments are:

Theorem 1 *Let $n > 3$ then $\mathcal{PE}_s \subsetneq \mathcal{PE}_{as} \subsetneq \mathcal{PE}_w$;*

Definition 6 A *tournament* on n vertices is a directed graph which is an orientation of the complete simple undirected graph. That is a tournament is a loopless digraph in which any two distinct vertices are connected by exactly one arc.

Let $T_{t,k}$ denote the digraph whose vertex set is $V = \{v_1, v_2, \ldots, v_k\}$ and whose arc set is $A = \{(v_i, v_j) | 1 \le i < j \le k\}$. That is, $T_{t,k}$ is a transitive tournament (a cycle free tournament) on k vertices.

For digraphs A and B let $A \Rightarrow B$ denote the digraph consisting of the vertices of A together with the vertices of B, the arc set consisting of the arc set of A together with the arc set of B and arcs from every vertex of A to every vertex of B.

Proposition 2 $\mathcal{PD} \subsetneq \mathcal{PE}_w$.

Proof Let $D \in \mathcal{PD}$ and let $\mathbf{p}$ be a directed path in D with incident vertex set V'. Let w be any vertex not in V'. Since $D \in \mathcal{PD}$ there is a path $\mathbf{p}'$ in D with vertex set $V' \cup \{w\}$. Thus, $D \in \mathcal{PE}_w$.

For $n \ge 5$ and $2 \le k \le n-2$ and w a vertex, the digraph $D = T_{t,k} \Rightarrow w \Rightarrow T_{t,(n-k-1)}$ is in $\mathcal{PE}_w$, but not in $\mathcal{PD}$ since $\{v_1, v_2, v_n\}$ is not the vertex set of any path in D but any path in D is weakly extendable. ■

Proposition 3 $\mathcal{PE}_s \subsetneq \mathcal{PE}_w$.

Proof Clearly, if a path is strongly extendable, it is weakly extendable. Thus, $\mathcal{PE}_s \subseteq \mathcal{PE}_w$. But, the path $v_2 \to v_3 \to \cdots \to v_n$ in $T_{t,n}$ is extendable to $v_1 \to v_2 \to v_3 \to \cdots \to v_n$ which is the only possible extension, and the initial vertex must change, thus it is not strongly extendable, and consequently $T_{t,n} \in \mathcal{PE}_w$ but $T_{t,n} \notin \mathcal{PE}_s$. That is $\mathcal{PE}_s \neq \mathcal{PE}_w$. ■

Let $T_{s,n}$ denote the tournament on vertices $V = \{v_1, v_2, \ldots, v_n\}$ and arc set $A = \{(v_i, v_j) | 1 \leq i < j \leq n\} \cup \{(v_n, v_1)\} \setminus \{(v_1, v_n)\}$. That is $T_{s,n}$ is the tournament $T_{t,n}$ with the one arc (v_1, v_n) reversed.

Proposition 4 $\mathcal{PE}_s \subsetneq \mathcal{CE}$.

Proof Let D be a digraph in $\mathcal{PE}_s$ and C a non Hamiltonian cycle in D. By reordering the vertices we may assume that the cycle is $v_1 \to v_2 \to \cdots \to v_k \to v_1$, so that $v_1 \to v_2 \to \cdots \to v_k$ is a non Hamiltonian path in D. Since $D \in \mathcal{PE}_s$ there is a path $v_1 \to u_2 \to \cdots \to u_k \to v_k$ in D for some vertices u_i, $i = 2, \ldots, k$. But then $v_1 \to u_2 \to \cdots \to u_k \to v_k \to v_1$ is a cycle extending C. That is $D \in \mathcal{CE}$.

Note that $v_1 \to v_2 \to \cdots \to v_{n-1}$ is a path that is not strongly extendable in $T_{s,n}$, thus $T_{s,n}$ is in $\mathcal{CE}$ but not in $\mathcal{PE}_s$. That is $\mathcal{CE} \neq \mathcal{PE}_s$. ■

Proposition 5 *If a directed graph dominates a directed Hamilton cycle then it is strongly connected.*

Proof Let D be a digraph that dominates a Hamilton cycle and let u and v be any two vertices in D. Then, u and v partition the cycle into two arc disjoint paths, one from u to v the other from v to u. ■

Corollary 1 *All the digraphs in $\mathcal{CE}_H$, $\mathcal{PE}_s$. and $\mathcal{HC}_s$ are strongly connected.*

Proposition 6 *If a digraph dominates a directed Hamilton path then it is path connected.*

Proof Let D be a digraph that dominates a Hamilton path and let u and v be any two vertices in D. Then, that path is incident with both u and v so there is either a path from u to v or from v to u. ■

Corollary 2 *All of the digraphs in $\mathcal{CE}_H$, $\mathcal{PE}_s$. $\mathcal{PE}_w$. $\mathcal{PE}_{as}$, $\mathcal{PD}$, $\mathcal{HC}_w$, and $\mathcal{HC}_s$ are path connected.*

Proposition 7 $\mathcal{PE}_{as} \subsetneq \mathcal{HC}_w$.

Proof Let D be a digraph in $\mathcal{PE}_{as}$ and u and v any two vertices. Then, D dominates a Hamilton path, and since u and v must be on this path, there is either a path from u to v or from v to u. In either case there is Hamilton path from u to v or from v to u by the definition of $\mathcal{PE}_{as}$. ■

Example 2 Let D be the digraph in $\mathcal{D}_n$ with arc set $\{(v_1, v_2), (v_2, v_3.)(v_3, v_1), (v_3, v_4), (v_4, v_3), (v_4, v_5), (v_5, v_4) \cdots, (v_{n-1}, v_n), (v_n, v_{n-1})\}$. D is a directed three cycle appended to a path of digons from vertex 3 to n. Then the path $v_n \to v_{n-1} \to \cdots \to v_3 \to v_1$ is extendable but only to one beginning at v_1 and ending at v_n, not the other way. Thus, $\mathcal{PE}_{as} \neq \mathcal{PE}_s$.

Sets	why	why $\neq$	Sets	why	why $\neq$
$\mathcal{CE}_H \subsetneq \mathcal{SC}$	Cor 1	$C_{n,4}$	$\mathcal{CE}_H \subsetneq \mathcal{CE}$	def	$T_{t,n}$
$\mathcal{PE}_s \subsetneq \mathcal{SC}$	trans	$C_{n,4}$	$\mathcal{PE}_{as} \subsetneq \mathcal{CE}$	Prop 4	P_n(vac)
$\mathcal{PE}_{as} \subsetneq \mathcal{SC}$	trans	$C_{n,4}$	$\mathcal{PE}_w \subsetneq \mathcal{PE}_{as}$	Def.	$C_{n,4}$
$\mathcal{P}an_w \subsetneq \mathcal{SC}$	trans	$C^{[2,d]}$	$\mathcal{PE}_s \subsetneq \mathcal{CE}_H$	Prop 4	P_n(vac)
$\mathcal{HC}_s \subsetneq \mathcal{SC}$	Cor 1	$C_{n,4}$	$\mathcal{PE}_s \subsetneq \mathcal{PE}_{as}$	Def.	Ex. 2
$\mathcal{SC} \subsetneq \mathcal{PC}$	Def.	P_n	$\mathcal{PE}_s \subsetneq \mathcal{PE}_w$	Def.	$C_{n,4}$
$\mathcal{CE}_H \subsetneq \mathcal{PC}$	trans	$C_{n,4}$	$\mathcal{PD} \subsetneq \mathcal{PE}_w$	Prop 2	Prop 2
$\mathcal{PE}_s \subsetneq \mathcal{PC}$	trans	C_n	$\mathcal{PE}_{as} \subsetneq \mathcal{HC}_w$	Prop 7	$Ch_{8,d}$
$\mathcal{PE}_{as} \subsetneq \mathcal{PC}$	trans	C_n	$\mathcal{PE}_s \subsetneq \mathcal{HC}_w$	trans	$Ch_{8,d}$
$\mathcal{PE}_w \subsetneq \mathcal{PC}$	trans	dbl-star	$\mathcal{HC}_s \subsetneq \mathcal{HC}_w$	Def.	$Ch_{8,d}$
$\mathcal{HC}_w \subsetneq \mathcal{PC}$	trans	$C_{n,4}$	$\mathcal{P}an_s \subsetneq \mathcal{PE}_s$	Def.	$C^{[2,d]}$
$\mathcal{HC}_s \subsetneq \mathcal{PC}$	trans	$C_{n,4}$	$\mathcal{P}an_s \subsetneq \mathcal{HC}_s$	Prop 1	$\overrightarrow{Ch_8}$
$\mathcal{PD} \subsetneq \mathcal{PC}$	Cor 2	C_n	$\mathcal{P}an_w \subsetneq \mathcal{PC}$	Def.	C_n

Fig. 3 Subset containment table

4.4 Summary

Figure 3 shows the set containments that we have established in the previous sections.

In Fig. 6 we have a table of directed graphs across the top and sets along the side. A check indicates that the digraph in that column is a member of the set in that row. An **x** indicates that is is not. The subscript "vac" indicates that the inclusion is vacuously true since there are no cycles to extend. The subscript "2-cyc" indicates that the only non extendable cycles are digons. the sets across the top are:

$C_{n,4}$, $n \geq 7$, the directed n-cycle $(v_1 \rightarrow v_2 \rightarrow v_3 \rightarrow \cdots \rightarrow v_{n-1} \rightarrow v_n \rightarrow v_1)$ plus the arc $v_4 \rightarrow v_n$;

$T_{t,n}$, the transitive n-tournament;

$T_{s,n}$, the strong n-tournament;

Ch_8 and $Ch_{8,d}$, See Fig. 2;

C_n, a directed n-cycle and $C_{n,d}$ an n-cycle with every edge replaced by an arc in both directions;

P_n, a directed Hamilton path;

$K^{[+]}$, K_{n-1} on vertices $\{v_1, v_2, \ldots, v_{n-1}\}$ plus a 2-path through vertex v_n;

Star, a double star, or the union of an out-star and an in-star both centered at the same vertex;

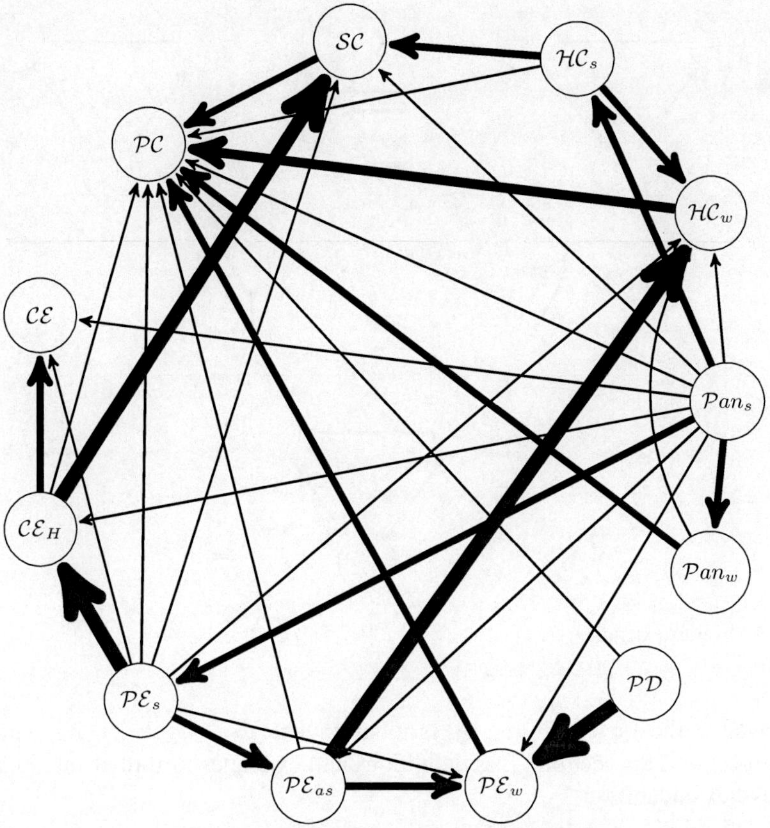

Heavy arrows are justified in the text.

Medium arrows are easily established, some justified in the text.

Light arrows follow by transitivity.

Each arrow represents $\subsetneq$.

Fig. 4 Set containment graph

$C^{[2]}$ and $C^{[2,d]}$, See Example 2.1.

The sets in the first column are those defined above.

5 Set Connectivity and Extendability

Some graphs or digraphs are not cycle extendable, but each cycle is extendable to cycles of length one or two more. Example 3 is an example of this. It is not cycle

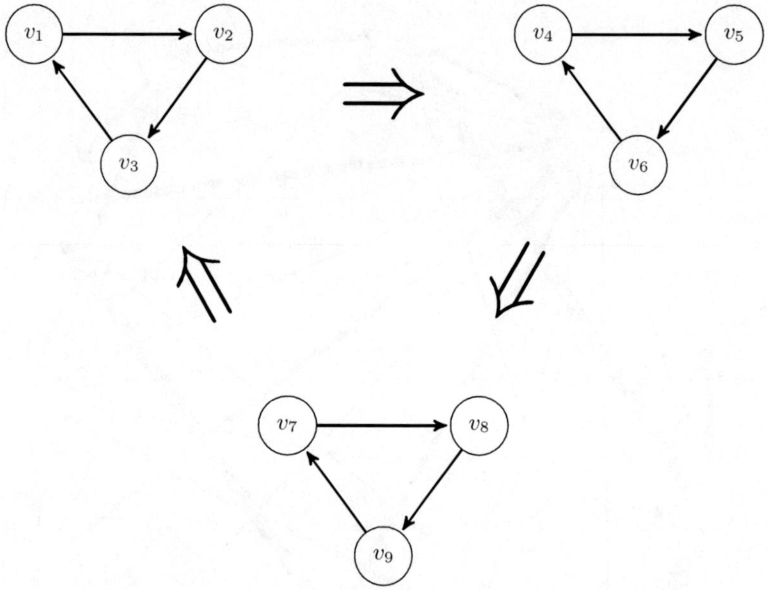

Fig. 5 9 vertex example

extendable, the 3-cycle (v_1, v_2, v_3) is not extendable to a 4-cycle, but it is extendable to a 5-cycle. This section gives definitions and examples to further investigate this type of extendability.

5.1 Definitions—S-Path- and S-Cycle-Extendability

We can further refine the concept of connectedness by limiting the length of the path connecting two vertices:

1. Let S be a subset of $\{1, 2, \ldots, n-1\}$. A digraph is *weakly-S-path connected* if given any two distinct vertices u and v there is a path whose length is in S connecting either u to v or v to u.
2. Let S be a subset of $\{1, 2, \ldots, n-1\}$. A digraph is *strongly-S-path connected* if given any two distinct vertices u and v there is are paths whose lengths are in S, one connecting u to v and the other connecting v to u.
   Note that being weakly/strongly-Hamilton-connected is equivalent to being weakly/strongly-$\{n\}$-path-connected.
3. Let $S \subseteq \{2, 3, \ldots, n\}$. A digraph $D \in \mathcal{D}_n$ is said to be *S-path dense* if given any set of vertices $V' \subseteq V$ with $|V'| \in S$ there is a directed path in D whose set of incident vertices is V'.
   Note that if $S = \{k, k+1, \ldots, n\}$ then S-path dense is the same as k-path dense.

subset	$C_{n,4}$	$T_{t,n}$	$T_{s.n}$	Ch_8	$Ch_{8,d}$	C_n	$C_{n,d}$	P_n	$K^{[+]}$	Star	$C^{[2]}$	$C^{[2,d]}$
$\mathcal{SC}$	√	x	√	√	√	√	√	x	√	x	√	√
$\mathcal{PC}$	√	√	√	√	√	√	√	√	√	x	√	√
$\mathcal{CE}$	x	$√_{vac}$	√	x	x	$√_{vac}$	x_{2cyc}	$√_{vac}$	√	x_{2cyc}	√	√
$\mathcal{CE}_H$	x	x	√	x	x	$√_{vac}$	x	x	√	x	√	√
$\mathcal{PE}_s$	x	x	x	x	x	x	x	x	x	x	√	√
$\mathcal{PE}_{as}$	x	x	x	x	x	x	x	x	√	x	√	√
$\mathcal{PE}_w$	√	√	√	√	√	√	√	√	√	x	√	√
$\mathcal{PD}$	x	√	√	x	x	x	x	x	√	x	x	x
$\mathcal{HC}_w$	x	x	x	√	√	x	x	x	√	x	√	√
$\mathcal{HC}_s$	x	x	x	x	x	x	x	x	x	x	√	√
$\mathcal{P}an_w$	x	x	x	x	x	x	x	x	x	x	√	√
$\mathcal{P}an_s$	x	x	x	x	x	x	x	x	x	x	√	x

Note 1: A check indicates that the test graph is in the set, and a bold x that it is not.

Note 2: The subscript ``2cyc" indicates that a digon is not extendable to a three cycle but otherwise is extendable.

Note 3: The subscrpt ``vac" refers to the fact that there are no cycles in the graph and hence are cycle extendable.

Fig. 6 Inclusion of test graphs

4. Let $S \subseteq \{1, 2, \dots, n\}$. A path in D whose incidence vertex set is V' is *weakly S-extendable* if there is some subset $W \subseteq (V \setminus V')$ with $|W| \in S$ such that D contains a path whose set of incident vertices is $V' \cup W$.
5. Let $S \subseteq \{1, 2, \dots, n\}$. A digraph in $\mathcal{D}_n$ is said to be *weakly-S-path-extendable* if every path of length 1 through $n - a$ is weakly S-extendable where a is the smallest element of S.
6. Let $S \subseteq \{1, 2, \dots, n\}$. A path in D whose incidence vertex set is V' with initial vertex v_I and terminal vertex v_T is *strongly S-extendable* if there is some subset $W \subseteq (V \setminus V')$ such that D contains a path whose set of incident vertices is $V' \cup W$ and whose initial vertex is v_I and whose terminal vertex is v_T.

7. Let $S \subseteq \{1, 2, \ldots, n\}$. A digraph in $\mathcal{D}_n$ is said to be *strongly-S-path-extendable* if every path of length 1 through $n - a$ is strongly S-extendable where a is the smallest element of S.
8. Let $S \subseteq \{1, 2, \ldots, n\}$. A path in D whose incidence vertex set is V' with initial vertex v_I and terminal vertex v_T is *almost strongly S-extendable* if there is some subset $W \subseteq (V \setminus V')$ such that D contains a path whose incident vertex set is $V' \cup W$ and whose initial vertex is v_I or v_T and whose terminal vertex is v_T or v_I, respectively.
9. Let $S \subseteq \{1, 2, \ldots, n\}$. A digraph in $\mathcal{D}_n$ is said to be *almost strongly-S-path-extendable* if every path of length 1 through $n - a$ is almost strongly S-extendable where a is the smallest element of S.
10. Let $S \subseteq \{1, 2, \ldots, n\}$. A cycle in D whose incidence vertex set is V' is *S-extendable* if there is some subset $W \subseteq (V \setminus V')$ with $|W| \in S$ such that D contains a cycle whose set of incident vertices is $V' \cup W$.
11. Let $S \subseteq \{1, 2, \ldots, n\}$. A digraph in $\mathcal{D}_n$ is said to be *S-cycle-extendable* if every cycle of length 2 through $n - a$ is S-extendable where a is the smallest element of S.

In [13, Theorem 1] J. W. Moon showed that any strongly connected orientation of the complete loopless graph is $\{1, 2\}$-cycle extendable. The digraph in Fig. 5 is a strongly connected orientation of the complete graph on nine vertices, and hence shows that Moon's theorem can not be improved to cycle extendable. In fact, the digraph in Fig. 5 is not only Hamiltonian, and Hamilton connected, but also weakly-pan-connected.

5.2 Sets Defined by Set-Continuity and Set-Extendability

1. Let $S \subseteq \{1, 2, \ldots, n\}$ and let $\mathcal{PD}_{S,n}$ denote the set of all digraphs in $\mathcal{D}_n$ that are S-path dense.
2. Let $S \subseteq \{1, 2, \ldots, n\}$ and let $\mathcal{PE}_{w,S,n}$ denote the set of all digraphs in $\mathcal{D}_n$ that are weakly S-path extendable.
3. Let $S \subseteq \{1, 2, \ldots, n\}$ and let $\mathcal{PE}_{s,S,n}$ denote the set of all digraphs in $\mathcal{D}_n$ that are strongly S-path extendable.
4. Let $S \subseteq \{1, 2, \ldots, n\}$ and let $\mathcal{PE}_{as,S,n}$ denote the set of all digraphs in $\mathcal{D}_n$ that are almost strongly S-path extendable.
5. Let $\mathcal{CE}_{S,n}$ denote the set of all digraphs in $\mathcal{D}_n$ that are S-cycle-extendable.

Note that the empty graph $\overline{K_n}$ is (vacuously) a member of all of the above sets except $\mathcal{PD}_{S,n}$.

Example 3 (See Fig. 5.) Let D be the digraph on 9 vertices, $v_1, v_2, \ldots, v_9$ such that v_1, v_2, v_3 induces a 3-cycle, v_4, v_5, v_6 induces a 3-cycle, v_7, v_8, v_9 induces a 3-cycle, and there is an arc from each vertex of the first 3-cycle to each vertex of the second

3-cycle, an arc from each vertex of the second 3-cycle to each vertex of the third 3-cycle, and an arc from each vertex of the third 3-cycle to each vertex of the first 3-cycle. Then the 3-cycle on v_1, v_2, v_3 cannot be extended to a 4-cycle containing v_1, v_2, v_3, but every cycle in D is $\{1, 2\}$-extendable. It should be noted that this graph is path dense, weakly-pan-connected, and strongly Hamilton connected.

Note that in the above example, each 3-cycle can be replaced with any strongly connected digraph on at least 3 vertices. The analysis will remain the same.

We end with some obvious containments.

Theorem 2 *Let* $n > 3$ *and* $T \subsetneqq S \subseteq \{1, 2, \ldots, n-1\}$,

1. $\mathcal{PD}_n \subsetneqq \mathcal{PD}_{S,n}$.
2. $\mathcal{PE}_{s,S,n} \subsetneqq \mathcal{PE}_{as,S,n} \subsetneqq \mathcal{PE}_{w,S,n}$.
3. $\mathcal{PE}_s \subseteq \mathcal{PE}_{s,T,n} \subsetneqq \mathcal{PE}_{s,S,n}$.
4. $\mathcal{PE}_{as} \subseteq \mathcal{PE}_{as,T,n} \subsetneqq \mathcal{PE}_{as,S,n}$.
5. $\mathcal{PE}_w \subseteq \mathcal{PE}_{w,T,n} \subsetneqq \mathcal{PE}_{w,S,n}$.
6. $\mathcal{CE}_n \subsetneqq \mathcal{CE}_{T,n} \subsetneqq \mathcal{PE}_{S,n}$.

Most containments involving only S and not T or only T and not S are parallel to the containments above and are not verified here. Containments involving $T \subsetneqq S$ follow solely from that containment. Further note that $\mathcal{CE}_n = \mathcal{CE}_{\{1\},n}$, etc.

References

1. Alavi, Y., Williamson, J. E., "Panconnected graphs", Studia Scientiarum Mathematicarum Hungarica, 10 (1975) Nos. 1–2: 19–22.
2. Arangno, D. C., "Hamiltonicity, Pancyclicity, and Cycle Extendability in Graphs", Utah State University (2014). All Graduate Theses and Dissertations. 3295. https://digitalcommons.usu.edu/etd/3295
3. Beasley, L. B. and Brown, D. E., "On cycle and bi-cycle extendability in chordal and chordal bipartite graphs", Congressus Numerantium. 174(2005)41–47.
4. Beasley, L. B. and Brown, D. E., "Cycle extendability in graphs and digraphs". Linear Algebra Appl. 435 (2011), no. 7, 1513–1519.
5. Beasley, L. B., Merz, S. K., Brown, D. E., and Thomas, B. J., "Cycle extendability in tournamendts". Congressus Numerantium. 233 (2019), pp.95-101.
6. Chartrand, G. and Lesniak, L., *Graphs and Digraphs*, 2nd *Ed.*, Wadswoth & Brooks/Cole, Monterey, CA, 1986
7. G. Chen, R.J. Faudree, R.J. Gould, M.S. Jacobson, "Cycle extendability of Hamiltonian interval graphs", SIAM J. Discrete Math. 20(3) (2006) 682–689.
8. Hendry, G.R.T., "Extending Cycles in Directed Graphs", J. Comb. Th. B., 46(1989) 162–172.
9. Hendry, G. R. T., "Extending cycles in graphs", Discrete Math. 85 (1990) no. 1, 59 – 72.
10. Hendry, G. R. T., "Path extendable graphs", Per. Math. Hungarica, 21(3, (1990), 205 – 219.
11. Lafond, M. and Seamone, B., "Hamiltonian chordal graphs are not cycle extendible", SIAM J. Discrete Math. 29(2), (2015) 877–887.
12. J. W. Moon, *Topics in Tournaments*, Holt, New York, 1968.
13. J. W. Moon, "On cycles in Tournaments", Matematicky Časopsis, 19(1969, No. 2, 121–125.
14. Zhang, Z.-B., Zhang, X., Broersma, H., and Lou, D., "Extremal and degree conditions for path extendability in digraphs". SIAM J. Discrete Math. 31 (2017), no. 3, 1990–2014.

On the Extraconnectivity of Arrangement Graphs

Eddie Cheng, László Lipták, and Daniel Tian

Abstract Extraconnectivity generalizes the concept of connectivity of a graph but it is more difficult to compute. In this note, we compute the g-extraconnectivity of the arrangement graph for small g (with $g \leq 6$) with the help of a computer program. In addition, we provide an asymptotic result for general g.

1 Introduction

The study of multiprocessor systems is an important aspect of parallel computing. The underlying topology of such a multiprocessor system is an interconnection network. Such an interconnection network is usually described and studied in terms of graph theory. One can view the vertices as processors in which the resulting system is a multiprocessor supercomputer (with edges being the links between processors), or they can be viewed as computers (with edges being the links between computers) in which the resulting system is a computer network. Using the example of a multiprocessor supercomputer, since processors and/or links can fail, it is important to come up with fault resiliency measurements.

The *(vertex) connectivity* of a connected non-complete graph is the minimum number of vertices whose deletion disconnects the graph. The vertex connectivity of a complete graph on n vertices is defined to be $n - 1$. Moreover, a connected non-complete graph is *k(-vertex)-connected* if with at most $k - 1$ vertices being deleted, the resulting graph is connected, that is, the vertex connectivity is at least k. There is

Daniel Tian started this project at the 2019 Oakland University Summer Mathematics Institute.

E. Cheng (✉) · L. Lipták
Department of Mathematics and Statistics, Oakland University, Rochester, MI 48309, USA
e-mail: echeng@oakland.edu

L. Lipták
e-mail: liptak@oakland.edu

D. Tian
Northville High School, 45700 Six Mile Rd, Northville, MI 48168, USA

F. Hoffman (ed.), *Combinatorics, Graph Theory and Computing*,
Springer Proceedings in Mathematics & Statistics 388,
https://doi.org/10.1007/978-3-031-05375-7_20

a corresponding edge version if edges are deleted. However, such measures are quite simplistic, thus researchers have proposed a number of more advanced parameters.

A set of vertices T in a connected non-complete graph G is called a *restricted vertex-cut of order m* or an *$(m-1)$-extra-vertex-cut* (or $(m-1)$-extra-cut for short) if $G-T$ is disconnected and every component in $G-T$ has at least m vertices. The *restricted vertex connectivity of order m* or the *$(m-1)$-extraconnectivity* is the size of a smallest restricted vertex-cut of order m. Thus a restricted vertex-cut of order 1 is a vertex-cut and the restricted vertex connectivity of order 1 (or 0-extraconnectivity) is the vertex connectivity. A similar definition can be made for the case when deleting edges. We remark that the term restricted connectivity have been used to mean different concepts by different authors.

Another way to generalize the concept of connectivity is the following. A graph G is *super m-vertex-connected of order q* if with at most m vertices being deleted, the resulting graph is either connected or it has one large component and the small components collectively have at most q vertices in total, that is, the resulting graph has a component of size at least $|V(G-T)|-q$, where T is the set of deleted vertices. Although this measurement is not as refined and somewhat raw, it is flexible. There is a connection between the two concepts as shown in the next result.

Proposition 1 *If G is super p-vertex-connected of order q, then the restricted vertex connectivity of order $q+1$ is at least $p+1$, that is the q-extraconnectivity of G is at least $p+1$.*

Proof By contradiction, assume that the q-extraconnectivity of G is at most p. So there exists a set of vertices F with $|F| \leq p$ such that $G-F$ is disconnected and each of its components has at least $q+1$ vertices. This is a contradiction as $G-F$ has one large component, and its small components have at most q vertices in total. □

The arrangement graph, denoted by $A_{n,k}$, is defined for positive integers n and k such that $n > k \geq 1$. The vertex set of the graph is all permutations of k elements of the set $\{1, 2, \ldots, n\}$. Two vertices corresponding to the permutations $[a_1, a_2, \ldots, a_k]$ and $[b_1, b_2, \ldots, b_k]$ are adjacent if and only if there exists *exactly one* integer $1 \leq i \leq k$ such that $a_i \neq b_i$. Figure 1 shows $A_{4,2}$. (For convenience, we write the (n, k)-permutation $[i, j]$ as ij in this figure, for example [1] as 14.) There have been much research on this class of interconnection networks including embeddings, Hamiltonicity and surface area. See [1] for a list of references. It is easy to see that the connectivity of $A_{n,k}$ is $k(n-k)$.

Let H_i be the set of vertices representing permutations whose kth element is i for $1 \leq i \leq n$, and let T denote a set of vertices to be deleted. Define $T \cap H_i = T_i$ and $|T_i| = t_i$ for $1 \leq i \leq n$. Clearly $A_{n,k}$ is $k(n-k)$-regular because for any vertex, all of its neighbors differ in one of the k positions in the permutation, and for each position there are $n-k$ other choices for the number in that position. Let us first note some other preliminary facts about $A_{n,k}$, which are easy to check.

1. H_i is isomorphic to $A_{n-1,k-1}$ when $n > k \geq 2$. This is because removing i from all the permutations in H_i results in permutations of $k-1$ elements from

Fig. 1 $A_{4,2}$

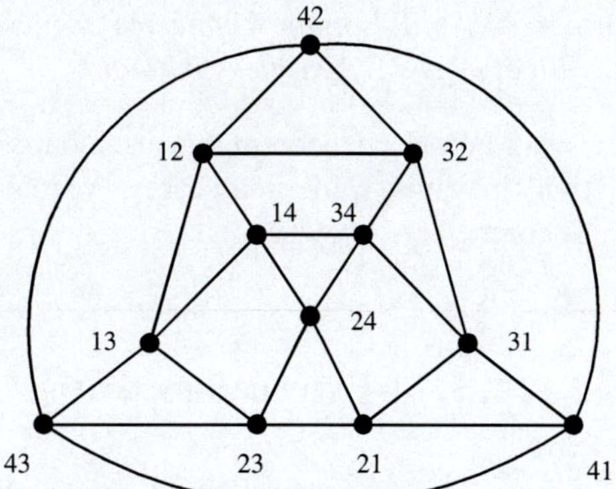

$\{1, 2, \ldots, n\} - \{i\}$. This fact is highly useful in the inductive proofs of the paper, as we can often use the induction hypothesis on H_i.

2. $A_{n,k}$ has $\frac{n!}{(n-k)!}$ vertices, which is the number of permutations of k elements from an n-element set. It follows that H_i has $\frac{(n-1)!}{(n-k)!}$ vertices for all $1 \leq i \leq n$.
3. For any j vertices in H_i, there are exactly $j(n-k)$ distinct vertices outside H_i that are adjacent to at least one of the j vertices. This follows from the fact that each vertex in H_i has $n-k$ neighbors outside H_i and that no two vertices share a common neighbor outside H_i.
4. For each pair H_i and H_j with $i \neq j$, there are exactly $\frac{(n-2)!}{(n-k-1)!}$ independent edges (that is, edges such that no two are incident to a common vertex) between them. Note that every edge between H_i and H_j must be between vertices whose permutations differ in their kth element. Thus the number of edges between H_i and H_j is just the number of permutations of $k-1$ elements from $\{1, 2, \ldots, n\} - \{i, j\}$.

The diagnosability of interconnection networks is an important concept and [2] gave diagnosability results for the arrangement graphs. In the process, they established the following results.

Theorem 1 ([2]) *Let $k \geq 3$ and $n \geq k+2$. Then the 1-extraconnectivity of $A_{n,k}$ is $(2k-1)(n-k)-1$.*

Theorem 2 ([2]) *Let $k \geq 4$ and $n \geq k+2$. Then the 2-extraconnectivity of $A_{n,k}$ is $(3k-2)(n-k)-3$.*

Theorem 3 ([2]) *Suppose either $k \geq 4$ and $n \geq k+2$, or $k \geq 3$ and $n \geq k+3$. Then the 3-extraconnectivity of $A_{n,k}$ is $(4k-4)(n-k)-4$.*

In this note, we make use of existing results together with Proposition 1 to turn finding g-extraconnectivity of the arrangement graph into an automated process. We need the following result.

Theorem 4 ([3]) *Let n, k, s be positive integers such that $n-1 > k \geq 2$, $s \geq 1$. If T is a subset of the vertices of $A_{n,k}$ such that $|T| \leq \left(s(k-2)+2-\frac{s^2}{2}\right)(n-k)$,* *then $A_{n,k} - T$ is either connected or has a large component and small components with at most $s - 1$ vertices in total.*

We note that Theorem 4 is an asymptotic result, thus our result for extraconnectivity may miss some small cases. We remark that for $s \leq 3$ more precise results than Theorem 4 are known.

2 {4, 5, 6}-Extraconnectivities

As expected, our process produces results already given in Theorems 1–3. Based on existing results, our guess is that in most cases, there exists a minimum r-extra-vertex-cut F of $A_{n,k}$ such that the resulting graph has exactly two components, one large component and one small component with exactly $r + 1$ vertices. If our guess is correct, then one can search for a connected subgraph of $A_{n,k}$ with the smallest neighbor set. This procedure gives the following result.

Theorem 5 *Let $k \geq 23$ and $n - k \geq 3$. Then the 4-extraconnectivity of $A_{n,k}$ is $(5k - 5)(n - k) - 7$.*

Theorem 6 *Let $k \geq 30$ and $n - k \geq 3$. Then the 5-extraconnectivity of $A_{n,k}$ is $(6k - 7)(n - k) - 9$.*

Theorem 7 *Let $k \geq 37$ and $n - k \geq 3$. Then the 6-extraconnectivity of $A_{n,k}$ is $(7k - 9)(n - k) - 11$.*

To prove Theorem 5, one may first want to show that the 4-extraconnectivity of $A_{n,k}$ is at least $(5k - 5)(n - k) - 7$. Suppose not. Then there exists a set of vertices F with $|F| < (5k - 5)(n - k) - 7$ such that every component in $G - F$ has at least 5 vertices. We let $s = 6$ in Theorem 4. Then we have that $A_{n,k}$ is super $\left(6(k - 2) + 2 - \frac{6^2}{2}\right)(n - k)$-vertex connected of order 5. Note that for $k \geq 23$, $(6(k - 2) + 2 - \frac{6^2}{2})(n - k) = (6k - 28)(n - k) \geq (5k - 5)(n - k) - 7$. Thus we can conclude that for such k the graph $G - F$ has one large component and a total of at most 5 vertices in the small components. But we also know that every component has at least 5 vertices. Therefore, $G - F$ has exactly two components, one large component and a small component with exactly 5 vertices, and hence F contains the neighborhood of these 5 vertices. This will give a contradiction if we can show that the neighborhood of every connected subgraph of 5 vertices has at least $(5k - 5)(n - k) - 7$ vertices. If one can find such a subgraph whose neighborhood is of size exactly $(5k - 5)(n - k) - 7$ and its deletion gives every component of size at least 5 (or perhaps even two components), then this shows that $(5k - 5)(n - k) - 7$ is also an upper bound. Indeed, if we can find a connected subgraph on 5 vertices whose neighborhood F is of size exactly $(5k - 5)(n - k) - 7$, then we know that $G - F$ has exactly two components since $A_{n,k}$ is super $((5k - 5)(n - k) - 7)$-vertex-connected of order 5 for $k \geq 23$. Thus we have reduced the problem to looking for all connected subgraphs on 5 vertices in $A_{n,k}$. The analysis for 5-extraconnectivity and 6-extraconnectivity can be done in a similar way and we have the following results.

Proposition 2 *Let H be a connected subgraph of $A_{n,k}$ and $n - k > 0$.*

1. *If $k \geq 23$, H has 5 vertices, and $|N(V(H))| \leq (5k-5)(n-k)-7$, then $N(V(H))$ is a 4-extra-cut.*
2. *If $k \geq 30$, H has 6 vertices, and $|N(V(H))| \leq (6k-7)(n-k)-9$, then $N(V(H))$ is a 5-extra-cut.*
3. *If $k \geq 37$, H has 7 vertices, and $|N(V(H))| \leq (7k-9)(n-k)-11$, then $N(V(H))$ is a 6-extra-cut.*

Proof The idea of this proof was given above as part of the overall scheme but we will reproduce and formalize it here for easy reference. We will prove (1) and the others can be done similarly. Clearly $A_{n,k} - N(V(H)$ is disconnected, so we need to show every component has at least 5 vertices. We let $s = 6$ in Theorem 4. Then we have that $A_{n,k}$ is super $\left(6(k-2)+2-\frac{6^2}{2}\right)(n-k)$-vertex connected of order 5. Note that for $k \geq 23$, $(6(k-2)+2-\frac{6^2}{2})(n-k) = (6k-28)(n-k) \geq (5k-5)(n-k)-7$. Thus we can conclude that $A_{n,k} - N(V(H)$ has one large component and a total of at most 5 vertices in the small components. But H is connected so it must be a component in $A_{n,k} - N(V(H)$. Thus $A_{n,k} - N(V(H)$ has exactly two components, one of which is H, and both components have at least 5 vertices. □

We remark that exhibiting one such admissible neighborhood of size $(5k-5)(n-k)-7$ for (1) shows that it is an upper bound, but we need to check all such neighborhoods to establish it as a lower bound. We further remark that there is a weaker lower bound that we have immediately: Letting $s = 5$ in Theorem 4, we have that $A_{n,k}$ is super $(5(k-2)+2-\frac{5^2}{2})(n-k)$-vertex-connected of order 4. Since $5(k-2)+2-\frac{5^2}{2} = 5k-20.5$, we can apply Proposition 1 to conclude that the 4-extraconnectivity of $A_{n,k}$ is at least $(5k-20.5)(n-k)+1$. So by exhibiting just one such 4-extra-vertex cut, we know that the 4-extraconnectivity of $A_{n,k}$ is asymptotically equal to $5k(n-k)$ as k and $n-k$ tend to infinity. The other two theorems can be treated in a similar way. So we have the following corollary.

Corollary 1 *As k and $n-k$ tend to infinity, the r-extraconnectivity of $A_{n,k}$ is asymptotically $(r+1)k(n-k)$ for $r = 4, 5, 6$.*

Proposition 3 *Let $k \geq 23$ and $n-k \geq 3$. Let*

$$S = \{1234B, 5234B, 1634B, 1274B, 5634B\}.$$

Then $N(S)$ is a 4-extra-cut in $A_{n,k}$ of size $(5k-5)(n-k)-7$, where B is a fixed permutation of length $k-4$ and $1234B$ is a vertex of $A_{n,k}$.

Proof We remark that we require $n - k \geq 3$ as we require 7 symbols $1, 2, 3, 4, 5, 6, 7$ and a block of length 4. Now the subgraph induced by S is connected, and it is a 4-cycle with a leaf-edge. More precisely, the 4-cycle is given by $1234B - 5234B - 5634B - 1634B - 1234B$ and the leaf edge is given by $1234B - 1274B$. Now each vertex of S has $k(n-k)$ neighbors, giving a total of $5k(n-k)$ vertices. However, three of the neighbors of $1234B$ are already in S, each of $5234B, 5634B, 1634B$ already has two neighbors in S, and $1274B$ has one neighbor in S. Thus we have counted $5k(n-k) - 10$ vertices. However some of the vertices have been counted multiple times. We will systemically consider them

1. $1234B$ and $1634B$ share $(n-k) - 1$ common neighbors outside of S.
2. $1234B$ and $1274B$ share $(n-k) - 1$ common neighbors outside of S.
3. $1234B$ and $5234B$ share $(n-k) - 1$ common neighbors outside of S.
4. $1234B$ and $5634B$ share 0 common neighbor outside of S.
5. $5234B$ and $1634B$ share 0 common neighbor outside of S.
6. $5234B$ and $1274B$ share 1 common neighbor outside of S.
7. $5234B$ and $5634B$ share $(n-k) - 1$ common neighbors outside of S.
8. $1634B$ and $1274B$ share 1 common neighbor outside of S.
9. $1634B$ and $5634B$ share $(n-k) - 1$ common neighbors outside of S.
10. $1274B$ and $5634B$ share 0 common neighbor outside of S. (In fact, they share no common neighbor in $A_{n,k}$.)

Thus $|N(S)| = 5k(n-k) - 10 - 5((n-k) - 1) - 2 = (5k-5)(n-k) - 7$. It now follows from Proposition 2 that $N(S)$ is a 4-extra-cut. □

Thus we have now proved that the 4-extraconnectivity of $A_{n,k}$ is in the interval $[5k - 20.5)(n-k) + 1, (5k-5)(n-k) - 7]$. To show that it is $(5k-5)(n-k) - 7$, we will need to show that $(5k-5)(n-k) - 7$ is the smallest neighbor set of a connected subgraph of size 5, which we will use a computer for. The following propositions can be proved in a similar way.

Proposition 4 *Let $k \geq 30$ and $n - k \geq 3$. Let*

$$S = \{12345B, 62345B, 17345B, 12845B, 67345B, 62845B\}.$$

Then $N(S)$ is a 5-extra-cut in $A_{n,k}$ of size $(6k-7)(n-k) - 9$, where B is a fixed permutation of length $k - 5$ and $12345B$ is a vertex of $A_{n,k}$.

Proposition 5 *Let $k \geq 37$ and $n - k \geq 3$. Let*

$$S = \{123456B, 723456B, 183456B, 129456B, 783456B, 729456B, 189456B\}.$$

Then $N(S)$ is a 6-extra-cut in $A_{n,k}$ of size $(7k-9)(n-k) - 11$, where B is a fixed permutation of length $k - 6$ and $12346B$ is a vertex of $A_{n,k}$.

Hence Corollary 1 is proved. We remark that finding such a "good" extra-cut is not necessary to prove Corollary 1. Let r be fixed and suppose $n - k$ is large.

Then consider the $(r+1)$-clique generated by $i23\ldots k$ for all $i \in \{1, k+1, k+2, \ldots, k+r\}$. This is part of an $(n-k+1)$-clique. Each vertex generates $(k-1)(n-k)$ neighbors via positions 2 to k, giving $(r+1)(k-1)(n-k)$ neighbors, which are all distinct. In addition, they share the remaining $(n-k+1)-(r+1) = n-k-r$ vertices in the $(n-k+1)$-clique via position 1 as common neighbors. Thus the neighbor set F of this $(r+1)$-clique has size $(r+1)(k-1)(n-k) + (n-k-r) = ((r+1)k-r)(n-k)-r$. Letting $s = r+2$ in Theorem 4, we have that $A_{n,k}$ is super $((r+2)(k-2)+2-\frac{(r+2)^2}{2})(n-k)$-vertex-connected of order $r+1$. But $((r+2)(k-2)+2-\frac{(r+2)^2}{2})(n-k) \geq ((r+1)k-r)(n-k)-r$ if k and $n-k$ are sufficiently large. Thus for sufficiently large k and $n-k$, $A_{n,k}-F$ has exactly two components, a large component and a small component that is an $(r+1)$-clique. So the r-extraconnectivity of $A_{n,k}$ is at most $((r+1)k-r)(n-k)-r$. We now let $s = r+1$ in Theorem 4, and we have that $A_{n,k}$ is super $((r+1)(k-2)+2-\frac{(r+1)^2}{2})(n-k)$-vertex-connected of order r. Thus we can apply Proposition 1 to conclude that the r-extraconnectivity of $A_{n,k}$ is at least $((r+1)(k-2)+2-\frac{(r+1)^2}{2})(n-k)+1$. Therefore the r-extraconnectivity of $A_{n,k}$ is in the interval $[((r+1)(k-2)+2-\frac{(r+1)^2}{2})(n-k)+1, ((r+1)k-r)(n-k)-r]$ for large k and $n-k$. (Yes! This is the same type of argument that we have used before.) Thus we have the following result that generalizes Corollary 1.

Proposition 6 *Let $r \geq 1$. As k and $n-k$ tend to infinity, the r-extraconnectivity of $A_{n,k}$ is asymptotically $(r+1)k(n-k)$.*

To finish the proof of Theorems 5–7, we need to show that the extra-cuts given in the above results are the best among all such cuts generated by a connected subgraph. For this, we use a computer search. Suppose we are looking for an r-extra-cut where $r \in \{4, 5, 6\}$. We note that since $A_{n,k}$ is vertex-transitive, we may assume that $123\ldots k$ is in the desired subgraph. Since this subgraph has exactly $r+1$ vertices, the distance between two of its vertices is at most r, thus they can differ in at most r positions. Thus we can assume that every vertex in the desired subgraph is of the form $a_1a_2\ldots a_rB$, where B is fixed. We note that when two vertices differ in exactly one position, then they are part of an $(n-k+1)$-clique, and hence they share $(n-k)-1$ common neighbors. If they differ in exactly two positions, then they share exactly 2 common neighbors. If they differ in more than two positions, then they share no common neighbors. Using these observations, we can grow a search tree from $123\ldots n$. Use $r = 4$ as an example. We can decide how many neighbors of $1234B$ should be included in the subgraph. Suppose the answer is 4, then we need to decide how many of the first 4 positions will be used to generate such 4 neighbors. Suppose the answer is 2. Then without loss of generality, we may assume that it is via the first two positions. Now suppose 1 neighbor is via the first position and 3 neighbors are via the second position. Then we may assume that the 4 neighbors are either $5234B, 1534B, 1634B, 1734B$ or $5234B, 1634B, 1734B, 1834B$. We note that there are two cases rather than a number involving n and k. This type of case analysis is suitable for a computer program. The program shows that the extra-cuts given in the above propositions are optimal. A sample code is given in the Appendix.

On a typical computer it took a few seconds to get the answer for $r = 4$ and a few minutes to get the answer for $r = 6$.

Acknowledgements We thank the anonymous referee for a number of helpful comments and suggestions.

A. Computer Code

```
import java.util.ArrayList;

public class Arrangement {
  public static String[] ver;
  public static int R;
  public static ArrayList<Integer> nk1ans;
  public static ArrayList<Integer> consans;
  public static ArrayList<String> ex;
   //We use R as number of vertices and K - as if K>R,
//WLOG let the last
//K-R be the same for all vertices
  public static void main(String[]args) {
      R = 5;
      ver = new String[R];
      nk1ans = new ArrayList<Integer>();
      consans = new ArrayList<Integer>();
      ex = new ArrayList<String>();
      String a = "";
      String b = "";
      a+=(char)('A'+0);
      b+=(char)('A'+R);
      for(int i = 1;i<R;i++) {
          a+=(char)('A'+i);
          b+=(char)('A'+i);
      }
      ver[0] = a;
      ver[1] = b;
      solve(2,R+1,0);
      for(int i = 0; i<nk1ans.size();i++) {
          System.out.println("("+R+"nk-"+nk1ans.get(i)+")
(n-k)-"+(nk1ans.get(i)+consans.get(i))+", EX: "+ex.get(i));
      }
  }

  //Recursive function for brute force solution
  public static void solve(int point, int nodl,int largchg) {
//nodl: number of different letters used,largchg is the
//largest index such that the for some vert at that
//index that char is different from ABCDEF.
      if(point!=R) {
          ArrayList<String> newVerts = new ArrayList<String>();
          for(int i = 0; i<=point-1;i++) {
              newVerts.add(ver[i]);//This is such that we\\
//don't add the previous vertices as new vertices - we
```

```
//will ignore the first few
        }
        for(int i = 0; i<=point-1;i++) {
            for(int j = 0; j<=nodl;j++) {
                char cur = (char)('A'+j);
                if(ver[i].indexOf(cur)!=-1) {
                    continue;
                }
                else {
                    for(int k = 0; k<=largchg+1;k++) {
                        String temp =
ver[i].substring(0,k)+cur+ver[i].substring(k+1);
                        if(!newVerts.contains(temp)) {
                            newVerts.add(temp);
                            ver[point] = temp;
                            solve(point+1,Math.max(nodl, j+1),
Math.max(largchg, k));
                        }

                    }
                }
            }
        }
    }
    else {
        String ans = calc();
        StringTokenizer st = new StringTokenizer(ans);
        int nk1 = Integer.parseInt(st.nextToken());
        int cons = Integer.parseInt(st.nextToken());
        if(!nk1ans.contains(nk1)) {
            nk1ans.add(nk1);
            consans.add(cons);
            String exa = "";
            for(int i = 0;i<R;i++) {
                exa+=ver[i]+" ";
            }
            ex.add(exa);
        }
        else {
            int poi = nk1ans.indexOf(nk1);
            if(consans.get(poi)<cons) {
                consans.remove(poi);
                consans.add(poi,cons);
                String exa = "";
                for(int i = 0;i<R;i++) {
                    exa+=ver[i]+" ";
                }
                ex.remove(poi);
                ex.add(poi,exa);
            }
        }
    }
}
```

```

//Calculates neighbor set of given set of vertices
public static String calc() {
    int nk1coef = 0;
    int cons = 0;

    for(int i = 1; i<R;i++) {
        String cur = ver[i];
            ArrayList<String> dcverts = new ArrayList<String>();
            ArrayList<Integer> chgs = new ArrayList<Integer>();

            boolean[] isShared = new boolean[R];
            int isSharednum = 0;
            for(int j = 0; j<i;j++) {
                String cur2 = ver[j];
                int differs = 0;
                int diff1 = 0;
                int diff2 = 0;
                for(int k = 0; k<R;k++) {
                    if(cur.charAt(k)!=cur2.charAt(k)) {
                        if(differs == 0) {
                            diff1 = k;
                            differs++;

                        }
                        else if(differs == 1) {
                        diff2=k;
                        differs++;

                    }
                    else if(differs==2) {
                        differs++;
                        break;
                    }
                }
            }
            if(differs == 1) {
                if(!isShared[diff1]) {
                    isShared[diff1] = true;
                        isSharednum++;
                    nk1coef++;
                }
            }
            if(differs == 2) {
                if(diff1>diff2){
                    int temp = diff1;
                    diff1 = diff2;
                    diff2 = temp;
            }
                if((cur.charAt(diff1)!=cur2.charAt(diff2))) {

                    String v = cur.substring(0,diff1)+
cur2.charAt(diff1)+cur.substring(diff1+1);

                    if(!dcverts.contains(v)) {
                        dcverts.add(v);
```

```
                        chgs.add(diff1);
                    }
                }
                if((cur.charAt(diff2)!=cur2.charAt(diff1))) {
                    String v = cur.substring(0,
diff2)+cur2.charAt(diff2)+cur.substring(diff2+1);
                    if(!dcverts.contains(v)) {
                        dcverts.add(v);
                        chgs.add(diff2);
                    }
                }
            }
        }
        for(int n = 0; n<chgs.size();n++) {
            if(isShared[chgs.get(n)]) {
                chgs.remove(n);
                dcverts.remove(0);
                n--;
            }
            else if(dcverts.get(n).indexOf(dcverts.get(n).
charAt(chgs.get(n)),chgs.get(n)+1)!=-1) {
                chgs.remove(n);
                dcverts.remove(0);
                n--;
            }
        }
        if(i==2) {
            for(int n = 0;n<3;n++) {
            }
        }
        cons+=dcverts.size();
        cons= (cons - isSharednum)+1;
    }
    return nk1coef+" "+cons;
 }
}
```

References

1. Cheng, E., Qiu, K., Shen, Z.: On the restricted connectivity of the arrangement graph. Journal of Supercomputing 73, 3669–3682 (2017).
2. Xu, L., Lin, L., Zhou, S., Hsieh, S.-Y.: The extra connectivity, extra conditional diagnosability, and t/m-diagnosability of arrangement graphs. IEEE Transactions on reliability 65, 1248–1262 (2016).
3. Cheng, E., Lipták, L., Yuen, A.: Linearly many faults in arrangement graphs. Networks 61, 281–289 (2013).

k-Paths of k-Trees

Allan Bickle

Abstract A k-tree is a graph that can be formed by starting with K_{k+1} and iterating the operation of making a new vertex adjacent to all the vertices of a k-clique of the existing graph. When the order $n > k + 1$, a k-path graph is a k-tree with exactly two vertices of degree k. We state a forbidden subgraph characterization for k-paths as k-trees. We characterize k-trees with diameter $d \geq 2$ based on the k-paths they contain.

Keywords k-Tree · k-Path · Diameter

1 Introduction

In this paper, we seek to describe the structure of k-trees using k-paths, particularly focusing on the diameter of k-trees. Undefined notation and terminology will follow [2].

This work builds on previous papers on the Wiener index of maximal k-degenerate graphs [4] (with Zhongyuan Che) and on maximal k-degenerate graphs with diameter 2 [3].

Definition 1 A k**-tree** is a graph that can be formed by starting with K_{k+1} and iterating the operation of making a new vertex adjacent to all the vertices of a k-clique of the existing graph. The clique used to start the construction is called the **root** of the k-tree.

A k**-leaf** is a degree k vertex of a k-tree.

A k**-path graph** G is an alternating sequence of distinct k- and $k+1$-cliques $e_0, t_1, e_1, t_2, \ldots, t_p, e_p$, starting and ending with a k-clique and such that t_i contains exactly two k-cliques e_{i-1} and e_i.

A. Bickle (✉)
Penn State Altoona, Altoona, PA, USA
e-mail: remove@mail.com
URL: https://allanbickle.wordpress.com/

F. Hoffman (ed.), *Combinatorics, Graph Theory and Computing*,
Springer Proceedings in Mathematics & Statistics 388,
https://doi.org/10.1007/978-3-031-05375-7_21

An example of a 2-path (which is also a 2-tree) is shown below left. A 2-tree that is not a 2-path (the triangular grid Tr_2) is below right.

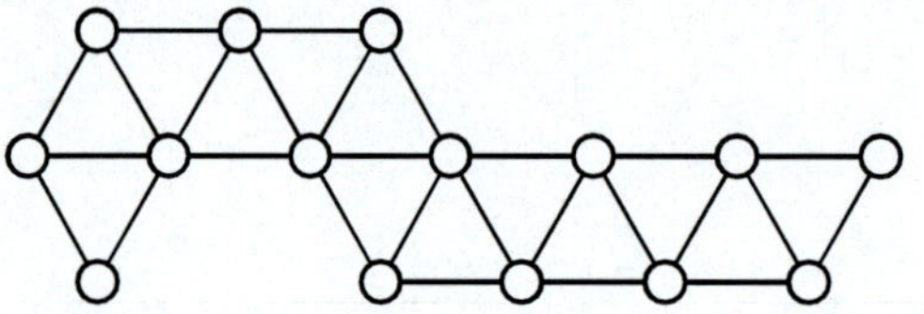

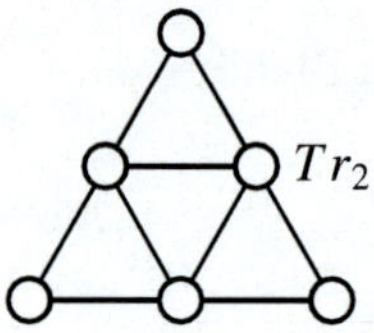

Tr_2

Note that k-paths are also known as linear k-trees [1]. They are closely related to pathwidth [6]; in particular, they are the maximal graphs with proper pathwidth k. There is a simple characterization of these graphs.

Theorem 1 [5] *Let G be a k-tree with $n > k + 1$ vertices. Then G is a k-path graph if and only if G has exactly two k-leaves.*

This leads to a forbidden subgraph characterization for k-paths as k-trees.

Theorem 2 *A k-tree is a k-path if and only if it does not contain $K_k + \overline{K}_3$ or for $k \geq 2$, $Tr_2 + K_{k-2}$.*

Proof ($\Rightarrow$) (contrapositive) These graphs contain three k-leaves, so they are not k-paths.

($\Leftarrow$) (contrapositive) A k-tree that is not a k-path must have at least three k-leaves. Then it must contain a subgraph G that is minimal with respect to this property. It will have exactly three k-leaves, and deleting any of them results in a k-path. Let H be the graph formed by deleting all k-leaves from G. If H is not a clique, then it has two k-leaves, one of which has only one k-leaf of G neighboring it, so G is not minimal.

If $H = K_k$, $G = K_k + \overline{K}_3$. If $H = K_{k+1}$, each of its vertices are adjacent to a k-leaf of G. If two k-leaves of G have the same neighborhood, then G is not minimal. Thus there are $k - 2$ vertices of H adjacent to all three k-leaves of G, and deleting them produces Tr_2. □

2 Diameter of k-Trees

A tree is minimal with respect to diameter d if and only if it is P_{d+1}. In [3], I found a characterization of k-trees minimal with respect to diameter 3.

Definition 2 A **dominating vertex** of a graph is a vertex adjacent to all other vertices.

Algorithm 1 *Let P be a $k - 2$-path, $k \geq 3$, of order $n - 4$ with k-leaves w and x. Join dominating vertices y and z to P, forming $P + K_2$. Add u with neighborhood $N(w) \cup \{w, y\}$, and v with neighborhood $N(x) \cup \{x, z\}$. Let $\mathbb{G}_k$ be the class of all graphs formed this way.*

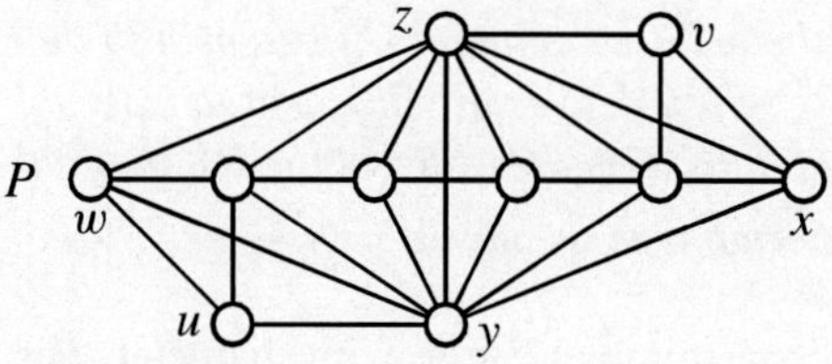

Theorem 3 [3] *A graph G is a k-tree minimal with respect to diameter 3 if and only if $G \in \mathbb{G}_k$.*

Equivalently, a k-tree has diameter at most 2 if and only if it does not contain any graph in $\mathbb{G}_k$.

The graphs in $\mathbb{G}_k$ are all k-paths. A generalization also holds.

Lemma 1 *A k-tree minimal with respect to diameter $d \geq 2$ is a k-path.*

Proof A k-tree with diameter at least d must contain a pair of vertices distance d apart. Now adding a vertex to a k-tree cannot change any existing distances. Thus in a minimal k-tree with diameter d, the vertices at distance d must be k-leaves, and no other vertices are k-leaves. □

The 2-paths with diameter d cannot be characterized solely by their degree sequences, as there are two 2-paths with degree sequence 5, 4, 4, 3, 3, 3, 2, 2 which have diameters 3 and 4 (see below). A characterization based on the arrangement of the degree 4 vertices is possible.

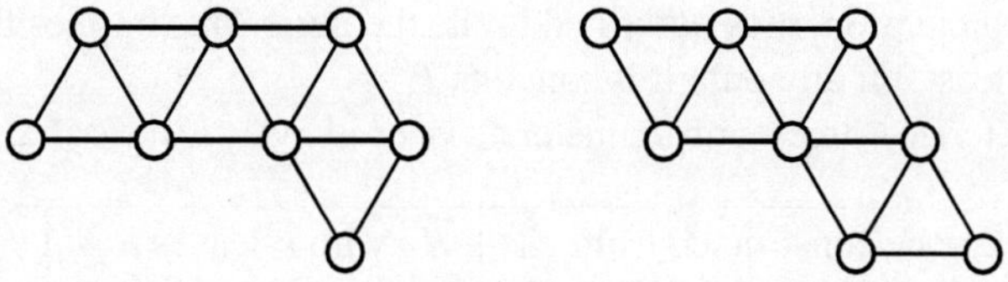

Definition 3 A **hub** is a vertex of degree at least 5 of a 2-path. A **truss** is a subgraph induced by vertices of degree 4 in a 2-path. An **external truss** has a vertex neighboring a 2-leaf, an **internal truss** does not.

In the 2-path below, the black vertex is an internal truss and the gray vertices induce an external truss.

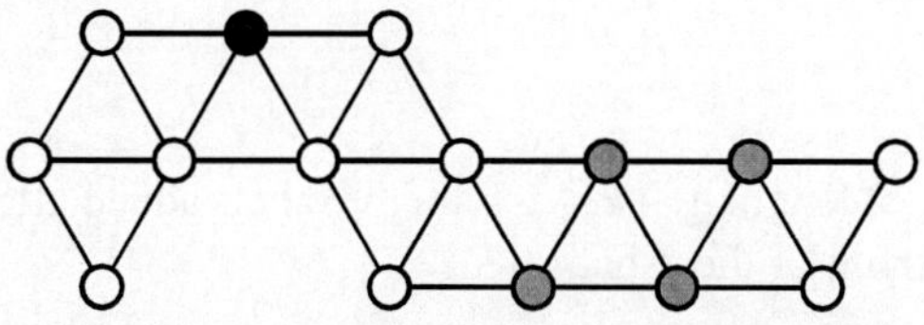

Theorem 4 *Let G be a 2-tree minimal with respect to diameter d. Then G is a 2-path, and if $G \neq P_{2d}^2$, the 2-leaves are adjacent to external trusses with odd order. If h is the number of hubs, t_i is the order of the ith internal truss, and t' and t'' are the orders of the external trusses, then* $d = h + \sum \lfloor \frac{t_i}{2} \rfloor + \lceil \frac{t'}{2} \rceil + \lceil \frac{t''}{2} \rceil + 1$.

Proof By Lemma 1, a minimal 2-tree with diameter d is a 2-path. To show the formula holds, we use induction on n. Since $G \neq P_{2d}^2$, it contains a hub. We start with the fan induced by its closed neighborhood. This has $h = 1, d = 2$, and all other quantities 0. We add vertices one at a time, checking that the formula holds in each case.

There are only two choices how to add a new 2-leaf next to an existing 2-leaf. In one choice, the other neighbor had degree at least 4. If it is already a hub, the diameter does not increase. If it is part of a truss of odd order, one vertex of the truss becomes a hub, the rest of the truss (if any) becomes internal, the sum does not change, and the diameter does not increase. If it is part of a truss of positive even order, one vertex of the truss becomes a hub, the rest of the truss becomes internal, the sum does not change, and the diameter does not increase.

In the other choice, the other neighbor had degree 3, so we create an external truss or add one vertex to an existing external truss. If the truss had odd order, adding this vertex does not change the diameter. If the truss is new or had even order, adding this vertex increases the diameter by 1.

Since only the last case increases the diameter, in a 2-path minimal with respect to d, the 2-leaves are adjacent to external trusses with odd order. □

Thus a 2-tree with order $n \geq 5$ has diameter at least d if any only if it contains a 2-path with the properties described in the theorem. This implies that a 2-tree has diameter at least 3 if any only if it contains P_6^2.

To characterize k-trees with diameter d, we need a way to describe the construction of k-paths.

A k-path can be constructed from $K_k + \overline{K}_2$ with k-leaves u and v by maintaining u as a k-leaf and adding a new k-leaf adjacent to v and $k - 1$ of its k neighbors. Label the k neighbors of u 1 through k (in any way). Each time a k-leaf x is added adjacent to (old) k-leaf w, label w with the label of its neighbor that does not neighbor x.

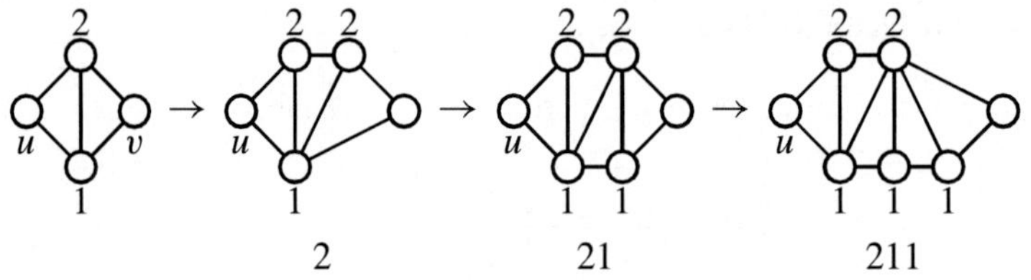

Define a string of length $n - k - 2$ with the labels added after the first k. Call this a **construction string** of the k-path.

Definition 4 A string of numbers **contains a pattern** if the numbers in the pattern occur in order (not necessarily consecutively) in the string.

For example, the pattern 321 is contained in 312213 but not 132233.

Theorem 5 *A k-tree has diameter $d \geq 2$ if and only if it contains a k-path whose construction string contains at least $d - 2$ consecutive permutations of* $\{1, \ldots, k\}$.

Proof By Lemma 1, a k-tree with diameter d contains a k-path with diameter d. Let G be a k-path with diameter d and k-leaves (say) u and v. We show that the number of consecutive permutations of $\{1, \ldots, k\}$ in the string is always $d - 2$. Certainly this is true for $K_k + \overline{K}_2$, which is minimal with diameter 2 and has an empty string.

Let H be a minimal k-path contained in G with k-leaves u and w. The vertices in $N(w)$ have labels 1, ..., k. Each vertex added to form G removes one vertex from the neighborhood of the k-leaf it replaces, so at most one vertex from $N_H(w)$. To increase the diameter, each vertex in $N_H(w)$ must be removed, and each will be replaced with another vertex with the same label. The diameter increases by one exactly when the string contains one more permutation of $\{1, \ldots, k\}$. □

References

1. A. Abiad, B. Brimkov, A. Erey, L. Leshock, X. Martinez-Rivera, S. O, S. Song, J. Williford (2017) On the Wiener index, distance cospectrality and transmission regular graphs, Discrete Appl. Math. 230:1–10.
2. A. Bickle (2020) Fundamentals of Graph Theory, AMS.
3. A. Bickle (2021) Maximal k-degenerate Graphs with Diameter 2, International Journal of Mathematical Combinatorics 2:67–78.
4. A. Bickle, Z. Che (2021) Wiener Indices of Maximal k-Degenerate Graphs, Graphs and Combinatorics, 37 2:581–589.
5. L. Markenzon, C. M. Justel, N. Paciornik (2006) Subclasses of k-trees: Characterization and recognition, Discrete Appl. Math. 154 5:818–825.
6. A. Proskurowski, J. Telle (1999) Classes of graphs with restricted interval models, Discrete Math. Theoret. Comput. Sci. 3:167–176.

Rearrangements of the Simple Random Walk

Marina Skyers and Lee J. Stanley

Abstract In this paper we will look at representations of the simple random walk, S_n, and show how to effectively rearrange the sequence of terms $\frac{S_n}{\sqrt{n}}$ in order to achieve almost sure convergence to the standard normal on the open interval $(0, 1)$. This is done via a suitable choice of permutation $F : \{0, 1\}^n \to \{0, 1\}^n$. We are interested in how much rearranging of the simple random walk is optimal. We will describe how to minimize the graph-theoretic complexity of these permutations and also show that they satisfy some additional nice properties.

Keywords Simple random walk · Permutations · Complexity

1 Introduction

Let S_n be the random walk on $(0, 1)$. In 1733, de Moivre postulated the first version of the central limit theorem for independent random variables that take on values ± 1. It is an important special case of the central limit theorem that the S_n converge in distribution to the standard normal on $(0, 1)$. Well-known results ([4, 6–8]) show this cannot possibly be improved to almost sure convergence. The random walk has been the subject of intense study (see the work of Erdös and Revesz [5] and Shi and Toth [9]). Indeed, the definition of the S_n is immediately accessible and intuitive and each S_n is readily representable as the sum of an i.i.d. family (of size n) of irreducibly simpler random variables. While immediately intuitive, the S_n are quite disorderly. This disorder is mirrored by the fact that, for almost all x, $\left\{ \frac{S_n(x)}{\sqrt{n}} \middle| n \in \mathbb{N}^+ \right\}$ diverges.

M. Skyers (✉)
Penn State Brandywine, Media, PA, USA
e-mail: mus61@psu.edu

L. J. Stanley
Lehigh University, Bethlehem, PA, USA
e-mail: ljs4@lehigh.edu

F. Hoffman (ed.), *Combinatorics, Graph Theory and Computing*,
Springer Proceedings in Mathematics & Statistics 388,
https://doi.org/10.1007/978-3-031-05375-7_22

Definition 1 Let λ will denote Lebesgue measure on $[0, 1]$ (or on one of the variants with either endpoint or both excluded). As usual, a *probability space* is a triple $(\Omega, \mathscr{S}, P)$, where Ω is the set of points, $\mathscr{S}$ is the σ-algebra of Borel subsets of Ω, and $P : \mathscr{S} \to [0, 1]$ is the (σ-additive) probability measure. In this paper we will have $\Omega = [0, 1)$, $\mathscr{S}$ will be the σ-algebra of Borel subsets of Ω, and P will be the restriction of Lebesgue measure to the Borel sets.

We will use card (x) to denote the cardinality of the set x. C will denote Cantor space, $\{0, 1\}^{\mathbb{N}^+}$.

Definition 2 For $x \in C := \{0, 1\}^{\mathbb{N}^+}$ excluding the two constant sequences, identify x with $\sum_{i=1}^{\infty} \frac{x_i}{2^i} \in (0, 1)$. For dyadic rationals, choose the representation with a tale of zeros. For $x \in C$ and for finite binary sequences $\boldsymbol{r}$ of length n, we will use the notation $x \supseteq \boldsymbol{r}$ to mean x extends $\boldsymbol{r}$, i.e., x agrees with $\boldsymbol{r}$ for the first n terms of its dyadic expansion. Define for $1 \le i \le n$, $R_i(x) := (-1)^{1+x_i}$ and $S_n(x) := \sum_{i=1}^{n} R_i(x)$. Define $\text{Weight}_n(x)$ as the sum of the first n coordinates of x. Notice that $S_n(x) = -n + 2\text{Weight}_n(x)$. Obviously, $S_n(x)$ and $\text{Weight}_n(x)$ depend only on the first n coordinates of x. So, for binary sequences $\boldsymbol{r}$ of length n, we can define

$$S_n(\boldsymbol{r}) := S_n(x) \text{ for any } x \in C \text{ such that } x \supseteq \boldsymbol{r}.$$

$$\text{Weight}(\boldsymbol{r}) := \text{Weight}_n(x) \text{ for any } x \in C \text{ such that } x \supseteq \boldsymbol{r}.$$

Observe that $S_n(\boldsymbol{r}) = -n + 2\text{Weight}(\boldsymbol{r})$. We can see this in the graph of $S_n(x)$, at each level n. The graphs of $S_n(x)$, for $n = 5, 6, 7$, will be illustrated below.

We will see that the quantile of S_n turns out to be a very orderly, non-decreasing step function, which we will call S_n^*, and it can be explicitly defined as follows. Define steps $A_{n,i}$, $i = 0, \ldots, n$, where

$$A_{n,i} = \left(\frac{1}{2^n} \sum_{j=0}^{i-1} \binom{n}{j}, \frac{1}{2^n} \sum_{j=0}^{i} \binom{n}{j} \right].$$

For such i, and for all $x \in A_{n,i}$ we define $S_n^* = -n + 2i$. For $n \in \mathbb{N}^+$, let $\kappa = \kappa_n = \kappa_n(x)$ be the following integer: $\kappa = \sum_{i=1}^{n} x_i 2^{n-i}$. Then $x \in \left[\frac{\kappa_n(x)}{2^n}, \frac{\kappa_n(x)+1}{2^n} \right)$. So we can compute $S_n^*(x)$ by identifying the step $A_{n,i}$ that includes the interval $\kappa_n(x)$. Note that, for each $n \in \mathbb{N}^+$ and for $\kappa \in [0, 2^n) \cap \mathbb{N}$,

$$-n \le S_n(\kappa), S_n^*(\kappa) \le n$$

and S_n, S_n^* satisfy the dualization equations

$$S_n(\kappa) = -S_n\left(2^n - 1 - \kappa\right),$$

$$S_n^*(\kappa) = -S_n^*(2^n - 1 - \kappa).$$

Below are the graphs for S_n and $S_n^*(x)$ when $n = 5, 6, 7$.

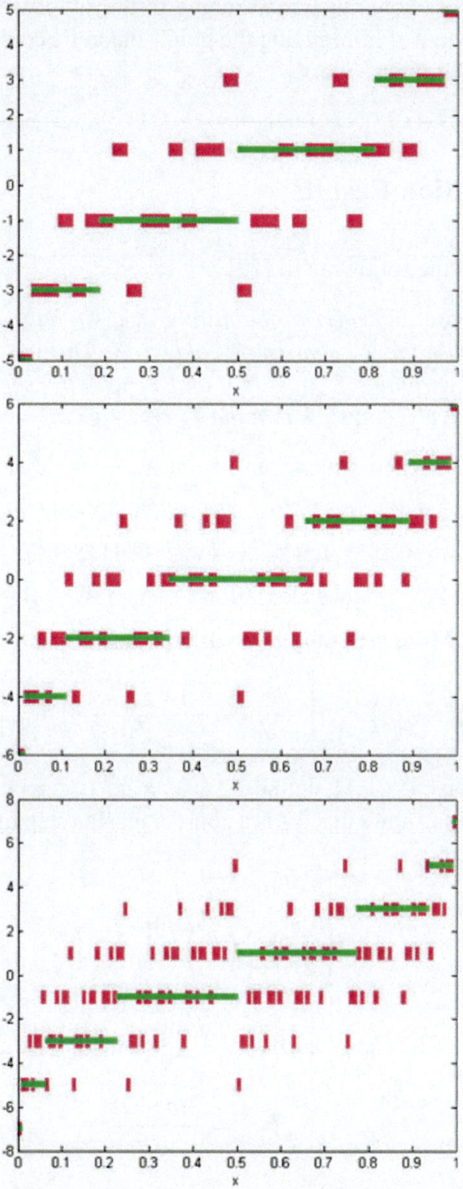

In this paper we will investigate representations of S_n^* that are as close as possible to the canonical representation for S_n, via permutations $F : \{0, 1\}^n \to \{0, 1\}^n$ such that $S_n^* = S_n \circ F$. In fact, it turns out that $\left(F_n | n \in \mathbb{N}^+\right)$ is uniformly primitive recursive ([1, 3]). Our results on the representability of S_n^* are proved in Sect. 2. In fact, somewhat surprisingly, Theorem 6 shows there are a large number of such representations of each S_n^*. In Sect. 3, we provide an explicit, highly effective construction of a preferred sequence of such representations, uniformly and highly effectively, in n. In Sect. 4, we discuss how much rearranging of the simple random walk is optimal from the point of view of minimizing the graph-theoretic complexity of the function F, and look ahead to future work.

2 Representation Results

Skorokhod proved the following in [10].

Theorem 3 *Suppose that on a probability space, we have random variables X_n, $n \in \mathbb{N}^+$, and suppose the X_n converge weakly to X. Then on $([0,1], B([0,1]), \lambda)$, there are random variables Y_n, $n \in \mathbb{N}^+$, and Y, with the same distributions as the X_n and X, respectively, and such that the Y_n converge almost surely to Y.*

If in Skorokhod's Theorem, we start from $X_n = \frac{S_n}{\sqrt{n}}$, then, the Y_n that result are exactly $\frac{S_n^*}{\sqrt{n}}$. Now we will look closely at Skorokhod's construction so as to obtain an explicit characterization of S_n^*. Let $A_t := \{y \in (0,1) \mid S_n(y) \le t\sqrt{n}\}$. So $\lambda(A(t)) = P\left(\frac{S_n}{\sqrt{n}} \le t\right) = P(X_n \le t)$ (see Definition 1). Then $A_t = \emptyset$ for $t < -\sqrt{n}$, and $A_t = (0,1)$ for $t \ge \sqrt{n}$. More generally, A_t will be constant on these intervals of t:

$$(-\infty, -\sqrt{n}), \left[-\sqrt{n}, \frac{2-n}{\sqrt{n}}\right), \ldots, \left[\frac{-n+2k}{\sqrt{n}}, \frac{-n+2(k+1)}{\sqrt{n}}\right), \ldots, \left[\frac{n-2}{\sqrt{n}}, \sqrt{n}\right), [\sqrt{n}, \infty),$$

for $0 \le k < n$. For $x \in (0,1]$, define $X_n^*(x) := \inf\{t \in \mathbb{R} \mid \lambda(A_t) \ge x\}$. A straightforward computation shows that X_n^* is a non-decreasing step function with steps $A_{n,i}$, $i = 0, \ldots, n$, where

$$A_{n,i} = \left(\frac{1}{2^n}\sum_{j=0}^{i-1}\binom{n}{j}, \frac{1}{2^n}\sum_{j=0}^{i}\binom{n}{j}\right].$$

Definition 4 For such i, and for all $x \in A_{n,i}$, we define

$$X_n^*(x) := \frac{-n+2i}{\sqrt{n}},$$

and

$$S_n^*(x) := -n + 2i.$$

This sequence of definitions, culminating in the definition of S_n^*, carries out Skorokhod's construction starting from the sequence $\left(\frac{S_n}{\sqrt{n}} \middle| n \in \mathbb{N}^+\right)$. Therefore the "Skorokhod sequence" $\left(\frac{S_n^*}{\sqrt{n}} \middle| n \in \mathbb{N}^+\right)$ converges almost surely to the standard normal, this time on (0, 1], but the fact that $S_n^*(1)$ happens to be defined turns out to be more of an annoyance than a feature, so we'll view S_n^* as defined only on (0, 1). Note that the definition of $S_n^*(x)$ requires only that we identify the step $A_{n,i}$ to which x belongs. This depends only on the first n coordinates of x, and so the same holds for $S_n^*(x)$ (as indeed it does for $S_n(x)$). This, in turn, means that we can view S_n^* as being defined on $\{0, 1\}^n$ just as we did for S_n in Definition 2:

$$S_n^*(\boldsymbol{r}) := S_n^*(x) \text{ for any } x \in C' \text{ such that } x \supseteq \boldsymbol{r}.$$

So we have shown that if in Skorokhod's Theorem, we start from $X_n = \frac{S_n}{\sqrt{n}}$, then, the Y_n that result are exactly $\frac{S_n^*}{\sqrt{n}}$. So for each $n \in \mathbb{N}^+$, $\frac{S_n^*}{\sqrt{n}}$ has the same distribution as $\frac{S_n}{\sqrt{n}}$ and, more importantly, the $\frac{S_n^*}{\sqrt{n}}$ converge almost surely to the standard normal on (0, 1). An important question that arises here is, are there representations of S_n^* similar to the canonical representation for S_n? And if so, how close can they be to the canonical representation for S_n? We can answer these questions as follows. (For additional work related to the following results, see [2].)

Theorem 5 *For any n, there is a canonical one to one correspondence between permutations* $F : \{0, 1\}^n \to \{0, 1\}^n$ *such that* $S_n^* = S_n \circ F$, *and representations* $S_n^* = \sum_{i=1}^n R_{n,i}^*$, *where* $\left(R_{n,i}^* \middle| 1 \le i \le n\right)$ *is an i.i.d. family of random variables on* (0, 1) *such that each* $R_{n,i}^*$ *depends only on the first n coordinates of x and takes on values* $-1, 1$ *with equal probability.*

Proof Let balanced mean takes on values $-1, 1$ each with probability $\frac{1}{2}$. Suppose $S_n^* = S_n \circ F$. Define

$$R_{n,i}^*(x) := (-1)^{1+(F(x_1,\ldots,x_n))_i}.$$

Since $S_n(x) = \sum_{i=1}^n (-1)^{1+x_i}$, $S_n(F(x)) = \sum_{i=1}^n R_{n,i}^*(x)$. To show the $R_{n,i}^*$ are balanced, it suffices to show for all $i = 1, \ldots, n$ and $\varepsilon \in \{0, 1\}$,

$$\lambda\left(\left\{x \middle| (F(x_1, \ldots, x_n))_i = \varepsilon\right\}\right) = \frac{1}{2}.$$

Let $A = \left\{\boldsymbol{t} \in \{0, 1\}^n \middle| t_i = \varepsilon\right\}$. So $\operatorname{card}(A) = \frac{2^n}{2} = 2^{n-1}$. Since F is 1–1, card $\left(F^{-1}[A]\right) = 2^{n-1}$. Now, $F^{-1}[A] = \left\{\boldsymbol{r} \in \{0, 1\}^n \middle| (F(\boldsymbol{r}))_i = \varepsilon\right\}$ and $\{x| (F(x_1, \ldots, x_n))_i = \varepsilon\} = \bigsqcup_{\boldsymbol{r} \in F^{-1}[A]} N_{\boldsymbol{r}}$. So, $\lambda(\{x| (F(x_1, \ldots, x_n))_i = \varepsilon\}) = \lambda\left(\bigcup_{\boldsymbol{r} \in F^{-1}[A]} N_{\boldsymbol{r}}\right) = 2^{n-1} \cdot \frac{1}{2^n} = \frac{1}{2}$.

To show the $R_{n,i}^*$ are independent, it suffices to show for all $\boldsymbol{s} \in \{-1, 1\}^n$,

$$p(s_1, \ldots, s_n) = p_1(s_1) \cdot \ldots \cdot p_n(s_n),$$

where p is the joint pmf of the $R^*_{n,i}$ and p_i is the pmf of $R^*_{n,i}$ alone. We showed the right hand side is simply $\left(\frac{1}{2}\right)^n$, so it suffices to show $p(s_1, \ldots, s_n) = \frac{1}{2^n}$. Recall that $p(s_1, \ldots, s_n) = P\left(R^*_{n,1} = s_1, \ldots, R^*_{n,n} = s_n\right)$. Let $\boldsymbol{t} \in \{0,1\}^n$ be such that $t_i = \begin{cases} 0 & \text{if } s_i = -1 \\ 1 & \text{if } s_i = 1. \end{cases}$
F is one-to-one, so there is a unique $\boldsymbol{r} \in \{0,1\}^n$ such that $F(\boldsymbol{r}) = \boldsymbol{t}$. Then the probability of the event $E_s = \left(R^*_{n,1} = s_1, \ldots, R^*_{n,n} = s_n\right)$ is exactly

$$\begin{aligned}
\lambda\left(\{x \mid (F(x_1,\ldots,x_n))_1 = t_1, \ldots, (F(x_1,\ldots,x_n))_n = t_n\}\right) &= \lambda\left(\{x \mid F(x_1,\ldots,x_n) = \boldsymbol{t}\}\right) \\
&= \lambda\left(\{x \mid (x_1,\ldots,x_n) = \boldsymbol{r}\}\right) \\
&= \lambda(N_{\boldsymbol{r}}) \\
&= \frac{1}{2^n}.
\end{aligned}$$

Now suppose $\left(R^*_{n,i} \middle| 1 \le i \le n\right)$ is as above. Fix $\boldsymbol{r} \in \{0,1\}^n$. $\left(R^*_{n,1}(x), \ldots, R^*_{n,n}(x)\right)$ is constant on $N_{\boldsymbol{r}}$. Denote that constant value by $G(\boldsymbol{r})$. So $G : \{0,1\}^n \to \{-1,1\}^n$. G is one-to-one since if $\boldsymbol{u} \in \{0,1\}^n$, $\boldsymbol{u} \neq \boldsymbol{r}$, and $G(\boldsymbol{u}) = G(\boldsymbol{r})$, then

$$P\left(R^*_{n,1} = (G(\boldsymbol{r}))_1, \ldots, R^*_{n,n} = (G(\boldsymbol{r}))_n\right) \ge \lambda(N_{\boldsymbol{r}}) + \lambda(N_{\boldsymbol{u}}) = \frac{1}{2^{n-1}},$$

but by our hypotheses of balanced and independent, $P\left(R^*_{n,1} = (G(\boldsymbol{r}))_1, \ldots, R^*_{n,n} = (G(\boldsymbol{r}))_n\right) = \frac{1}{2^n}$. Since $G : \{0,1\}^n \to \{-1,1\}^n$, and since the domain and target of G are finite sets of the same cardinality, G is one-to-one if and only if it is onto. So we have that G is both one-to-one and onto. Define $F(\boldsymbol{r}) = \boldsymbol{t}$, where $t_i = \begin{cases} 0 & \text{if } (G(\boldsymbol{r}))_i = -1 \\ 1 & \text{if } (G(\boldsymbol{r}))_i = 1. \end{cases}$ Then $S_n(F(x)) = \sum_{i=1}^n (-1)^{1+t_i}$ $= \sum_{i=1}^n R^*_{n,i}(x) = S^*_n(x)$, i.e., F is as required. □

In addition, the following theorem shows there are many such permutations.

Theorem 6 *For each n, there are exactly* $\prod_{i=0}^{n}\left(\binom{n}{i}!\right)$ *permutations* $F : \{0,1\}^n \to \{0,1\}^n$ *such that* $S_n^* = S_n \circ F$.

Proof Recall that

$$A_{n,i} = \left\{ \boldsymbol{s} \in \{0,1\}^n \,\middle|\, S_n^*(x) = -n + 2i \text{ for all } x \supseteq \boldsymbol{s} \right\},$$

and let

$$B_{n,i} = \left\{ \boldsymbol{s} \in \{0,1\}^n \,\middle|\, S_n(\boldsymbol{s}) = -n + 2i \right\}.$$

Let f be a permutation of $\{0,1\}^n$. Then $S_n^* = S_n \circ f$ if and only if for all $0 \le i \le n$, $f\left[A_{n,i}\right] = B_{n,i}$, i.e., if and only if $f \restriction A_{n,i}$ is a bijection from $A_{n,i}$ to $B_{n,i}$, and of course there are $\binom{n}{i}!$ such bijections. Since $f = \bigcup_{i=0}^{n}\left(f \restriction A_{n,i}\right)$ and since the $A_{n,i}$ (respectively $B_{n,i}$) are pairwise disjoint, the conclusion is clear. □

Corollary 7 *For each n, there are exactly* $\prod_{i=0}^{n}\left(\binom{n}{i}!\right)$ *families* $\left(R_{n,i}^* \,\middle|\, i = 1, \ldots, n\right)$ *as above.*

Additional criteria make some of these permutations more natural than (and therefore preferable to) others. We say $\left(F_n \middle| n \in \mathbb{N}^+\right)$ is suitable if and only if for all n, F_n is a permutation of $\{0,1\}^n$ satisfying $S_n^* = S_n \circ F$ and such that:
(a) $\left(F_n \middle| n \in \mathbb{N}^+\right)$ is explicitly and naturally definable, uniformly and highly effectively in n,
(b) if $\boldsymbol{r} \in \{0,1\}^n$ and $S_n^*(\boldsymbol{r}) = S_n(\boldsymbol{r})$, then $F_n(\boldsymbol{r}) = \boldsymbol{r}$,
(c) F_n is as close as possible to being self-inverse (even for fairly small n (such as $n = 5, 6, 7$), it is impossible for F_n to literally be self-inverse).

3 Rearrangements of the Random Walk

We first look at a variant, $\left(G_n \middle| n \in \mathbb{N}^+\right)$, satisfying only the first two criteria, (a) and (b), as well as the composition equation, $S_n^* = S_n \circ G_n$. So each G_n will map Step to Weight ($\text{Step}_n(\kappa) = \text{Weight}_n(G_n(\kappa))$), and further, the mapping will be in an order-preserving fashion (except as ruled out by criterion (b)). This means that for all $0 \le \kappa < 2^n$,
(i) If $\text{Step}_n(\kappa) = \text{Weight}_n(\kappa)$, then $G_n(\kappa) = \kappa$,
(ii) If $\text{Step}_n(\kappa) \ne \text{Weight}_n(\kappa)$, and, if further, $\kappa < m < 2^n$ and $\text{Step}_n(\kappa) = \text{Step}_n(m) \ne \text{Weight}_n(m)$, then $G_n(\kappa) < G_n(m)$.

Lemma 8 *(i) and (ii) define a unique sequence* $\left(G_n \middle| n \in \mathbb{N}^+\right)$ *satisfying the composition equations* $S_n \circ G_n = S_n^*$.

Proof We have $A_{n,i} = \{\kappa | \text{Step}_n(\kappa) = i\}$ and we define $B_{n,i} := \{\kappa | \text{Weight}_n(\kappa) = i\}$. Further, let

$$A^1_{n,i} := A_{n,i} \setminus B_{n,i} = A_{n,i} \setminus (A_{n,i} \cap B_{n,i}),$$

$$B^1_{n,i} := B_{n,i} \setminus A_{n,i} = B_{n,i} \setminus (A_{n,i} \cap B_{n,i}).$$

These are the sets of things that are out of place on the ith step, or of the ith weight, respectively. We have the following equation:

$$\text{card}\left(A^1_{n,i}\right) = \binom{n}{i} - \text{card}\left(A_{n,i} \cap B_{n,i}\right) = \text{card}\left(B^1_{n,i}\right).$$

$G_n \restriction A^1_{n,i}$ is simply the order-preserving bijection between $A^1_{n,i}$ and $B^1_{n,i}$. □

In fact $(G_n | n \in \mathbb{N}^+)$ is uniformly primitive recursive in the following precise sense: there exists a single primitive recursive function $G(n, \kappa)$ such that for all n, $G(n, \cdot) \restriction \{0, \ldots, 2^n - 1\} = G_n$. Simply take $G(n, \kappa)$ to be equal to $G_n(\kappa)$, when $0 \le \kappa < 2^n$, and supply a suitable default value (e.g., $G(n, \kappa) = 0$, or $G(n, \kappa) = \kappa$), when $\kappa \ge 2^n$ or $n = 0$, then we have defined a unique function $G : \mathbb{N}^2 \to \mathbb{N}$. It is not very difficult to show that G primitive recursive. Each G_n satisfies the dualization equation $G_n(2^n - 1 - \kappa) = 2^n - 1 - G_n(\kappa)$.

For $n = 3, 4, 5, 6, 7$ and each κ such that Step $(n, \kappa) \ne$ Weight (κ) (i.e., κ is out of place at level n), the orbit of κ under G_n is given in the following table.

n	Orbits under G_n
3	{3, 4}
4	{7, 3, 8, 12}
5	{16, 7, 3, 8, 5}, {15, 24, 28, 23, 26}, {11, 17}, {13, 18}, {14, 20}
6	{32, 42, 15, 34, 49, 30, 19, 40, 56, 60, 55, 58, 47, 27, 13, 24, 11, 6}, {31, 21, 48, 29, 14, 33, 44, 23, 7, 3, 8, 5, 16, 36, 50, 39, 52, 57}
7	{64, 15, 34, 21, 48, 73, 46, 69, 39, 25, 68, 30, 11, 5, 16, 36, 22, 65, 23, 66, 27, 80, 57, 84, 99, 31, 13, 6, 32, 14, 33, 19, 40, 26, 72, 45, 67, 29, 7}, {63, 112, 93, 106, 79, 54, 81, 58, 88, 102, 59, 97, 116, 122, 111, 91, 105, 62, 104, 61, 100, 47, 70, 43, 28, 96, 114, 121, 95, 113, 94, 108, 87, 101, 55, 82, 60, 98, 120}, {3, 8}, {51, 74}, {53, 76}, {124, 119}

Our construction of $(F_n | n \in \mathbb{N}^+)$, which will satisfy all three criteria (a), (b) and (c), takes place within the general framework implicit in the construction of

$(G_n | n \in \mathbb{N}^+)$. While G was implicitly constructed in two stages, F will be built in three. As before, the first stage is that F is the identity on the κ's that are in place: $F_n(\kappa) = \kappa$ if $\text{Step}_n(\kappa) = \text{Weight}_n(\kappa)$. Then identify which κ's are part of a two-cycle and pair them up. After we have maximized two-cycles (this satisfies criterion (c)), removing those from $A^1_{n,i}$, $B^1_{n,i}$ leaves us with sets $A^2_{n,i}$, $B^2_{n,i}$ of equal cardinality and we map $A^2_{n,i} \to B^2_{n,i}$ in an order-preserving fashion.

Just as we noted for $(G_n | n \in \mathbb{N}^+)$, it is not very difficult to show $(F_n | n \in \mathbb{N}^+)$ is uniformly primitive recursive in the following precise sense: there exists a single primitive recursive function $F(n, \kappa)$ such that for all n, $F(n, \cdot) \restriction \{0, \ldots, 2^n - 1\} = F_n$. As before, if we simply take $F(n, \kappa)$ to be equal to $F_n(\kappa)$, when $0 \leq \kappa < 2^n$, and supply a suitable default value (e.g., $F(n, \kappa) = 0$, or $F(n, \kappa) = \kappa$), when $\kappa \geq 2^n$ or $n = 0$, then we have defined a unique function $F : \mathbb{N}^2 \to \mathbb{N}$. As before, each F_n satisfies $F_n(2^n - 1 - \kappa) = 2^n - 1 - F_n(\kappa)$.

As noted in the table above, the first time there are values of κ that are out of place is when $n = 3$. Below are the graphs of F_n for $n = 3$ and $n = 4$.

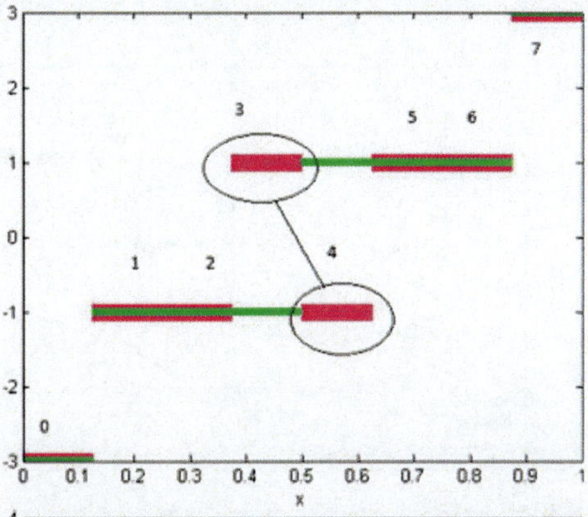

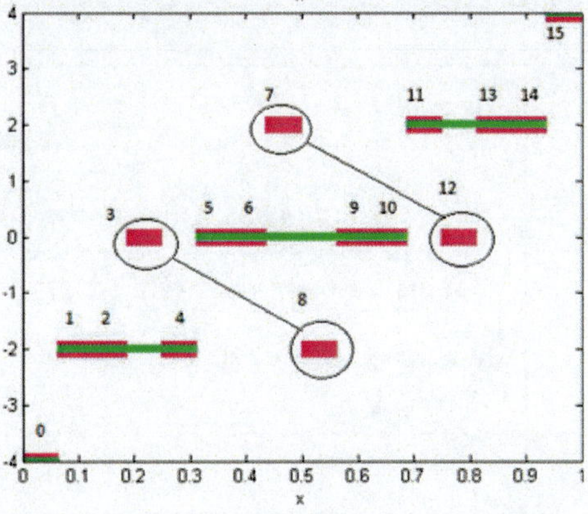

The first time there are values of κ that are not part of a two-cycle is when $n = 5$. For $n = 5, 6, 7$, the table below presents the orbits under F_n for those values of κ at level n.

n	Orbits under F_n
5	$\{16, 28, 15, 3\}$
6	$\{32, 56, 60, 31, 7, 3\}$
7	$\{64, 108, 31, 11, 3\}$, $\{13, 72, 113, 47\}$, $\{14, 80, 114, 55\}$, $\{63, 19, 96, 116, 124\}$

The resulting cycles of these values of κ, corresponding to each of the rows of the table, are illustrated in the graphs below. We have a single four-cycle at $n = 5$ and a single six-cycle at $n = 6$.

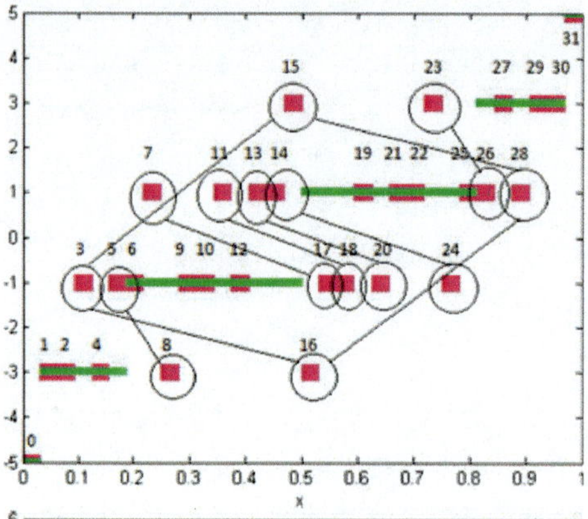

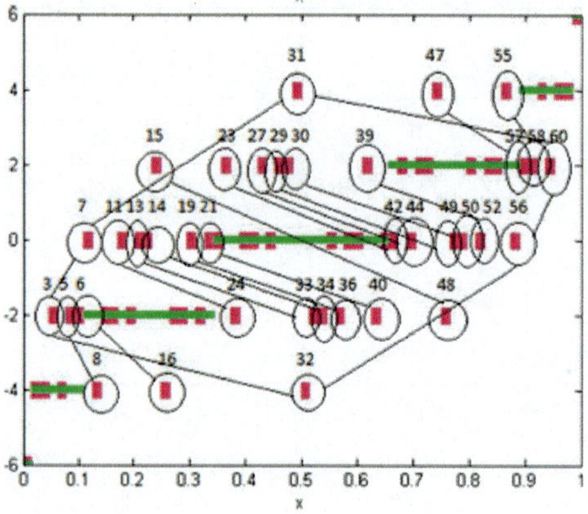

At $n = 7$ we have two four-cycles and two five-cycles. Because the graph of F_n is rather complicated by $n = 7$, we will leave out the two-cycles from the graph and only illustrate the four-cycles and five-cycles. The four-cycles are highlighted in the figure below.

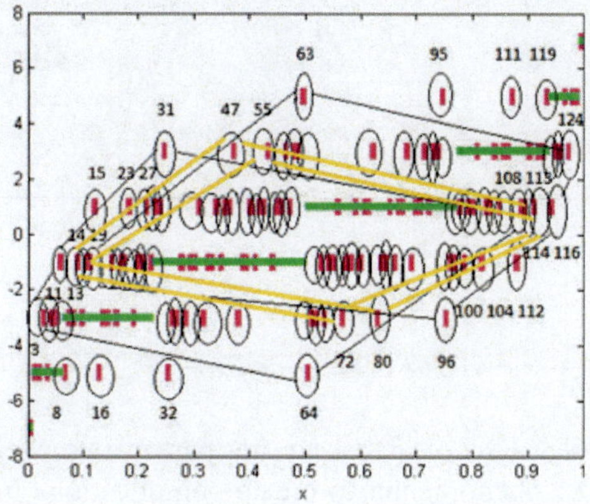

4 Graph-Theoretic Complexity of the Permutations

The results we have presented do indeed narrow the distance between the S_n and the S_n^* with respect to the important issue of representation. The form of the composition equation that we have used so far, $S_n^* = S_n \circ F$, emphasizes the point of view of providing suitable representations of the S_n^*. But this equation could just as well be written in the form $S_n = S_n^* \circ F^{-1}$, which would emphasize the point of view of seeking to tame the disorder of the S_n. This is related to the rearrangement idea that is illustrated in the above graphs of F_n: we rearrange S_n to get S_n^*, and thus achieve almost sure convergence. The question remains how much rearranging of the S_n is optimal. One direction involves attempting to minimize the graph-theoretic complexity of the function F. As described in the construction of F above, F maximizes the number of two-cycles (with the proper choice of the two-cycles), but then will act just as the function G on the remaining κ's which are not part of a two-cycle. Of course we know by Theorem 3 that there are many other possible variants for the function F. One may add some additional stages to the construction of F. In stage three (which might no longer be the terminal stage), we would seek to maximize the number of three-cycles just as we maximized the number of two-cycles in stage two, and fixed all the κ's which were in place (thereby maximizing the number of one-cycles) in stage one. If some κ remain outside the domain, proceed to stage four and continue. The goal would be to minimize lengths of cycles which could be viewed as one way of seeking to minimize the graph-theoretic complexity of the permutations. This idea is illustrated below for $n = 6$.

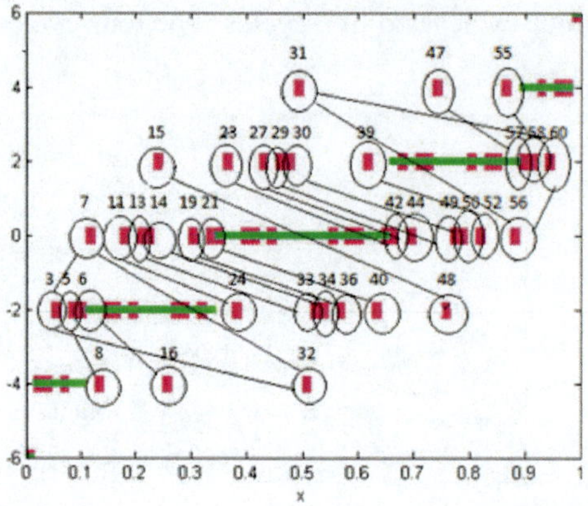

The existence of values of κ that are not part of two-cycles at level n, starting at $n = 5$, is the last phenomenon to create complications in the definition of the function F. It is conceivable that further interesting phenomenon (which do not create additional complications for the definition of F) first occur for some n larger than 5, and it is far from certain whether there are finitely many such n.

References

1. Nigel Cutland. Computability. Cambridge University Press (1980).
2. V. Dobric, P. Garmirian, M. Skyers and L. Stanley, Polynomial time relatively computable triangular arrays for almost sure convergence, Illinois J. Math. vol. 63, (2019), no. 2, pp 219-257.
3. Herbert B. Enderton. Computability Theory. Academic Press (2010).
4. P. Erdos. On the Law of the Iterated Logarithm. Annals of Mathematics, vol. 43 (1942) pp 419-436.
5. P. Erdös and P. Revesz. On the favourite points of a random walk. Mathematical Structures, Computational Mathematics, Mathematical Modelling 2, (1984) pp 152-157.
6. W. Feller. The general form of the so-called Law of the Iterated Logarithm. Transactions of the American Mathematical Society, vol. 54 (1943) pp 373-402.
7. A. Khintchine. Uber dyadische bruche. Math. Zeit., vol. 18 (1923) pp 109-116.
8. Kolmogoro, A. Über das gesetz des iterierten logarithmus, Math. Ann., 101 (1929) pp 126-135.
9. Z. Shi and B. Tóth. Favourite sites of simple random walk. Periodica Mathematica Hungarica, 41 (2000) pp 237–249.
10. A. V. Skorokhod. Limit theorems for stochastic processes. Theory of Probability and its Applications, vol. 1 (1956) pp 261-290.

On the Energy of Transposition Graphs

M. R. DeDeo

Abstract We analyze and compare properties of Cayley graphs of permutation graphs called transposition graphs as this family of graphs has better degree and diameter properties than other families of graphs. Cayley graphs are directly related to the properties of its generator set and thus Cayley graphs of permutation groups generated by transpositions inherit almost all of the properties of the hypercube. In particular, we study properties of the complete transportation, (transposition) star graph, bubble-sort graph, modified bubble-sort graph and the binary hypercube and use these properties to determine bounds on the energy of these graphs.

Keywords Transposition graphs · Permutation groups · Network computing

1 Introduction

1.1 Definitions

Parallel computing is largely dependent on the properties of the interconnection network that connects processors amongst themselves and/or to memory. The interconnections also affect the network operating system (OS) and the effectiveness of the system software. Many of the schemes used to model these interconnections can be classified into two types of networks: dynamic and static. In this paper we study the design and properties of a particular type of static network modeled by transposition graphs.

Static networks can be modeled by their corresponding graphs with at most two of the following properties: (1) if all processors are connected by the same number of edges, the network is called *regular* or *k-regular* where k is the number of edges emanating from each processor; (2) if not, the graph is not regular; (3) if the proces-

M. R. DeDeo (✉)
University of North Florida, Jacksonville 32224, USA
e-mail: mdedeo@unf.edu
URL: http://www.unf.edu/~mdedeo

F. Hoffman (ed.), *Combinatorics, Graph Theory and Computing*,
Springer Proceedings in Mathematics & Statistics 388,
https://doi.org/10.1007/978-3-031-05375-7_23

sors can be grouped into m subsets where each processor within each subset is not connected to another with the subset, but is connected to a processor in each other subset, the network is called *m-partite*. In particular, if the processors can be divided into two subsets where processors within each subset are not connected to another within its subset, but only to processors within the other subset (and vice versa), then the network is called *bipartite*. We now refer to each processor as a vertex and each connection as an *edge*.

The *degree* of each vertex is the number of edges emanating from it. As no processor will be connected to itself, and hence have no loops, the graph is called simple. The degree relates to the port capacity of the processors and thus relates to the hardware cost of the network. A *path* is the routing from one vertex to another. The length of a path is the number of edges a signal traverses from a given vertex to reach another vertex. The distribution of parallel paths is the number of paths of a given length and is crucial to the design of the routing table for an operating system. This also relates to the fault-tolerance of a network as the number of parallel paths between two vertices is limited by the degree of the network.

The *diameter* of the graph is the maximum eccentricity of any vertex in a graph. That is, it is the greatest distance between any pair of vertices. To find the diameter of a graph, first find the shortest path between each pair of vertices. The greatest length of any of these paths is the diameter of the graph. The diameter relates to the maximum communication delay and hence the running cost of the network.

1.2 Symmetry and Recursive Scalability

The study of several other, more complex properties is crucial for the understanding of the effectiveness of a network. These include symmetry and recursive scalability. Symmetry in graphs can be analyzed using graph theory and finite group theory. A symmetric graph is a graph that is both vertex-transitive and edge-transitive. In a *vertex-transitive graph*, its automorphism group acts transitively upon its vertices. In other words, the graph looks the same through the lens of any vertex. In addition, every symmetric graph without isolated vertices is vertex-transitive, and every vertex-transitive graph is regular. However, not all vertex-transitive graphs are symmetric (for example, the edges of the truncated tetrahedron), and not all regular graphs are vertex-transitive. In a vertex-transitive graph, the structure embedded in one region of the network can be readily translated into another region without affecting the quality of the original embedding. Vertex transitivity also enables the design of efficient distributed routing algorithms.

An *edge-transitive* graph is analogously defined. In particular, the number of vertex-disjoint paths between any two vertices is maximum and hence has a maximum fault-tolerance capacity. Recursive scalability refers to the ability to build larger networks from smaller subnetworks. These networks then possess naturally occurring symmetry that is often used in the design of routing tables, fault-tolerance and more.

A *distance-transitive* graph is a graph such that, given any two vertices at any distance, and any other two vertices at the same distance, there is an automorphism of the graph that carries the pairs of vertices to each other. Distance transitivity ensures good fault-tolerance, translations of embedded substructures from one region to another and decentralizes routing algorithms for packet communication. Thus distance transitivity is one of the most important of the symmetric properties.

A *Cayley graph* is a graph that encodes the abstract structure of a group using a specified, usually finite, set of generators for the group. Cayley graphs provide a unified framework for the design of interconnection networks for parallel computing. In particular, linear groups which are automorphism groups of finite dimensional vector spaces and semi-direct products of groups, such as degree 4 super-toroids and Borel Cayley graphs, provide classes of graphs with "good" routing algorithms where "good" is a multi-faceted decision problem based on the analysis of trade-offs in the symmetry and topology of the networks.

In this paper, we analyze and compare properties of Cayley graphs of permutation graphs called transposition graphs. The conjugacy class of its permutation group combined with its generators dictates the type of the symmetry possessed by their respective Cayley graphs. In addition, the family of transposition graphs has better degree and diameter properties than the hypercube. We note that the base-b ($b \geq 2$) hypercube of dimension n is a class of networks known to possess virtually every known notion of symmetry. Next, properties of the transposition graphs are given.

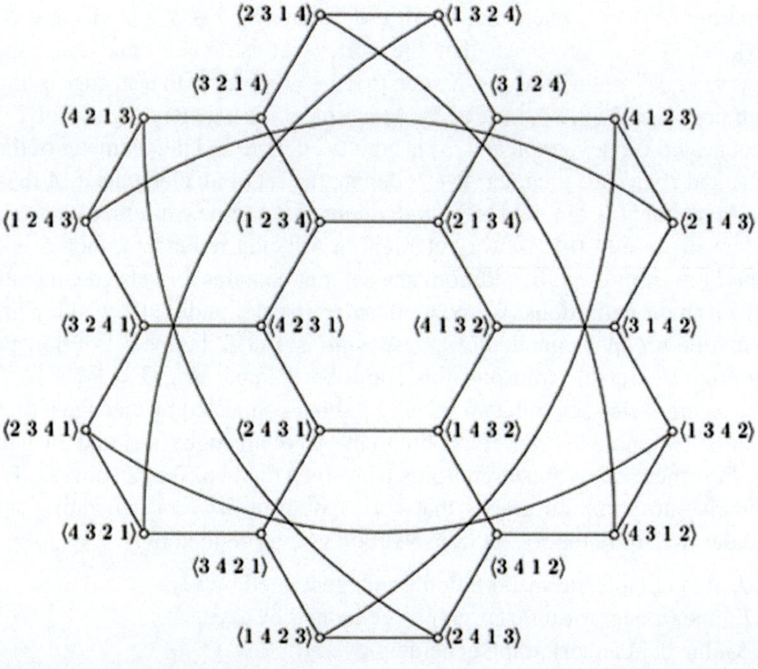

Fig. 1 Star transposition graph ST_4 [13, p. 65]

Lastly, the spectra of this class of graphs is discussed and formulas for the energy of these graphs is given as they are dependent on these properties (Fig. 1).

2 Transposition Graphs

A *permutation* of $\{1, 2, \ldots, n\}$ is a bijection onto itself. Let

$$p = \begin{pmatrix} 1 & 2 & \cdots & i & \cdots & n \\ p_1 & p_2 & \cdots & p_i & \cdots & p_n \end{pmatrix}$$

where $p_i \leq p_j$ for all i, j where p_i denotes the object at position i. For simplicity, we write $p = p_1 p_2 \cdots p_n$. Let S_n denote the set of all permutations of p. For any permutation $p \in S$, p can be represented as a product of k disjoint cycles and l invariants. For convenience, the invariants are deleted when p is represented in terms of its cycle structure. Cycles of length two are called *transpositions*.

Let T_{ij} denote the permutations that swap objects in positions i and j such that $T_{ij} = (i, j)$. For permutations p, if $p_i < p_j$ for $i < j$, then the pair p_i and p_i is said to be an inversion in p. Thus, a permutation is said to be an *odd (or even) permutation* if the number of inversions in p is odd (or even). We note that transpositions are odd permutations as the number of inversions is odd.

Let Γ be a finite group under multiplication with identity I. Let $S \subseteq \Gamma$ be a generating set for Γ such that (i) if $g \in S$, then $g^{-1} \in S$ and (ii) $I \notin S$. Given (Γ, S), let $G = (V, E)$ such that the vertex set is $V = \Gamma$ and the edge set is $E = \{(x, y)_g | x, y \in V \text{ and } g \in S \text{ such that } xg = y\}$. Note that an edge is undirected if both $(x, y)_g$ and $(y, x)_g^{-1}$ are in E. Also, since S is a generating set for Γ, then G is a connected Cayley graph and $|S|$ is both the degree and the diameter of the graph.

If A and B are sets then, let $A - B$ denote the set of all elements in A that are not in B. An element $h \in S$ said to be *redundant* if it can be expressed as a product of the elements in $S - \{h\}$. If every element in S is non-redundant, then S is called a *minimal generating set*. In addition, any set that contains S is also a generating set.

Given these definitions, we now consider simple, undirected Cayley graphs of permutation groups generated by transpositions. Let Ω be the set of transpositions generating Γ. Let the transposition graph be defined as $TG = (< n >, \Gamma)$ with $< n >$ as the vertex set with two vertices i and j connected by an edge if and only if $(i, j) \in \Gamma$. We note that Ω refers both to the set of all edges in G and the generating set of Γ as there exists an automorphism between the two. We can now study several important transportation graphs that correspond to different generating sets S. In particular, we study the sets of transposition graphs defined by:

1. CT_n, the complete transportation graph generated by Ω_0;
2. ST_n, the (transposition) star graph[1] generated by Ω_1 ;
3. BS_n, the bubble-sort graph generated by Ω_2;

[1] ST_n should not be confused with the Star Graph $S_k = K_{1,k}$.

4. MB_n, the modified bubble-sort graph generated by $\Omega_{2'}$; and
5. BC_n, the binary hypercube generated by Ω_3

where

$$\begin{aligned}
\Omega_0 &= \{(i\ j)|1 \le i \le j \le n\};\\
\Omega_1 &= \{(1\ i)|2 \le i \le n\};\\
\Omega_2 &= \{(i\ i+1)|1 \le i < n\};\\
\Omega_{2'} &= \Omega_2 \cup \{(1\ n)\};\ \text{and}\\
\Omega_3 &= \{(2i-1\ 2i)|1 \le i \le n\}.
\end{aligned}$$

We call this family of transposition graphs $\mathcal{T}$. It can be verified that both ST_n, the star graph, and BS_n, the bubble-sort graph, can be built recursively, but not MB_n, the modified bubble-sort graph. In addition, Ω_1, Ω_2 and Ω_3 are minimal generating sets, but Ω_0 and $\Omega_{2'}$ are not as they have redundant elements. In particular, the bubble-sort graph is the Cayley graph corresponding to the case where the transposition graph is the path graph on n vertices. The reason it is called the bubble-sort graph is that this Cayley graph is closely related to the (inefficient) bubble-sort algorithm for sorting an array. Many of the properties in Table 1 can be found in [1, 2, 6, 11, 14]. The values in Table 2 will not be presented here as they can be found using counting arguments on the generating sets (see [16] for a few of them).

Given a permutation in S_n, the array swaps elements in consecutive positions of the array. Observe that the minimum number of swaps of elements in consecutive positions required to sort a given array p is exactly the distance in the Cayley graph between the permutation p and the identity vertex e. The modified bubble-sort graph

Table 1 Symmetrical properties of transposition graphs in $\mathcal{T}$

	Vertex transitive	Edge transitive	Distance transitive	Shortest path distance	Hamiltonian cycle
CT_n	Yes	Yes	No	Yes	Known
ST_n	Yes	Yes	No	Yes	Known
BS_n	Yes	No[2]	No	Yes	Known
MB_n	Yes	Yes	No	Yes	Known
BC_n	Yes	Yes	Yes	Yes	Known

Some literature incorrectly assumes that the bubble-sort graph is edge transitive [14]

Table 2 Computational properties of transposition graphs in $\mathcal{T}$

	No. of vertices	Degree	Diameter	Bipartite	Recursive
CT_n	$n!$	$n(n-1)/2$	$n-1$	Yes	Yes
ST_n	$n!$	$n-1$	$\lfloor 3(n-1)/2 \rfloor$	Yes	Yes
BS_n	$n!$	$n-1$	$n(n-1)/2$	Yes	Yes
MB_n	$n!$	n	Unknown	Yes	No
BC_n	2^n	n	n	Yes	Yes

is obtained by modifying the bubble-sort graph by adding another generator (and hence, by adding extra edges) to the bubble-sort graph, thereby reducing its diameter. Additionally, Cayley graphs are directly related to the properties of its generator set. In particular, Cayley graphs of permutation groups generated by transpositions inherit almost all of the properties of the hypercube.

3 Energy and Spectra of Graphs

3.1 *Energy of Graphs*

In the 1940s, a close correspondence between the graph eigenvalues and the molecular orbital energy levels of π-electrons in conjugated hydrocarbons was realized [7, 9]. In particular,

$$E_\pi = n\alpha + \beta \sum_{i=1}^{n} |\lambda_i|$$

where n, α, β are constants. In addition, the energy is related to several other concepts in analysis, linear algebra and spectral graph theory. The general theory and chemical applications can be found in [8]. After it was recognized that spectral graph theory can be used more broadly than just in orbital theory, the notion of the energy of a graph was defined [8]. Given the eigenvalues of the adjacency matrix of a graph, λ_1, $\lambda_2, \ldots, \lambda_n$, the energy of a graph is defined to be

$$E(G) = \sum_{i=1}^{n} |\lambda_i|.$$

This is a natural extension of this property as crystallographic groups tell us about the structure of matter and graphs based on groups are used to model individual molecules as well as a variety of chemical systems. Broader study of the energy of graphs began in the 2000s and has expanded the field of spectral analysis. Not only does this newer graph invariant allow for a new relation on all graphs, the energy of a graph is related to several other concepts in analysis, linear algebra and spectral graph theory (Fig. 2).

3.2 *Spectra of $\mathcal{T}$*

Let G be a simple (no loops or repeated edges), undirected, and connected graph with vertices v_i for $i = 1, \ldots, n$ and m edges. Let $A(G)$ be the $n \times n$ adjacency

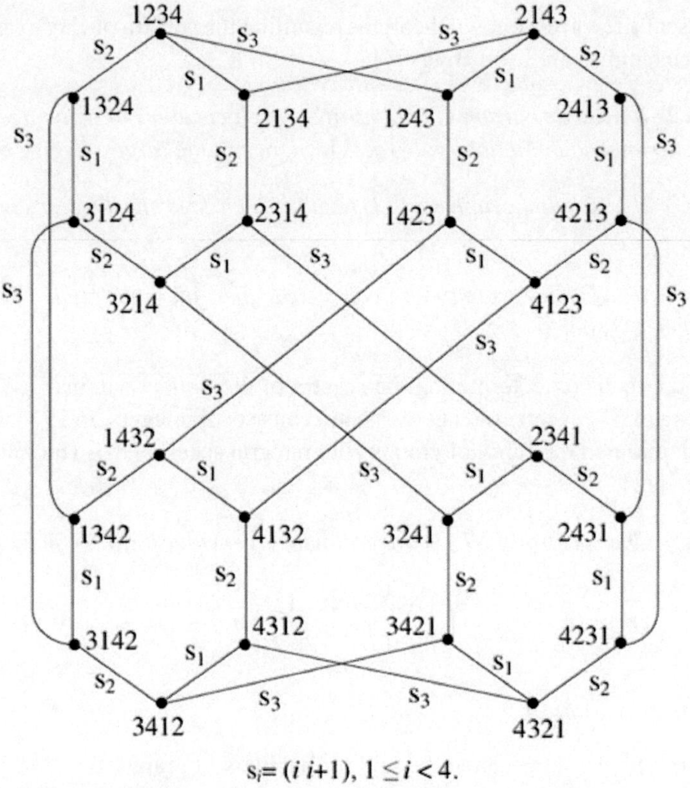

Fig. 2 Bubble sort graph BS_4 [5]

matrix associated with G such that

$$A(G) = \begin{cases} 1 & \text{if } v_i \text{ is connected to } v_j \text{ for } i \neq j \text{ by an edge;} \\ 0 & \text{otherwise.} \end{cases}$$

Let $\lambda_1, \lambda_1, \ldots, \lambda_n$ denote the n eigenvalues of $A(G)$ and $Spec(A) = \{\lambda_1, \lambda_2, \ldots, \lambda_n\}$. Let $m(\lambda_i)$ denote the multiplicity of λ_i.

Lemma 1 *All graphs in the family $\mathcal{T}$ of transposition graphs are regular and bipartite.*

Proof By counting arguments, all graphs in $\mathcal{T}$ are regular with fixed degree (see Table 1). Transposition graphs are odd permutations as the number of inversions is always odd. It can also be easily verified that Cayley graphs defined using permutations are necessarily bipartite graphs. Thus, all graphs in the family $\mathcal{T}$ of transposition graphs are necessarily bipartite. □

The spectra of these graphs are of interest for their own sake, as well as for various applications such as card shuffling and random walks on the symmetric group. We now present a few well-known theorems regarding the spectra of Cayley graphs. The next three lemmas are from Biggs [4]:

Lemma 2 *If a Cayley graph G is bipartite, the eigenvalues of its adjacency matrix* $A(G)$ *is symmetric in the interval* $[-k, k]$ *where k is the largest degree of the set.*

Lemma 3 *If a Cayley graph G is k-regular, then k is the largest eigenvalue in* $Spec(A)$.

Lemma 4 *If a Cayley graph G is connected, then the multiplicity of its largest eigenvalue, k, is one.*

We now focus on determining the spectra of the graphs within $\mathcal{T}$. We note that the spectra of ST_n is integral, i.e. its spectra consists of integers. In 1974, Harary and Schwenk initiated the study of graphs with integral spectra [10]. The following was proved in [11]:

Lemma 5 *The spectra of* ST_n *is integral with eigenvalues* $\pm(n-j)$ *with*

$$m(\pm(n-j)) \geq \binom{n-2}{j-1}\binom{n-1}{j} \text{ for } j \in \{1, \ldots, n-1\}$$

and $m(0) \geq \binom{n-1}{2}$ *for* $n > 3$.

Unfortunately, the same cannot be said for bubble-sort graphs:

Lemma 6 *The spectra of* BS_n *is not integral.*

Proof By counterexample, $Spec(BS_4)$ is in the subset of rational numbers in the interval $(\sqrt{2}, \sqrt{3})$ which is not integral. □

Moreover, BS_n is not a family of expander graphs. *Expanders* are sparse graphs (few edges relative to the number of vertices) that are highly connected. Thus expanders model efficient communication networks as they exhibit few edges while retaining high connectivity. We can prove a graph is an expander (or not) by finding its isoperimetric number which is a numerical measure of whether or not a graph has a "bottleneck". For a collection of vertices $V' \subseteq V(G)$, let $\partial V'$ denote the collection of all edges going from a vertex in V' to a vertex outside of V' (also called the edge boundary of V') and let $|V|$ denote the number of elements in the set. Then the isoperimetric number $h(G)$ is

$$h(G) := min\left\{\frac{|\partial V'|}{|V'|}|V' \subseteq V(G) \text{ and } 0 < |V'| \leq \frac{1}{2}|V(G)|\right\}$$

An expander family is one that satisfies $h(G_\alpha) \geq \varepsilon$ for a fixed ε and all α.

Lemma 7 *Bubble-sort graphs* BS_n *do not form a family of expander graphs.*

Proof For $n = 3$, let $S = \{p, p-1, (1\ 2)\} \subset S$. Consider the set $P_m = \{\pi \in S | 1 \leq \pi^{-1} \leq m\}$ where $m = \lfloor \frac{n}{2} \rfloor$. Then $|P_m| = m(n-1)!$. We know that $\rho \notin P_m$ and $\pi \in P_m$ are adjacent if and only if $\pi = \rho\theta = \rho \circ \theta$ for some $\theta \in P_m$. Assume $\rho \notin P_m$, $\pi \in P_m$, and $\pi = \rho\theta$. Then:

Case (i). Suppose $\theta = p$. Since $\rho \notin P_m$, the permutation ρ cannot map any element in the interval 1 through m to 1. Because $\pi \in P_m$, we must have $\rho(m+1) = 1$. There are $(n-1)!$ such π's. Thus there is one edge from p that leaves P_m for each such π.

Case (ii). Suppose $\theta = p^{-1}$. Then $\rho(n) = 1$ and $\pi(1) = 1$. Thus we also have $(n-1)!$ such π's giving $(n-1)!$ edges with one extreme in P_m and one in $BS_n - P_m$.

Case (iii). The case where $\theta = (1\ 2)$ cannot happen. Thus

$$h(BS_n) \leq \frac{|\partial P_m|}{|P_m|} = \frac{2(n-1)!}{m(n-1)!} = \frac{2}{m}.$$

By an analogous argument, for any change in size n, $h(BS_n)$ is never bounded below. Thus the set of bubble-sort graphs BS_n do not create an expander family. □

This observation could be helpful in the future to either determining the eigenvalues of BS_n or in creating a better upper bound on the energy of BS_n.

4 Bounds on the Energy of Transposition Graphs

Using the fact that all of the graphs in $\mathcal{T}$ are regular, bipartite, and connected, we have the following bounds for the energy of the graphs in $\mathcal{T}$:

Theorem 1 *An upper bound for the energy of CT_n is $n!n(n-1)$.*

Proof By its definition and Table 2, CT_n is regular, bipartite, and connected. The eigenvalues of CT_n occur in $[-n(n-1)/2, n(n-1)/2]$ where the multiplicity of both $\pm n(n-1)/2$ is one. Thus

$$E(CT_n) = \sum_{i=1}^{n!} |\lambda_i| < n!n(n-1).$$

□

Theorem 2 *The graph energy bounds of the family of graphs ST_n are*

$$2(n-1) + 2\sum_{i=1}^{n-1}(n-j)\binom{n-2}{j-1}\binom{n-1}{j} \leq E(ST_n) \leq 2n!(n-1).$$

Proof By its definition and Table 2, ST_n is regular, bipartite, and connected. The eigenvalues the eigenvalues of ST_n occur in $[-(n-1), (n-1)]$ where the multiplicity of both $\pm n(n-1)$ is one. Thus the upper bound is $2n!(n-1)$.

In addition, by Lemma 5 we have that the eigenvalues of ST_n are $\pm(n-j)$ with

$$m(\pm(n-j)) \geq \binom{n-2}{j-1}\binom{n-1}{j} \text{ for } j \in \{1, \ldots, n-1\}$$

and $m(0) \geq \binom{n-1}{2}$ for $n > 3$. Thus

$$2(n-1) + 2\sum_{i=1}^{n-1}(n-j)\binom{n-2}{j-1}\binom{n-1}{j} \leq E(ST_n).$$

□

Theorem 3 *An upper bound for the energy of BS_n is*

$$E(BS_n) \leq 2(n-1)\{(n+2)! + 1\}.$$

Proof By its definition and Table 2, BS_n is regular, bipartite, and connected. The eigenvalues of BS_n occur in $[-(n-1), (n-1)]$ where the multiplicity of both $\pm(n-1)$ is one. By [3], the second largest eigenvalue is at least $n-2$ with multiplicity $n-1$. Thus we have

$$E(BS_n) = \sum_{i=1}^{n!} |\lambda_i| < 2(n-1) + 2(n-2)!(n-1).$$

By combining like terms, the inequality is produced. □

Theorem 4 *An upper bound for the energy of MB_n is $n(n!)$.*

Proof By its definition and Table 2, MB_n is regular, bipartite, and connected. The eigenvalues of MB_n occur in $[-n, n]$ where the multiplicity of both $\pm n$ is one. Thus

$$E(MB_n) = \sum_{i=1}^{n!} |\lambda_i| < n(n!)$$

□

Theorem 5 *The $Spec(BC_n)$ is $(-n, -n+2, -n+4, \ldots, n-4, n-2, n)$ with the jth eigenvalue having multiplicity $\binom{n}{j}$.*

Proof For simplicity, let $BC_n = Q_n$. Let A_{Q_n} be the adjacency matrix of the binary hypercube Q_n. Then

$$A_{Q_n} = \begin{pmatrix} A_{Q_{n-1}} & I_{Q_{n-1}} \\ I_{Q_{n-1}} & A_{Q_{n-1}} \end{pmatrix}$$

where $I_{Q_{n-1}}$ is the $2^n \times 2^n$ identity matrix corresponding to Q_{n-1}. Its spectra follows recursively from the characteristic polynomial for the binary hypercube as

$$det(A_{Q_n} - \lambda I_{Q_n}) = det[(A_{Q_{n-1}} - \lambda I_{Q_{n-1}})^2 - I_{Q_{n-1}}]$$
$$= det(A_{Q_{n-1}} - \lambda I_{Q_{n-1}}) * (det(A_{Q_{n-1}} + \lambda I_{Q_{n-1}})$$
$$= det(A_{Q_{n-1}} - (\lambda + 1)I_{Q_{n-1}}) * det(A_{Q_{n-1}} - (\lambda - 1)I_{Q_{n-1}})).$$

The solutions for λ in this equation are $(-n, -n + 2, -n + 4, \ldots, n - 4, n - 2, n)$. □

Theorem 6 *The graph energy of the binary hypercube defined above is*

$$E(BC_n) = 2 \sum_{j=0}^{\lfloor n/2 \rfloor} \binom{n}{j} (n - 2j).$$

Proof From Theorem 5, the $Spec(BC_n)$ is $(-n, -n + 2, -n + 4, \ldots, n - 4, n - 2, n)$ with the jth eigenvalue having multiplicity $\binom{n}{j}$. Using counting arguments, the formula is attained. □

It is hoped that better bounds on the energy of this family of graphs are revealed with further study into the spectra of these graphs.

References

1. Akers, S.B., Harel, D., Krishnamurthy, B.: The star graph: An attractive alternative to the n-cube. Proc. Int. Conf. Parallel Process, 393–400 (1987).
2. Akers S. B., Krishnamurthy, B.: A group-theoretic model for symmetric interconnection networks, IEEE Trans. Comput. 38, 555–566 (1989). https://doi.org/10.1109/12.21148
3. Bacher, R.: Valeur propre minimale du laplacien de Coxeter pour le groupe symétrique. J. Algebra, 167(2), 460–472 (1994).
4. Biggs, N.: Algebraic Graph Theory. Cambridge University Press, England (2001).
5. Cheng, E., Liptak, L., Shawash, N.: Orienting Cayley graphs generated by transposition trees. Computers and Mathematics with App 55 (11) 2662–2672 (2008).
6. Day, K., Tripathi, A.: A comparative study of topological properties of hypercubes and star graphs. IEEE Trans. Comput., 5(1), 31–38 (1994).
7. Graovac, A., Gutman, I., Trinajsti, N., Topological Approach to the Chemistry of Conjugated Molecules, Springer, Berlin (1977).
8. Gutman, I., Polansky, O.E.: Mathematical Concepts in Organic Chemistry, Springer, Berlin (1986).
9. Gutman, I., Trinajstic N., Graph theory and molecular orbitals. Topics Curr. Chem. 42(49), (1973).
10. Harary, F., Schwenk, A.J.: Which graphs have integral spectra? In: Graphs and Combinatorics (Lecture Notes in Mathematics 406, ed. R. Bari, F. Harary), Springer-Verlag, Berlin-Heidelberg-New York pp. 45–51 (1974).
11. Heydemann, M.C., Ducourthial, B.: Cayley graphs and interconnection networks. In: G. Hahn, G. Sabidussi, Graph Symmetry, NATO Advanced Science Institutes Series C. In: Mathematica and Physical Sciences, Kluwer Academic Publishers, Dordrecht, 497, 167–224 (1997).
12. Krakovski R., Mohar B.: Spectrum of Cayley graphs on the symmetric group generated by transpositions. arXiv:1201.2167, 201

13. Labarre, A.: Combinatorial Aspects of Genome Rearrangements and Haplotype Networks/Aspects (Combinatoires Des Réarrangements Génomiques Et Des Réseaux D'haplotypes). Ph.D. Thesis, Université Libre De Bruxelles (2008).
14. Lakshmivarahan, S., Jwo, J.S., Dhall, S.: Symmetry in interconnection networks based on Cayley graphs of permutation groups: a survey. Parallel Computing, 19, 361–407 (1993).
15. Latifi, S., Srimani, P.K.: Transposition networks as a class of fault-tolerant robust networks, IEEE Trans. Comput. 230–238 (1996). https://doi.org/10.1109/12.485375
16. Shi, Ling-Sheng, and Peng Wu. Conditional Connectivity of Bubble Sort Graphs. Acta Mathematicae Applicatae Sinica, English Series, 33(4), 933–944 (2017). https://doi.org/10.1007/s10255-017-0708-8.

A Smaller Upper Bound for the $(4, 8^2)$ Lattice Site Percolation Threshold

John C. Wierman

Abstract The $(4, 8^2)$, or "bathroom tile," lattice is one of the eleven Archimedean lattices, which are infinite vertex-transitive graphs with edges from the tilings of plane by regular polygons. The site percolation model retains each vertex of an infinite graph independently with probability p, $0 \leq p \leq 1$. The site percolation threshold is the critical probability p_c^{site} above which the subgraph induced by retained vertices contains an infinite connected component almost surely, and below which all components are finite almost surely. Using computational improvements for the substitution method, the upper bound for the site percolation threshold of the $(4, 8^2)$ lattice is reduced from 0.785661 to 0.749002.

Keywords Site percolation · Percolation threshold · Set partitions · Non-crossing partitions

MSC Primary 60K35; Secondary 05C80, 05A18, 82B43

1 Introduction

Percolation theory studies connectivity of infinite random graph models, with particular emphasis on the existence or non-existence of an infinite connected component. Its popularity in the engineering and physical sciences is due to the behavior that occurs as the random graph becomes more richly connected, making a drastic qualitative change as an infinite cluster forms. As a result, applications of percolation models are widespread, concentrating on modeling critical phenomena where some type of phase transition occurs. Examples are thermal transitions of a liquid freezing into a solid, an infectious disease spreading only locally versus becoming epidemic, or a conductor-insulator alloy becoming a conductor as the proportion of conducting

---

J. C. Wierman (✉)
Department of Applied Mathematics and Statistics, Johns Hopkins University, Baltimore, MD 21218, USA
e-mail: jwierma1@jhu.edu

F. Hoffman (ed.), *Combinatorics, Graph Theory and Computing*,
Springer Proceedings in Mathematics & Statistics 388,
https://doi.org/10.1007/978-3-031-05375-7_24

atoms increases. Since the introduction of percolation models in the 1950s, they have provided explanations of many phenomena that could not be explained satisfactorily before.

Given an infinite graph G, the *site percolation model* on G creates a random subgraph by retaining each vertex independently with probability p, $0 \le p \le 1$, deleting it otherwise, and constructing the random subgraph G_p of G induced by the set of retained vertices. Retained vertices are often referred to as "open" or "occupied," and deleted vertices as "closed" or "vacant." (Another classic percolation model is the *bond percolation model* in which the edges of G are retained or deleted independently at random with probability p.)

Due to the interest in modeling a phase transition, the emergence of infinite connectivity as the parameter p increases is the principal focus, with particular interest on the critical value of p at which the transition occurs. For an infinite graph G, this quantity, the *site percolation threshold* $p_c^{site}(G)$, satisfies the following two conditions: (1) If $p < p_c^{site}(G)$, all connected components of G_p are finite, with probability one. (2) If $p > p_c^{site}(G)$, there exists an infinite connected component of G_p with probability one. (The fact that these events have probability one is a consequence of Kolmogorov's Zero-One Law and independent retention of the vertices of G.)

Although the main interest in percolation theory is in the phase transition point, little progress has been made toward determining the value of the site percolation threshold of common infinite lattice graphs, such as the square and hexagonal lattices. One highlight is that the site percolation threshold of the triangular lattice was proved by Kesten [4] to be $1/2$. The value of the site percolation threshold is highly dependent upon the structure of the infinite graph, but the nature of that dependence is not well understood. Most knowledge of the threshold values are from extensive simulation studies, which produce values claiming 5 or 6 digit accuracy. However, it is not very unusual for the interval estimates from different studies to be disjoint.

Due to the lack of exact solutions, it is of mathematical interest to provide rigorous bounds for the site percolation thresholds of common lattices that are as accurate as possible, and to develop bounding techniques which may eventually help determine exact solutions. Table 1 in Sect. 4 provides a compilation of rigorous bounds for the eleven Archimedean lattices, which are vertex-transitive graphs constructed with the vertices and edges of a tiling of the plane by regular polygons. (See the beautiful monograph by Grünbaum and Shephard [2] for discussion and illustrations.) Note that the bounds are typically rather poor, providing intervals of width 0.10–0.20 for most lattices.

This article substantially reduces the upper bound for the site percolation threshold of one of the Archimedean lattices, known as the $(4, 8^2)$ or "bathroom tile" lattice. The $(4, 8^2)$ lattice is illustrated in Fig. 1. The name reflects the fact that each vertex is incident to a square and two octagons.

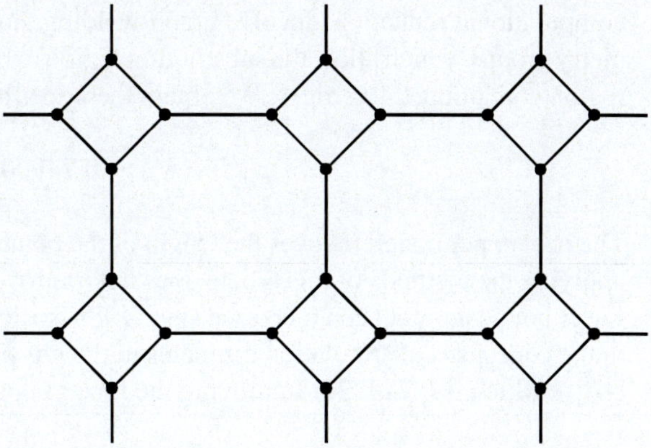

Fig. 1 A subgraph of the $(4, 8^2)$ or "bathroom tile" lattice

2 Bounds for the $(4, 8^2)$ Lattice Site Percolation Threshold

As for many infinite lattice graphs, there is a long history of improving rigorous bounds for the $(4, 8^2)$ lattice, yet the bounds are still not satisfyingly accurate.

In 1988, Luczak and Wierman [5] applied a grouping method to show that

$$0.707106 \approx \sqrt{p_c^{bond}(Square)} \leq p_c^{site}(4, 8^2) \leq \sqrt{p_c^{site}(Square)},$$

when there was only a very crude upper bound for the site percolation threshold of the square lattice.

In 1995, Wierman [13] adapted the substitution method to a site model for the first time, obtaining an upper bound of 0.679492 for the site percolation threshold of the square lattice, establishing

$$0.707106 \leq p_c^{site}(4, 8^2) \leq \sqrt{p_c^{site}(Square)} < 0.824313.$$

In 2001 a substitution method comparison of the $(4, 8^2)$ lattice to the line graph of the 2-subdivided square lattice, using a four-vertex substitution region, further improved the upper bound [14]:

$$p_c^{site}(4, 8^2) \leq 0.79970.$$

A different substitution method comparison, in 2019, produced the upper bound [17]

$$p_c^{site}(4, 8^2) \leq 0.785661.$$

In this article, we return to a comparison with the line graph of the 2-subdivided square lattice, adapting a collection of more efficient computational methods that were developed for bond percolation models for use on site percolation models. The computational reductions involve graph-welding, non-crossing partitions, and symmetry groups, which allow the substitution region to be applied to a larger substitution region containing 24 vertices. We obtain the upper bound

$$p_c^{site}(4, 8^2) \leq 0.749002.$$

The new upper bound reduces the length of the bounding interval by 46%. However, applying the methods of this article does not improve the lower bound. Note that the lower bound has not been increased since 1988, so that the upper bound is now closer to the consensus of simulation estimates in the physical sciences literature (See, e.g. [10].), which is 0.729724. Improving the lower bound remains a challenge.

3 Derivation of the Upper Bound

The new upper bound established in this article was derived using the substitution method, and was made possible by combining several previous computational advances, involving graph-welding, non-crossing partitions, symmetry reduction, and conversion to a network flow model. Descriptions of the general substitution method appear in [16, 18], while details of the computational reduction methods may be found in [7, 8]. The following description of the derivation of the bound focuses on specific items involved in the application to the $(4, 8^2)$ lattice and issues that had to be overcome, but only provides a sketch rather than complete details.

3.1 Substitution Method

The substitution method derives percolation threshold bounds for an unsolved percolation model by comparing it to a solved percolation model. It has produced most of the best current bounds for bond percolation thresholds. In particular, it derived bounds that determined the three leading digits of the bond percolation threshold of the $(3, 12^2)$ lattice [16] and the two leading digits of the bond percolation threshold of the kagome lattice [18], disproving long-standing conjectured exact values by Tsallis [11]. However, there are complications in adapting the substitution method to site percolation models, so bounds for site percolation thresholds are generally much less accurate than bounds for bond percolation thresholds.

3.2 Substitution Regions

To apply the substitution method to site percolation models, both the unsolved and solved lattices must be decomposed into vertex-disjoint isomorphic subgraphs, called *substitution regions* so that the random retentions and deletions associated with the sets of vertices in different regions are stochastically independent. Since no vertex of the original lattice can be on the boundary between two substitution regions, each edge that connects two substitution regions must be subdivided by inserting a new *boundary vertex* which is open with probability one. Since the substitution method compares the probabilities of connections between the boundary vertices of the substitution regions, the substitution regions of the two lattices must have the same number of boundary vertices. For bond percolation models, substitution method calculations have been completed for some substitution regions with eight boundary vertices, but not for nine or more.

3.3 The Comparison Lattice

The small number of exactly-solved site percolation models, and the constraints on the substitution regions, constrain the choice of comparison lattice. Yet, it is still more of an art than a science to choose a comparison lattice and a substitution region that provides an accurate bound and for which the necessary computations are manageable. A natural goal is to find a solved lattice graph that has relatively similar structure to the unsolved lattice.

The comparison lattice used here is obtained by two transformations from a solved bond percolation model. Kesten [3] proved that the bond percolation threshold of the square lattice is one-half. If every edge of the square lattice is subdivided into two edges, the bond percolation threshold of the resulting lattice is $\sqrt{1/2}$. Since a bond percolation model is equivalent to the site model on its line graph, the site percolation threshold of the line graph of the subdivided square lattice is exactly $\sqrt{1/2}$. For convenience, we will call this lattice the *reference lattice* and denote it by R. A substitution region of R with ten boundary vertices is illustrated in Fig. 2. Note

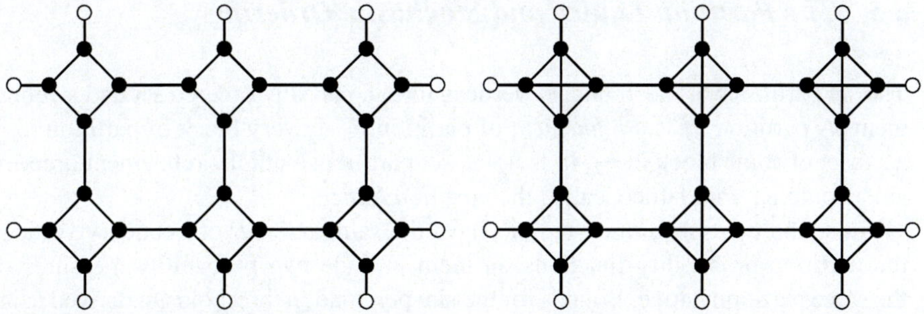

Fig. 2 Substitution regions with ten boundary vertices for the (4, 8^2) lattice (on the left) and the line graph of the subdivided square lattice (on the right). Filled circles represent vertices of the original lattice. Empty circles represent boundary vertices introduced by subdividing edges, and are always open

that R is a super-graph of the $(4, 8^2)$ lattice with diagonals inserted in every square face.

3.4 Set Partitions of the Boundary Vertices

The two percolation models are compared on the basis of the probabilities of connections between the boundary vertices of the substitution regions. A *configuration* is a designation of every vertex as open or closed. Note that for each substitution region, since there are 24 vertices with independent randomness, the probability of each configuration is a 24 degree polynomial function of p. Every configuration partitions the set of boundary vertices into blocks of boundary vertices which are in a common connected component of open vertices. For example, if the boundary vertices are labeled $1, 2, 3, \ldots, 10$, then $\{1, 4, 5, 6\}\{2, 3\}\{7, 8, 9, 10\}$ denotes the partition in which vertices 1, 4, 5, and 6 are in one connected component, 2 and 3 are in a different one, and 7, 8, 9, and 10 are in a third one. There may be many configurations which produce the same partition, so the probability of a partition is the sum of all configurations which produce it, which is also a 24 degree polynomial in the parameter p.

The number of set partitions of n objects is an extremely rapidly-increasing function of n, given by the Bell numbers. In our case, there are $Bell(10) = 115975$ partitions, and we must compute the probabilities of each partition for both models. While this could be done by calculating the probabilities of each of the 2^{24} configurations and summing to obtain partition probabilities, this is extremely inefficient. A more efficient graph-welding approach [6] applies the configuration approach to a small subgraph of the substitution region, then builds partition probability functions for increasingly larger graphs.

3.5 The Partition Lattice and Stochastic Ordering

The set partitions of the boundary vertices form a partially ordered set under refinement: A partition π_1 is a *refinement* of partition π_2 if every block of partition π_1 is a subset of some block in π_2. In fact, the set partitions with the refinement ordering are a combinatorial lattice, called the *partition lattice*.

Since the two substitution regions have the same number of boundary vertices, the partition probability functions on them provide two probability measures on the same partition lattice. Bounds for the site percolation threshold are derived using *stochastic ordering* of the two probability measures, defined in this context as follows: An *upset* is a set U of partitions such that if partition $\pi_1 \in U$ and $\pi_1 \leq \pi_2$, then partition $\pi_2 \in U$. For two probability measures P_1 and P_2 on the partition lattice, P_1 is *stochastically smaller* than P_2, denoted $P_1 \leq_{st} P_2$, if $P_1(U) \leq P_2(U)$ for every upset U. (In this case, we also say that P_2 is *stochastically larger* than P_1.)

The partition lattice on ten boundary vertices is a ranked poset, with the largest rank containing 42525 partitions. Since this subset is an anti-chain, any of its subsets (along with the top element) generates a different upset. Thus, a crude lower bound on the number of upset probability inequalities that need to be checked to determine stochastic ordering is 2^{42525}. Without major computational reductions, checking stochastic ordering would be impossible.

3.6 *"Finding Two Needles in a Haystack"*

We compare the probability measure $P_{(4,8^2),p}$ from the $(4, 8^2)$ lattice substitution region with parameter p to the reference probability measure P_R from the line graph of the subdivided square lattice with parameter $\sqrt{1/2}$, which is its site percolation threshold. For values of p for which $P_{(4,8^2),p} \geq_{st} P_R$, the parameter $p \geq p_c(4, 8^2)$, while if $P_{(4,8^2),p} \leq_{st} P_R$, then $p \leq p_c(4, 8^2)$.

Rather than treating upset inequalities, the inequalities can be converted to equations in p, with the lower and upper bounds for the percolation threshold of the $(4, 8^2)$ lattice site model being the smallest and largest solutions, respectively.

To use a colloquial expression, the derivation of the site percolation threshold bound is like the problem of finding a needle in a haystack, except that we need to find two needles! The needles are the largest and smallest solutions, while the haystack is the set of solutions to the more than 2^{42525} upset equations. Our strategy is to avoid searching large parts of the haystack, by proving that the needles are not there. The following summarizes the nature of the reductions employed.

3.7 *Non-crossing Partitions*

One reduction makes use of non-crossing partitions. Suppose that the boundary vertices are labeled from 1 to 10 clockwise around both substitution regions, starting from the left side of the top in Fig. 2. A partition is a *non-crossing partition* if $i < j < k < l$, with vertices i and k in the same block and vertices j and l in the same block, implies that all four vertices are in the same block. A partition that is not a non-crossing partition is called a *crossing partition*. Since the $(4, 8^2)$ lattice is planar, any crossing partition has probability zero. Although the reference lattice R is not planar, since two open paths crossing via diagonals of a square face implies that all four vertices of the face are open, any crossing partition in it also has probability zero. Thus, only non-crossing partitions contribute to the upset probability functions. Since the non-crossing partitions are counted by the Catalan numbers, there are only $Catalan(10) = 16796$ non-crossing partitions to be considered, rather than all $Bell(10) = 115975$, which greatly reduces the number of relevant upsets. For further details, see [8].

3.8 Symmetry Reduction

A second reduction uses rotational and reflection symmetry to find equivalence classes of partitions, and reduces the problem to checking stochastic ordering on a partially ordered set of classes. Partitions which are rotations or reflections of each other have identical probability functions, so can be combined into classes. The classes form a partially ordered set under refinement, where one class is a refinement of another if any of its partitions is refinement of any partition of the other. May and Wierman [7] show that the upper and lower bounds must be solutions of equations for upsets which are unions of classes. For the comparison in this article, the number of classes is 4388, a significant reduction from $Catalan(10) = 16796$.

3.9 Network Flow Model

The third major computational savings are from converting the problem to a network flow problem, based on a proof of Preston [9] of the equivalence of stochastic ordering and coupling for probability measures on a finite partially ordered set. Intuitively, a probability measure P_1 is stochastically larger than a probability measure P_2 if probability from P_1 can flow down in the partially ordered set to produce P_2. The network flow problem was solved symbolically in MATLAB using the augmented path algorithm, to obtain a rational number for the upper bound, avoiding any round-off error except in converting the final rational number to a decimal.

The entire computation, from calculating partition probablities through solving the network flow problem, took slightly less than two weeks on a 2.90 GHz Dell XPS 15 9570 laptop with 32 GB RAM, producing the upper bound $p_c(4, 8^2) \leq 0.749001747369766$. This is the first substitution method comparison that has been completed for substitution regions with ten boundary vertices, for either bond or site percolation models.

4 Future Research

Future research will focus on improving site percolation threshold bounds for other Archimedean lattices. Table 1 summarizes the current bounds, exact values, and consensus of simulation estimates. Although the site percolation threshold is exactly known for three of the lattices, only the triangular lattice solution was derived using the site model, with the other two obtained by transformations of bond model solutions: The kagome lattice is the line graph of the hexagonal lattice, so its site percolation threshold equals $1 - 2\sin(\pi/18)$, the bond percolation threshold of the hexagonal lattice. The $(3, 12^2)$ lattice is the line graph of the 2-subdivided hexagonal lattice, so its site percolation threshold is $\sqrt{1 - 2\sin(\pi/18)}$. The upper bound

Table 1 Site percolation threshold bounds and values for the archimedean lattices

Lattice name	Lower bound	Exact value or estimate	Upper bound
$(3, 12^2)$		= 0.807900...	
(4, 6, 12)	0.721730	0.747806	0.770935
$(4, 8^2)$	0.707106	0.729724	0.749002
Hexagonal	0.652703	0.697043	0.74335
(3, 4, 6, 4)	0.522394	0.621819	0.652704
Kagome		= 0.652703...	
Square	0.556000	0.592746	0.679492
$(3^4, 6)$	0.500000	0.579498	0.652704
$(3^3, 4^2)$	0.500000	0.550213	0.679492
$(3^2, 4, 3, 4)$	0.500000	0.550806	0.679492
Triangular		= 0.500000...	

for the $(4, 8^2)$ lattice site percolation threshold obtained in this article produces the shortest bounding interval for any of the unsolved lattices. The bounding interval lengths for three of the other lattices are almost 0.18.

The most important and challenging cases are the square and hexagonal lattices. For the square lattice, the lower bound was proved by van den Berg and Ermakov [1] in 1996, while the upper bound was derived in 1995 by Wierman [13], with no improvements since. For the hexagonal lattice, the lower bound is the bond percolation threshold of the hexagonal lattice, established [12] in 1981 while the upper bound was proved [8] in 2007 using the substitution method. Despite repeated attempts, these bounds have not been improved.

To make progress on these problems with the substitution method, substantial adaptations must be made to augment the partition lattice as in [17] and then adapt the graph-welding, non-crossing partition, symmetry reduction, and network flow model accordingly.

Acknowledgements This research was supported by the Aceheson J. Duncan Fund for the Advancement of Research in Statistics.

References

1. van den Berg, J. and Ermakov, A. (1996) A new lower bound for the critical probability of site percolation on the square lattice. *Random Structures & Algorithms* 8, 199–212. https://doi.org/10.1002/(SICI)1098-2418(199605)8:3<199::AID-RSA4>3.0.CO;2-T
2. Grünbaum, B. and Shephard, G. C. (1987) *Patterns and Tilings*, W. H. Freeman.
3. Kesten, H. (1980) The critical probability of bond percolation on the square lattice equals 1/2. *Communications in Mathematical Physics* 74, 41–59. https://doi.org/10.1007/BF01197577

4. Kesten, H. (1982) *Percolation Theory for Mathematicians,* Birkhäuser.
5. Łuczak, T. and Wierman, J. C. (1988) Critical probability bounds for two-dimensional site percolation models, *Journal of Physics A: Mathematical and General* 21, 3131–3138. https://doi.org/10.1088/0305-4470/21/14/014
6. May, W. D. and Wierman, J. C. (2003) Recent improvements to the substitution method for bounding percolation thresholds. *Congressus Numerantium* 162, 5–25.
7. May, W. D. and Wierman, J. C. (2005) Using symmetry to improve percolation threshold bounds. *Combinatorics, Probability and Computing* 14, 549–566. https://doi.org/10.1017/S0963548305006802
8. May, W. D. and Wierman, J. C. (2007) The application of non-crossing partitions to improving percolation threshold bounds. *Combinatorics, Probability and Computing* 17, 285–307. https://doi.org/10.1017/S0963548306007905
9. Preston, C. J. (1974) A generalization of the FKG inequalities. *Communications in Mathematical Physics* **36**, 233–241. https://doi.org/10.1007/BF01645981
10. Suding, P. N. and Ziff, R. M. (1999) Site percolation thresholds for Archimedean lattices. *Physical Review E* 60, 275–283. https://doi.org/10.1103/PhysRevE.60.275
11. Tsallis, C. (1982) Phase diagram of anisotropic planar Potts ferromagnets: A new conjecture. *Journal of Physics C* 15, L757–L764. https://doi.org/10.1088/0022-3719/15/23/007
12. Wierman, J. C. (1981) Bond percolation on honeycomb and triangular lattices, *Advances in Applied Probability* 13, 293–313. https://doi.org/10.2307/1426685
13. Wierman, J. C. (1995) Substitution method critical probability bounds for the square lattice site percolation model. *Combinatorics, Probability and Computing* 4, 181–188. doi: https://doi.org/10.1017/S0963548300001565
14. Wierman, J. C. (2001) Site percolation critical probability bounds for the $(4, 8^2)$ and $(4, 6, 12)$ lattices. *Congressus Numerantium* 150, 117–128.
15. Wierman, J. C. (2002) An improved upper bound for the hexagonal lattice site percolation critical probability. *Combinatorics, Probability and Computing* 11, 629–643. https://doi.org/10.1017/S0963548302005345
16. Wierman, J. C. (2016) Tight bounds for the bond percolation threshold of the $(3, 12^2)$ lattice, *Journal of Physics A: Mathematical and Theoretical* 49, 475002. https://doi.org/10.1088/1751-8113/49/47/475002
17. Wierman, J. C. (2019) An improved upper bound for the site percolation threshold of the $(4, 8^2)$ Archimedean lattice. *Congressus Numerantium* 232, 139–152.
18. Wierman, J. C.; Yu, G.; and Huang, T. (2015) A disproof of Tsallis' bond percolation threshold conjecture for the kagome lattice. *Electronic Journal of Combinatorics* 22, P2.52. https://doi.org/10.37236/5117